Molecularizing Biology and Medicine

Studies in the History of Science, Technology and Medicine
Edited by John Krige, CRHST, Paris, France

Studies in the History of Science, Technology and Medicine aims to stimulate research in the field, concentrating on the twentieth century. It seeks to contribute to our understanding of science, technology and medicine as they are embedded in society, exploring the links between the subjects on the one hand and the cultural, economic, political and institutional contexts of their genesis and development on the other. Within this framework, and while not favouring any particular methodological approach, the series welcomes studies which examine relations between science, technology, medicine and society in new ways e.g. the social construction of technologies, large technical systems.

Other Titles in the Series

Volume 1
Technological Change: Methods and Themes in the History of Technology
Edited by Robert Fox

Volume 2
Technology Transfer out of Germany after 1945
Edited by Matthias Judt and Burghard Ciesla

Volume 3
Entomology, Ecology and Agriculture: The Making of Scientific Careers in North America, 1885–1985
Paolo Palladino

Volume 4
The Historiography of Contemporary Science and Technology
Edited by Thomas Söderqvist

Volume 5
Science and Spectacle: The Work of Jodrell Bank in Post-war British Culture
Jon Agar

Volume 6
Molecularizing Biology and Medicine: New Practices and Alliances, 1910s–1970s
Edited by Soraya de Chadarevian and Harmke Kamminga

Other Volumes in Preparation

Making Isotopes Matter: F.W. Aston and the Culture of Physics
Jeff Hughes

Molecularizing Biology and Medicine

New Practices and Alliances, 1910s–1970s

Edited by
Soraya de Chadarevian and Harmke Kamminga
Wellcome Unit for the History of Medicine, Cambridge, UK

harwood academic publishers
Australia • Canada • China • France • Germany • India • Japan • Luxembourg
Malaysia • The Netherlands • Russia • Singapore • Switzerland • Thailand

Amsteldijk 166
1st Floor
1079 LH Amsterdam
The Netherlands

British Library Cataloguing in Publication Data

Molecularizing biology and medicine: new practices and
alliances, 1910s–1970s — (Studies in the history of
science, technology and medicine; v. 6)
1. Molecular biology — History — 20th century
I. Chadarevian, Soraya de II. Kamminga, Harmke
572.8′0904

ISBN 90-5702-293-1

Contents

List of Illustrations and Credits

List of Tables

Preface

The project of this book is part of a collaboration between the Cambridge Wellcome Unit for the History of Medicine and the INSERM Unit 158, Paris. Jean-Paul Gaudillière and Ilana Löwy from the Paris Unit have been involved in all stages of this project. We are very grateful for their important input and continuous advice.

Most of the chapters were first presented at a conference on "Molecularizing Biology and Medicine", held in Cambridge in July 1994. We thank all conference participants for their lively contributions to the discussions, and the speakers for their stimulating papers. We are also grateful to those authors in this volume who joined the project later, especially for their cooperation in meeting tighter deadlines.

We thank the British Council and the French Ministry of Education and Research for funding the collaboration between Cambridge and Paris. We are grateful for additional funding from the Wellcome Trust and for general support from our colleagues in the Cambridge Unit.

Soraya de Chadarevian
Harmke Kamminga

Notes on Contributors

Olga Amsterdamska is a member of the Department of Science and Technology Dynamics at the University of Amsterdam. Her research focuses on the development of the biomedical sciences and the interactions between the laboratory and the clinic in twentieth-century medicine. She has published on schools of thought in linguistics, on the history of bacteriology and biochemistry as well as on some theoretical issues in the social studies of science. She is editor of *Science, Technology, and Human Values*.

Soraya de Chadarevian is a research fellow at the Hamburg Institute for Social Research in Germany and associated with the Wellcome Unit for the History of Medicine in Cambridge. She is the author of *Between Discourses. Maurice Merleau-Ponty and the Sciences* (Königshausen and Neumann, 1990) and of several articles on the history of the biological and biomedical sciences in the nineteenth and twentieth century. She is currently working on a book on the history of molecular biology in postwar Britain.

Angela N.H. Creager is Assistant Professor in the Department of History and Program in History of Science at Princeton University, where she has been teaching history of biology since 1994. She is currently working on a book-length account of the importance of virus research to the growth of post-war biology in the US. Her other research interests are directed at understanding the impact of World War II on biomedical research and using comparative approaches to the history of molecular biology.

Paul J. Edelson is a physician and historian on the faculty of Cornell University Medical College with a special interest in the social impact of scientific medicine. He is currently working on an extended study of the origins of newborn genetic screening in the United States.

Jean-Paul Gaudillière is historian of biology and medicine at INSERM, Paris. He is the author of several articles on the history of molecular biology, cancer research and medical genetics. With Ilana Löwy he is the editor of *The Invisible Industrialist: Manufactures and the Construction of Scientific Knowledge* (Macmillan, 1997). He is currently working on a book on the reconstruction of biomedicine in post-war France.

Jordan Goodman is Senior Lecturer in History at the University of Manchester Institute of Science and Technology. He has published widely in the field of European economic history, the history of technology and medicine, and cultural history. He has published four books, the latest being an edited collection of essays, *Consuming Habits: Drugs in History and Anthropology*. He is currently involved in a joint project exploring the cultural history of an anticancer drug since the 1960s.

Harmke Kamminga works at the Wellcome Unit for the History of Medicine in Cambridge, where she heads the program in the history of modern biomedicine. She has published in the history and philosophy of biology and the history of the biomedical sciences, and is co-editor (with Steven French) of *Correspondence, Invariance and Heuristics* (Boston Studies Series in Philosophy of Science, Kluwer 1993) and (with Andrew Cunningham) of *The Science and Culture of Nutrition, 1840–1940* (Rodopi, 1995). She is currently writing a book on the history of vitamins.

Ilana Löwy is a senior researcher at INSERM, Paris. She was trained as an immunologist and historian of science. She is author of *The Polish School of Philosophy of Medicine, from Tytus Chalubinski to Ludwik Fleck* (Kluwer, 1990), of *Between Bench and Bedside: Healing and Interleukin-2 in a Cancer Ward* (Harvard University Press, 1996), and of numerous articles on the history of immunology, bacteriology and cancer research.

Lara Marks is a lecturer in the history of medicine in the Centre for the History of Science, Technology and Medicine, Imperial College, London. She has written extensively on the history of ethnicity and health and maternal and child health, including *Model Mothers: Jewish Mothers and Maternity Provision in East London, 1870–1939* (Oxford University Press, 1994) and *Metropolitan Maternity: Maternal and Infant Welfare Services in Early Twentieth Century London* (Rodopi, 1996). She is currently preparing a book on the international history of the contraceptive pill.

Diane B. Paul is Professor of Political Science and Co-director of the Program in Science, Technology, and Values at the University of Massachusetts at Boston. She writes primarily on the history and social impact of genetics. Her book *Controlling Human Heredity: 1865 to the Present* was recently published by Humanities Press.

Steve Sturdy is based in the Science Studies Unit at the University of Edinburgh, where he holds a Wellcome University Award in the History of Medicine. He has published several articles on the history of physiology, clinical medicine and health policy, and is currently completing a book on medicine, science and statecraft in Britain, 1900–1920.

Introduction

Soraya de Chadarevian and Harmke Kamminga

It is our aim in this volume to draw attention to the formation of particular strategic approaches in biology and medicine centred on molecules. These approaches became prominent in the interwar period and gained new momentum with the biomedical mobilization of World War II.[1] The identification, production, circulation and uses of molecules in biological research and in the explanation and treatment of diseases created new links between the laboratory, the clinic and industry. We introduce the term "molecularization" to describe the creation and transformation of these alliances. By following molecules through different biomedical contexts and networks, the contributors to this volume provide a novel historical perspective on innovation in the biological sciences and medical practice.

Nowadays a molecular vision of life, health and disease is linked specifically to the spectacular successes of molecular biology in the study of structure-function relationships of nucleic acids and proteins since the 1950s and 1960s, the development of recombinant DNA technologies in the 1970s, and the making of the Human Genome Project with its promises of a genetically based molecular medicine. Historians have used the terms "molecularization", "molecular vision", "molecular revolution", and "molecular politics" in this context (see Kay, 1993a and 1993b; Olby, 1990; Rheinberger, 1995; and Wright, 1994).[2]

The coupling between a molecular vision of life and molecular biology has been promoted vigorously by participant scientists and has been taken up in different forms by historians. Taking on board the molecular biologists' focus on nucleic acids and proteins, historians have tended to look for paths towards these ends in earlier work on macromolecules and in genetics (Judson, 1994; Kay, 1993a; Olby, 1990, 1995). Biochemists, on the other hand, have claimed their own place in this history by insisting that molecular approaches to life had been practised by biochemists all along, but that these encompassed a much broader range of molecules (Cohen, 1984; Fruton, 1992; Krebs, 1969).[3]

The notion of molecularization used in this volume opens up a new historiographic approach to the growth of a molecular culture in biomedicine, which differs from both these perspectives. Rather than treat work on macromolecules in historical isolation, we use molecularization to refer to small and large molecules alike. Yet we do not simply embrace the biochemists' outlook which places the accumulation of knowledge about the role of molecules in biological processes at centre-stage. It is neither our intention to trace the "origins" of molecular biology, nor to restrict molecularization to the growing understanding of molecules as biological and medical objects. Instead, we use molecularization to refer to practices centred on molecules, and focus on the interactions between different social groups in the creation and transformation of these practices. Building on a variety of approaches, from network theory to social interactionism and interest theory, with a focus on laboratory practices and alliances, the authors make manifest the powerful role that molecules have played in forging links between the laboratory, the clinic, industry and wider social interests.[4]

The studies in this volume bring out particularly sharply that the relationships between these domains have been much more complex and varied than is suggested by the still common notion that "basic knowledge" is acquired in the research laboratory and is then "applied" in the clinic, with industry becoming involved in the large-scale production and commercialization of drugs. Particular alliances between research scientists, clinicians, industrialists and policy makers have confounded the distinction between "basic" and "applied" science. The very introduction of the terms "biomedical sciences" and "biomedical complex" in the 1960s made explicit the interconnections which, as the contributors show, had been formed and reformed in new configurations for many decades.

The studies presented in this volume cover the period from the 1910s to the 1970s. We have chosen to start in the 1910s, because strategies of molecularization were by then in the making in different fields of research and production, in several countries. In closing the volume in the 1970s, we consciously exclude the more recent developments linking molecular biology, biotechnology and genetic medicine, which are already receiving considerable attention in the literature (see especially Kevles and Hood, 1992; Krimsky, 1991). The studies collected here nevertheless offer tools for a historical re-evaluation of these more recent strategies of molecularization.

In the following sections, we provide, in outline, a historical framework for the chapters in this volume and then point to a number of common and intersecting themes running through them. In the epilogue to this volume, Steve Sturdy reflects on the history of molecularization, focussing on the central role of standardization in the creation of a "molecular economy" in biology and medicine.

MOBILIZING MOLECULES

The nineteenth century saw the "laboratory revolution in medicine" (Cunningham and Williams, 1993), with a large-scale move of medical men into the research laboratory, the creation of new sciences allied to medicine, such as experimental physiology, bacteriology, and physiological chemistry, and the beginnings of the pharmaceutical industry. Links were formed between the research laboratory, the clinic and industry, especially with the development and production of antisera and vaccines in the 1890s. The early successes of serum therapy for diphtheria created demands for new products on a massive and international scale, which could not be met by the few institutions devoted to research and production in this field, notably the Institut Pasteur in Paris and Robert Koch's Institute of Infectious Diseases in Berlin (Moulin, 1991; Weindling, 1992). Industry became involved in scaling up production of antisera against many bacterial diseases, simultaneously entering new medical markets for their products. In Germany, the state, too, became a significant player in this network, when the Ministry of Health assumed control over the standardization of antisera (Liebenau, 1990; Sturdy, this volume).

Early forms of molecularization of medicine in the twentieth century built on these developments, but with two significant and interlinked innovations: a shift towards medical intervention at the level of specific molecules, and the active participation of industry, not only in the production, but in the identification of therapeutic molecules. Both features are exemplified by Paul Ehrlich's program of chemotherapy, which was initiated around the turn of the century in Frankfurt and was aimed at identifying synthetic dye stuff derivatives as therapeutic agents against parasitic diseases (see especially Lenoir, 1988; Liebenau, 1990).

Ehrlich tested hundreds of dye stuff derivatives, synthesized at his request by industrial chemists in the Hoechst Chemical Works, for specific therapeutic properties. Most of them turned out to be inactive, toxic, or both. Those showing promising activity in experimental animals were tested on patients with the cooperation of local hospitals and physicians. The active collaboration of the chemical industry, which made Ehrlich's research possible, was spurred by economic interests in finding uses for waste products of dye synthesis and in entering new markets opened up earlier by the vaccine and antiserum business. Their investment in this work, and Ehrlich's research strategy, was vindicated spectacularly by the most famous product of the chemotherapeutic program, "compound 606", the anti-syphilitic drug which was marketed by Hoechst under the name of Salvarsan in 1911. With the successful development of chemically well-defined synthetic products with therapeutic properties that were clinically tested, standardized and used widely for treatment, molecules in medicine became big business.

Ehrlich explained the action of dye stuff derivatives in terms of specific binding to (non-defined) chemical groups attached to the cellular protoplasm (Cambrosio *et al.*, 1993). His notion of protoplasmic receptors did not, however, provide molecular explanations and could not guide the design of drugs, nor could Ehrlich explain the etiology of the diseases in question in molecular terms. Nevertheless, Ehrlich's program of chemotherapy, together with its products and the new collaborations on which it relied, is an early example of molecularization in our sense. Furthermore, the success of the program in yielding therapeutically active products directed attention to molecular explanations in pathology and stimulated further searches for specific molecular drugs.

The chemotherapeutic tradition remained strong in Germany, as illustrated by the industrial development of sulphonamides in the 1930s. Chemotherapy was not, however, the only area in which molecularization was pursued. During the 1910s, chemical understanding was sought and clinical uses were explored in the case of other classes of substances, for example hormones and vitamins. In these instances, however, the active involvement of industry and treatment with well-defined molecules came at a later stage, after World War I (Bliss, 1982; Horrocks, 1995; Kamminga, this volume; Oudshoorn, 1993). Wartime pressures gave molecularization a new impetus, on a broader geographic canvas.

With the disruption of trade links during World War I, many overseas licences were revoked, and Salvarsan production and standardization, for instance, were taken up outside Germany. From quite different directions, vitamin research was intensified in the light of concerns about the nutritional value of army rations and civilian diets, under the threat of shortages in the food supply. Especially in Britain, scientists who were investigating the detailed chemical composition of healthy diets became involved in the formulation of food policy at the highest levels (Teich, 1995).

The experience gained and the networks established during the war gave momentum to the search for means of medical intervention at the level of specific molecules more widely. The articulation of research strategies towards this end increasingly involved industry and the state as active partners, as well as clinicians and scientists. The involvement of industries, with their interest in finding cheaper synthetic ways of production, stimulated research on the structure of vitamins and hormones. The need for structural research on molecules was also stressed by biochemists, for their own purposes, in the promotion of their subject as the fundamental science of life and as the basis for a scientific medicine (Kamminga and Weatherall, 1996).

Much of this structural research was directed at vitamins and hormones, in view of their wide preventive and therapeutic use. In some instances, the structure determination of these molecules was not achieved until decades after their industrial production and clinical use began. A striking example is insulin, first produced and used therapeutically in the 1920s, whose three-dimensional molecular structure was determined only in the late 1960s.

In the interwar period, biochemical research was supported strongly by the policies of major funding agencies, such as the Medical Research Council (MRC) in Britain and the Rockefeller Foundation based in the United States (Abir-Am, 1987; Austoker and Bryder, 1989; Kohler, 1978, 1991). Both these bodies devoted a great deal of their funds to research on the structure and functions of biologically active molecules, from the 1920s onwards. The MRC also assumed a prominent role in setting international biological standards, especially in relation to vitamins and hormones (Liebenau, 1989).

By the late 1930s, multiple links between the research laboratory, the clinic, state and industry were in place, based on molecules of biological, medical and economic interest. Strategies of molecularization pursued in the interwar period were built upon and put to new uses during World War II, in which scientists were enlisted on an unprecedented scale.

The biomedical mobilization of World War II was in important ways centred on molecules, a point to which this book draws attention for the first time. Research scientists, clinicians, industry, workers, and the state machinery of the nations at war were mobilized for the investigation and large-scale production of a series of strategic molecules. These included above all antibacterial drugs, especially sulphonamides and the antibiotic penicillin, antimalarial drugs, and blood fractionation products for use in transfusion. The penicillin work in Britain and the United States provided the allied armies and physicians with a powerful weapon in the "war against disease," spurring searches for other therapeutic molecules, or "magic bullets" (Bäumler, 1965; Neushul, 1993; Swann, 1983).[5]

In the postwar reconstruction, the penicillin project was made into a model, both to argue for state support of fundamental research and to set in place new managerial research strategies in the biomedical sciences (Swann, 1988). The mobilization model was also put to use in conducting large-scale clinical trials for drugs and in introducing screening programs for diseases defined in terms of abnormalities at the level of molecules.

Physical approaches to the study of biological materials promised fundamental knowledge of life processes and of disease causation, and attracted important funding in the postwar years. This support, together with the commercialization of a range of instruments for the analysis of large biological molecules and subcellular structures, encouraged many scientists to engage in this area of study (Elzen, 1986; Kay, 1988; Rasmussen, 1996). When a group of researchers at the California Institute of Technology, applying the new technologies, identified a new type of hemoglobin in the blood of sickle cell anemia patients, they announced their findings by heralding sickle cell anemia as the first "molecular disease". Sickle cell hemoglobin and its simple mechanism of inheritance stimulated hopes of finding molecular explanations for other genetic disorders. This focus on molecular causes of inherited diseases, however, was not necessarily followed by therapeutic breakthroughs.

More recently, new protein and nucleic acid technologies have opened up novel prospects of intervention at the molecular level. The biotechnology industry has been a major player in these developments since it was set in place in the late 1970s. The production of genetically engineered drugs such as insulin and growth hormone has been supplemented by the development of research tools and diagnostic technologies for use in the laboratory and the clinic. These new strategies and alliances have had, and continue to have, far-reaching and novel consequences. From the perspective of this book, however, they represent but one particular articulation of multiple practices centred on molecules.

FRAGILITIES AND POWERS OF MOLECULARIZATION

At the same time as drawing attention to processes of molecularization, the book shows up the contingencies and fragilities of these processes. As many contributors show, molecular strategies were frequently contested and often aborted. Both the powers and fragilities of molecularization become particularly clear when strategies of molecularization are followed over a long period, as is done, in different ways, in the opening and closing chapters of the volume.

Following the variegated career of one molecule, colchicine, Jordan Goodman shows that the successful use of colchicine as a molecular probe in cellular processes was bound to a particular context and could co-exist with, or be shelved in favour of, non-molecular uses of the same substance by other people in other places. Ilana Löwy shows that multiple efforts in the immunotherapy of cancer, from the 1910s onwards, did not lead to the development of effective anti-cancer drugs. Interferons, regulatory molecules of the cell which were hailed as "miracle drugs" in the 1980s, were no exception. According to Löwy, the ideal of molecularization nonetheless survived and found its articulation in a powerful network which linked scientists, industrialists and oncologists.

Molecularization has rarely converged successfully on etiology, diagnosis and therapy conjointly. Insight into the etiology of a disease in molecular terms, or its diagnosis at the molecular level, does not necessarily imply that a molecular cure is available, feasible, economic, or indeed desirable. On these grounds, and for reasons of professional autonomy, clinicians have often resisted the introduction of molecular diagnostics. In other cases, the therapeutic effect of chemically well-defined molecules is exploited without their mode of action being well understood at the molecular level. Salicylic acid, sold under the brandname of Aspirin since 1899, is a striking example of a substance which is widely used for the prevention and treatment of a variety of symptoms, but whose molecular mechanism of action remains elusive. Even when the mechanism of molecular therapies is believed to be well understood, as in the case of insulin treatment of diabetes or a phenylalanine-free diet for phenylketonuria patients, unwanted "side-effects" and the varying susceptibility of patients to the same treatment indicate that the explanations are not complete (Paul and Edelson, this volume). Furthermore, the successful use of molecular drugs always depends on complex cultural set-ups and routines, as shown by Marks in this volume for the case of the contraceptive pill.

Especially in the clinic, molecular practices have existed side by side with physiological or organismic approaches, often developed in response to the obvious limitations of molecular or, more generally, reductionist practices. Holistic approaches do not necessarily negate the achievements of reductionist explanations, but are often integrated with the latter into more comprehensive frameworks of medical knowledge and practice (Lawrence and Weisz, in press).

Molecularization, then, has not been the only strategy pursued in biomedicine, nor has it been uniformly successful in its achievements. Even strong material, technological and political support provided no guarantee for success defined in terms of criteria dictated by the molecular strategies themselves, at least in the short and medium term. For instance, Nixon's cancer campaign of the 1970s, which injected large sums of money into molecular strategies, did not produce a "cure for cancer", and gene therapy for molecular disorders such as sickle cell anemia continues to be a promissory note instead of clinical reality (Gaudillière; de Chadarevian, this volume). Furthermore, there have been trade-offs between successes of molecular approaches and losses on other levels. Funds and commitments channelled into screening for the inherited metabolic disorder phenylketonuria, for example, were at the same time channelled away from social programs for help with the practical problems faced daily by mental patients and their families (Paul and Edelson, this volume). The apparent power and success of molecular strategies today, then, depend on the criteria of evaluation which society has set for biomedical research and public health policies.

STRATEGIES OF MOLECULARIZATION

We now outline some major themes of molecularization, as they are presented in this volume. In treating these themes, it is our aim not only to highlight particular links between different chapters, but to delineate further the concept of molecularization as a historiographic tool. The manifold connections between the different themes are important: far from taking the laboratory, the clinic, industry, politics (or, indeed, molecules) as given, fixed entities, we want to indicate how they and their interrelations were created and transformed around molecular practices.

Instruments and Technologies

Technologies played a crucial role in the representation, definition, measurement, analysis, production and circulation of molecules, big and small. Not surprisingly, therefore, instruments and technologies form a pervasive theme of the chapters of this volume.

The development of technologies designed to measure and monitor molecules in the body became a central concern for those who reduced bodily functions to the interplay of molecules and understood diseases as disturbances of chemical processes. For Olga Amsterdamska the study of these technologies opens the way to investigating the interactions between biochemists and clinicians in a particular institution, the Rockefeller Hospital, created at the beginning of the century with the explicit aim of fostering a new scientific medicine, or a medicine based on physicochemical approaches to diseases. Van Slyke's apparatus for the measurement of bicarbonate concentration in the blood of diabetic patients, developed at this hospital, was based on a gasometric principle. Yet the choice of bicarbonate as an indicator of diabetes as well as the attempts to demonstrate the utility of the apparatus were based on and propagated a biochemical definition of the disease. While his name has been attached to the gasometric apparatus, Van Slyke, during his career at the Rockefeller Hospital, developed a whole series of instruments aimed at monitoring the concentration of metabolites in the blood and urine of patients. These same parameters were used as a means of classification for pathological conditions. A crucial requirement for these instruments was that they were simple and could be applied as routine diagnostic procedures.

Angela Creager stresses the importance of technologies in enhancing the authority of molecular approaches in the acrimonious debate between "molecularists" and "colloidalists" in the 1920s and 1930s. Where, as in the case of proteins, classical chemical methods for establishing molecular purity often failed, instruments such as ultracentrifuges provided important new criteria for classifying substances as molecules. Ironically, the ultracentrifuge and the Tiselius apparatus, which played a crucial role in shifting the balance of influence between the two groups of scientists involved in the debate, were designed by colloid chemists with the aim of studying the properties of colloidal substances. The same instruments were later employed to define quality control standards for therapeutic molecules.

In the interwar years only a few prototypes of these instruments existed, conferring a privileged status on those who had access to them and experience and skill in handling them. The production of simpler commercial models which set in after World War II made the technologies more generally available. With governments and foundations prepared to allocate funding for costly technologies, a large number of laboratories acquired these instruments. By the 1960s, they became part of the standard armamentarium of all biochemistry and molecular biology laboratories.

Jean-Paul Gaudillière emphasizes the impact of wartime biomedical mobilization on the establishment of certain research practices around big technologies. The ultracentrifuge and the electron microscope were deployed as instruments for purifying and visualizing viruses in the wartime project on influenza. The same instrumental practices, Gaudillière argues, were deployed in the construction of cancer viruses and later in scaling up cancer virus research. The availability of commercial instruments was crucial for the standardization of such biomedical practices.

As in the case of viruses, so the construction of abnormal hemoglobins and the diseases associated with them was inextricably linked to the availability of technologies aimed at a molecular level of analysis. Soraya de Chadarevian follows the career of abnormal hemoglobins from Pauling's first representation of sickle cell hemoglobin in the Tiselius apparatus to the identification of new hemoglobin variants through chromatographic and electrophoretic fingerprinting techniques. Like Amsterdamska, she points to the importance of simple commercial instruments for the applicability of new diagnostic techniques in the clinic as well as in surveys in the field. She also discusses the role of molecular models as a technology for the representation and analysis of molecular structure-function relationships.

De Chadarevian, like Goodman in his biography of colchicine, describes how molecules (or collections of molecules) themselves are turned into laboratory or diagnostic tools. In the course of its career, colchicine turned from a therapeutic agent to a tool for the study of mitosis and later into one for tracing microtubuli in the cell. Similarly, hemoglobin and its variants were used as tools for genetic analysis on a molecular level and more generally for structure determinations.

Finally, several authors deal with large-scale trials, surveys and screening programs as technologies aimed at tracing the presence of specific molecules in the body, or assessing the therapeutic efficacy of molecules. Gaudillière discusses the systematic screening for mammalian and human cancer viruses under the contract system established by the U.S. National Cancer Institute. Löwy explicitly links the cooperative organization of large-scale clinical trials, such as those for new cancer drugs organized by the Cancer Chemotherapy National Service Center in the U.S., to the experience of wartime research. In her chapter Löwy notes that "war research 'acclimatized' physicians with large-scale projects, coordinated actions and the need to suspend part of their expertise-based authority to achieve collective goals". This climate favored the development of cooperative enterprises such as the clinical trials of penicillin and streptomycin.

Also Lehmann's first experience with large-scale surveys dated, as de Chadarevian points out, from his service in the Royal Army Medical Corps in India during the war, where he studied the causes of anemia which commonly affected the troops. He acquired further experience in large-scale blood sampling as a colonial medical officer in Uganda, where he continued to study the causes of anemia and became interested in the distribution of sickle cell hemoglobin in African populations.

Phenylketonuria (PKU) screening marked a shift from social to biological, in this case biochemical, intervention in mental disorders. As Diane Paul and Paul Edelson show in their chapter, this shift was supported and welcomed by government agencies as well as by parents' associations. Despite many unassessed problems, such as the high incidence of false positive results, the unclear effects of dietetic treatment in healthy children and open questions relating to the prevention of the disease through phenylalanine-free diets, PKU screening was heralded as a model for the control of molecular diseases.

Lara Marks demonstrates the problems involved in applying a reductionistic model of women as hormonal bodies in the first clinical trials of the pill in Puerto Rico. Unlike experimental animals, the women participating in the trial could not be kept in a cage and monitored for 24 hours per day. Instead, the organizers of the trials had to rely on the cooperation of women in following certain procedures. Despite the obvious limitations of the reductionistic approach, the trials did support the efficacy of the contraceptive pill in preventing pregnancy in women.

The Laboratory and the Clinic

Molecules created new alliances between the clinic and the laboratory. Several chapters in this volume take this as their central theme.

Laboratory techniques are not simply transferred or applied to the clinic. This point is made most forcefully by Amsterdamska, but also by other authors. Amsterdamska argues that the biochemists working at the Rockefeller Hospital did not merely take up "questions" posed by clinicians in their research. Clinical practices and concerns shaped all aspects of the biochemical work in this institution, as Amsterdamska illustrates in particular with respect to the development of monitoring techniques of bodily chemical functions. The construction as well as the meaning and the testing of these technologies relied on the close collaboration between clinicians and biochemists. Co-production rather than transfer of technology was therefore at work.

Focussing on hemoglobin as a model molecule for structural and genetic work, de Chadarevian takes up a similar point. She argues that, long before the advent of new therapies based on molecular genetic technologies, laboratory scientists derived material resources, functional knowledge and legitimation for their work from the clinic. Like Löwy in her chapter, she argues that molecular technologies (in protein research, genetics, or immunology) often failed to be translated into new therapeutic practices. Through the use of these technologies for diagnostic procedures or for monitoring patients, however, clinicians and laboratory scientists came to share a common molecular culture and provided mutual legitimation for their respective practices. Interferon or hemoglobin can thus be seen as "boundary objects" or molecules which, despite their different uses, link clinicians and laboratory scientists.

Following colchicine's life of multiple identities, Goodmann shows how molecules can move in and out of the clinic in a series of moves over time. Starting life as a therapeutic substance for the treatment of gout and rheumatic disorders, colchicine, after an uneven career as a research tool, came back to the clinic as a drug with a known, if disputed, molecular mechanism of action to which many disorders respond.

Together, these contributions render problematic the simplistic view of a one-way transfer of scientific advances from the laboratory to the clinic, and help to sharpen our view of how a molecular culture was constructed by interactions between both sites.

Molecules and Industry

Molecularization, in the laboratory and the clinic, also had important industrial dimensions. Industries were crucially involved in the production and circulation of standardized molecular reagents and drugs, as well as in the commercialization of instruments for the representation, measurement and analysis of molecules. Commercial interests often promoted molecular approaches. At the same time, industrial procedures of production were also set in place in laboratories, and academic laboratories could keep a control function on industry through patenting procedures or monitoring industrial quality standards. The links between laboratory and industry were thus manifold and complex.

The chapters in this volume follow the industrial trail from the role of pharmaceutical industries in molecularization in the 1920s through the experiences of World War II to the rise of biotechnology industries in the 1970s. In her contribution focussing on early vitamin research, Harmke Kamminga shows how industrial interests in selling vitamin preparations and vitamin-enriched foods promoted structural investigations of vitamins in the interwar period. In this light, she discusses the transformation of a modest British food company, through its participation in vitamin research, into the well-known pharmaceutical company Glaxo.

Under the special conditions of World War II, the links between research laboratories and industries became tighter. This has been shown in the literature in relation to the production of penicillin. Angela Creager focusses on a much less well known military biomedical project, Edwin Cohn's Plasma Fractionation Project at Harvard. She emphasizes not only the close links between Cohn and the collaborating pharmaceutical industries, but the industrial style in which Cohn ran his own laboratory. The laboratory built a pilot plant for plasma fractionation which became the model for subsequent plants. The Harvard pilot plant served as the central production site for serum albumin as a substitute for full blood until commercial production was well underway in 1943. To cover the demands for serum albumin posed by the military, Harvard researchers had to operate the plant 24 hours per day. The management style Cohn imposed on the laboratory paralleled industrial as well as military structures. Following the fate of Cohn's empire after the war, Creager shows that Cohn chose to continue his close association with pharmaceutical industries, although professing his ideal of 'fundamental' research. Cohn exercised a control function over the industrial production of blood fractionation products by patenting his procedures and imposing biochemical quality standards which could only be met in his laboratory.

In his chapter on the virus cancer program in postwar America, Gaudillière stresses that the success of the virus theory of cancer causation relied on an extended network of contractors associated with the National Cancer Institute. These contract industries organized and guaranteed the production, circulation and control of biochemical and immunological tools for tracing viral cancer genes. They thus played a crucial role in the standardization of research practices, linking the laboratory as well as the clinic to industry. The contract system, Gaudillière argues, imposed a managerial style on scientific research which was modelled on industrial research practices.

Löwy traces new developments in the complex field of immunotherapy of cancer to the entrance of the biotechnology industry as a new player in the game. Mass-produced genetically engineered interferon could not fulfill the high-pitched promises of a cure for cancer, but was nevertheless integrated into routine treatment of cancer patients. A further point implicit in this and other chapters is that the industrial production of certain molecules rather than others both steers and limits choices of molecules for use in the clinic and the laboratory.

Policies and Politics

At important junctures, molecular approaches in research and to problems of health and disease were actively promoted by government agencies, foundations and associations representing patients. This support was often crucial in directing the course of molecular strategies, and in overturning resistances to them.

As discussed by Amsterdamska, in founding the Rockefeller Hospital in 1910, the American patrons intended to promote "scientific medicine" based on a chemical notion of the body and its malfunctions. Clinical practice of this kind required a closer interaction of biochemists and clinicians, which the Hospital was to foster. The Hospital remained a privileged place for the kind of collaborations envisaged by its founders, but it also provided the training ground for many future professors in the most important medical schools in the U.S. Training and educational policies more generally were obviously crucial in promoting molecular approaches in biomedicine.

Kamminga shows that the Royal Society Food (War) Committee set up in Britain during World War I and new government funds controlled by the Medical Research Council (MRC) were instrumental in boosting research on the structure and function of molecules. The role of the MRC in promoting new biophysical research after World War II is discussed by de Chadarevian.

As shown most explicitly by Creager, World War II, by means of government interventions and military interests, presented novel opportunities for the circulation of molecules. Cohn could build up his empire for blood products only through the special conditions of war. Once the war was over, he could maintain the central position of his laboratory in this field only by complying, at least in words if not in deeds, to the new government policy of "fundamental research".

Gaudillière and Paul and Edelson also stress the role of government programs and health policies in promoting molecular approaches in biomedicine in postwar America. Nixon launched a large cancer virus program rather than supporting other approaches to the cancer problem or to public health more generally. Behind the decision of the program stood a "policy-making community" which encompassed the American Cancer Society, the National Cancer Institute, big pharmaceutical companies and the U.S. Congress, linked together by experts circulating between these institutions. The molecularization of cancer etiology, Gaudillière concludes, had important political content and was carried by large biomedical networks formed in the aftermath of World War II.

Government organizations, together with parents' associations, were also instrumental in the shift from social to biochemical approaches to mental disorders. Paul and Edelson remind us that, before reliable tests for PKU or other metabolic disorders were available, a government report stressed the importance of metabolic screening. When Guthrie developed his bacteriological test to measure blood phenylalanine levels in newborn babies, he negotiated directly with the National Association for Retarded Children and with the Children's Bureau of the Department of Health, Education, and Welfare. Before the first field trials ended, and notwithstanding existing doubts concerning the diagnosis and treatment of PKU, the Bureau was already committed to a national screening program.

As the chapters collected in this volume indicate, molecularization had many meanings and involved many actors. Van Slyke's attempt to design instruments to measure and monitor chemical functions in the body was not the same as Pauling's or Lehmann's search for molecular causes of diseases. The isolation, purification and analysis of molecules in the laboratory was not the same as the mass-production and commercialization of molecular drugs. Molecularization involved research laboratories, industries, the clinic and the policy arena. Molecular strategies created new links, or radically changed existing links, between these domains, affecting society at large. The concept of molecularization allows us to link these changes together and to emphasize that the molecular-genetic culture in biomedicine known today rests on more than advances in molecular biology and biotechnology: it was preceded by, and embedded in, technically and socially much more diverse biomedical practices and links centred on strategic molecules.

NOTES

1 The term "biomedical sciences" was introduced in the 1960s to justify the National Institutes of Health's diversion of funds, allocated for the study of human health and disease, into basic molecular biology (Feinstein, 1995, p. 289). For convenience, we occasionally use the term with reference to earlier periods.

2 The authors differ in the connotations they give to these terms, either implicitly or explicitly. Lily Kay (1993b) has used "molecularization" to refer to practices centred on macromolecules, especially proteins, from the 1930s. Robert Olby (1990) locates the beginnings of the "molecular revolution" in biology around the same period, also restricting this to studies of macromolecules. Susan Wright's (1994) term "molecular politics" refers to the recombinant DNA debates of the 1970s. Hans-Jörg Rheinberger (1995) has used "molecularization" specifically in relation to the new genetic medicine.

3 On the dispute between molecular biologists and biochemists, see Abir-Am (1992) and de Chadarevian and Gaudillière (1996).

4 The pioneering study of laboratory practices in the biomedical field is Latour and Woolgar (1979); see further especially Clarke and Fujimura (1992). Alliances between laboratory science and medicine are examined in Pickstone (1992) and Löwy (1993), while industrial connections are at centre-stage in Gaudillière and Löwy (1997), Liebenau (1987) and Swann (1988). Studies of science policy in relation to the life sciences and medicine

are presented in Abir-Am (1982), Austoker and Bryder (1989) and Kohler (1991).

5 The term "magic bullet" was first used by Ehrlich in the context of serum therapy. He assumed that the protective substances, being produced by the human body itself, were both highly specific and non-toxic. Synthetic molecules as used in chemotherapy, however, would not have perfect affinity for the cellular protoplasm and would inevitably be toxic to some degree (Marquardt, Chapter 12). The term "magic bullet" was revived in relation to penicillin, even if it did not conform strictly to Ehrlich's criteria. It was later used more loosely for "wonder drugs" in general, and retrospectively also for the synthetic drug Salvarsan (Bäumler, 1965).

REFERENCES

P. Abir-Am, "The Discourse of Physical Power and Biological Knowledge in the 1930s: A Reappraisal of the Rockefeller Foundation's 'Policy' in Molecular Biology" *Social Studies of Science* (1982), 12: 341–382. (See also "Responses and Replies" *Social Studies of Science* (1984), 14: 225–263).

P. Abir-Am, "The Biotheoretical Gathering, Trans-disciplinary Authority and the Incipient Legitimation of Molecular Biology in the 1930s: New Perspectives on the Historical Sociology of Science" *History of Science* (1987), 25: 1–70.

P. Abir-Am, "The Politics of Macromolecules: Molecular Biologists, Biochemists and Rhetoric" *Osiris* (1992), 7: 210–237.

J. Austoker and L. Bryder (eds), *Historical Perspectives on the MRC* (Oxford: Oxford University Press, 1989).

E. Bäumler, *In Search of the Magic Bullet: Great Adventures in Modern Drug Research* (London: Thames and Hudson, 1965).

M. Bliss, *The Discovery of Insulin* (Toronto: McLelland and Stewart, 1982).

A. Cambrosio, D. Jacobi and P. Keating, "Ehrlich's 'Beautiful Pictures' and the Controversial Beginnings of Immunological Imagery" *Isis* (1993), 84: 662–699.

S. de Chadarevian and J.P. Gaudillière, "The Tools of the Discipline: Biochemists and Molecular Biologists" *Journal of the History of Biology* (1996), 29: 327–330; and all contributions in this special issue.

A.E. Clarke and J.H. Fujimura (eds), *The Right Tools for the Job* (Princeton, NJ: Princeton University Press, 1992).

S.S. Cohen, "The Biochemical Origins of Molecular Biology: Introduction" *Trends in Biochemical Sciences* (1984), 9: 334–336.

A. Cunningham and P. Williams (eds), *The Laboratory Revolution in Medicine* (Cambridge: Cambridge University Press, 1992).

B. Elzen, "Two Ultracentrifuges: A Comparative Study of the Social Construction of Artefacts" *Social Studies of Science* (1986), 16: 621–662.

A.R. Feinstein, "The Crisis in Clinical Research" *Bulletin of the History of Medicine* (1995), 69: 288–291.

J.S. Fruton, *A Skeptical Biochemist* (Cambridge, MA: Harvard University Press, 1992).

J.P. Gaudillière and I. Löwy (eds), *The Invisible Industrialist* (London: MacMillan, in press).

S.M. Horrocks, "The Business of Vitamins: Nutrition Science and the Food Industry in Inter-war Britain" in H. Kamminga and A. Cunningham (eds), *The Science and Culture of Nutrition, 1840–1940* (Amsterdam: Rodopi, 1995), pp. 235–258.

H.F. Judson, *The Eighth Day of Creation* (London: Jonathan Cape, 1979; repr. Penguin, 1994).

H. Kamminga and M.W. Weatherall, "The Making of a Biochemist. I: Frederick Gowland Hopkins' Construction of Dynamic Biochemistry" *Medical History* (1996), 40: 269–292.

L. Kay, "Laboratory Technology and Biological Knowledge: The Tiselius Electrophoresis Apparatus, 1930 to 1945" *History and Philosophy of the Life Sciences* (1988), 10: 51–72.

L. Kay, *The Molecular Vision of Life: Caltech, the Rockefeller Foundation, and the Rise of the New Biology* (Oxford: Oxford University Press, 1993a).

L. Kay, "Life as Technology: Representing, Intervening, and Molecularizing" *Rivista di Storia della Scienza* (1993b), series II, 1: 85–103.

D.J. Kevles and L. Hood (eds), *The Code of Codes: Scientific and Social Issues in the Human Genome Project* (Cambridge, MA: Harvard University Press, 1992).

R.E. Kohler, "Walter Fletcher, F.G. Hopkins and the Dunn Institute of Biochemistry" *Isis* (1978), 69: 331–355.

R.E. Kohler, *Partners in Science: Foundations and Natural Scientists 1900–1945* (Chicago: University of Chicago Press, 1991).

H. Krebs, *Biochemistry, "Molecular Biology" and the Biological Sciences* (London: Biochemical Society, 1969).

S. Krimsky, *Biotechnics and Society. The Rise of Industrial Genetics* (New York: Praeger, 1991).

B. Latour and S. Woolgar, *Laboratory Life: The Construction of Scientific Facts* (London: Sage, 1979).

C. Lawrence and G. Weisz (eds), *Greater than the Parts: Holism in Biomedicine, 1920–1950* (New York: Oxford University Press, in press).

T. Lenoir, "A Magic Bullet: Research for Profit and the Growth of Knowledge in Germany Around 1900" *Minerva* (1988), 26: 66–88.

J. Liebenau, *Medical Science and Medical Industry: The Formation of the American Pharmaceutical Industry* (London: MacMillan, 1987).

J. Liebenau, "The MRC and the Pharmaceutical Industry: The Model of Insulin" in J. Austoker and L. Bryder (eds), *Historical Perspectives on the MRC* (Oxford: Oxford University Press, 1989), pp. 163–180.

J. Liebenau, "Paul Ehrlich as a Commercial Scientist and Research Administrator" *Medical History* (1990), 34: 65–78.

I. Löwy (ed.), *Medicine and Change: Historical and Sociological Studies of Medical Innovation* (Paris: John Libbey, 1993).

M. Marquardt, *Paul Ehrlich* (London: Heinemann, 1949).

A.M. Moulin, *Le dernier langage de la médecine: Histoire de l'immunologie de Pasteur au Sida* (Paris: Presses Universitaire de France, 1991).

P. Neushul, "Science, Government and the Mass Production of Penicillin" *Journal of the History of Medicine and Allied Sciences* (1993), 48: 371–395.

R.C. Olby, "The Molecular Revolution in Biology" in R.C. Olby, G.N. Cantor, J.R.R. Christie and J.M.S. Hodge (eds), *Companion to the History of Modern Science* (London/New York: Routledge, 1990), pp. 503–520.

R.C. Olby, *The Path to the Double Helix* (Seattle: University of Washington Press, 1974; rev. edn Dover, 1995).

N. Oudshoorn, *Beyond the Natural Body: An Archeology of Sex Hormones* (London: Routledge, 1995).

J.V. Pickstone (ed.), *Medical Innovations in Historical Perspective* (London: MacMillan, 1992).

H.-J. Rheinberger, "Beyond Nature and Culture: A Note on Medicine in the Age of Molecular Biology" *Science in Context* (1995), 8: 249–263.

N. Rasmussen, "Making a Machine Instrumental: RCA and the Wartime Origins of Biological Electron Microscopy in America, 1940–1945" *Studies in History and Philosophy of Science* (1996), 27: 311–349.

J.P. Swann, "The Search for Synthetic Penicillin during World War II" *British Journal for the History of Science* (1983), 16: 154–190.

J.P. Swann, *Academic Scientists and the Pharmaceutical Industry. Cooperative Research in Twentieth Century America* (Baltimore, MD: Johns Hopkins University Press, 1988).

M. Teich, "Science and Food During the Great War: Britain and Germany" in H. Kamminga and A. Cunningham (eds), *The Science and Culture of Nutrition, 1840–1940* (Amsterdam: Rodopi, 1995), pp. 213–234.

M.W. Weatherall and H. Kamminga, "The Making of a Biochemist. II: The Construction of Frederick Gowland Hopkins' Reputation" *Medical History* (1996), 40: 415–436.

P. Weindling, "Scientific Elites and Laboratory Organisation in *Fin de Siècle* Paris and Berlin: The Pasteur Institute and Robert Koch's Institute for Infectious Diseases Compared" in A. Cunningham and P. Williams (eds), *The Laboratory Revolution in Medicine* (Cambridge: Cambridge University Press, 1992), pp. 170–188.

S. Wright, *Molecular Politics: Developing American and British Regulatory Policy for Genetic Engineering, 1972–1982* (Chicago: University of Chicago Press, 1994).

CHAPTER 1

Plants, Cells and Bodies: The Molecular Biography of Colchicine, 1930–1975

Jordan Goodman

It is rarely appreciated that biologically active molecules used in therapeutic regimes often have important strategic uses outside the body. These uses may precede, coincide with or proceed from their therapeutic function (Goodman, 1998). Or, as in the case of colchicine, the molecule that is the subject of this chapter, the non-therapeutic uses may form a complex, non-linear relationship with the therapeutic ones.

This chapter is organized in the form of a biography covering the period from the 1930s until the 1970s. Colchicine was known and used therapeutically for centuries but during this period it also became a powerful biochemical research tool in a variety of cognitive and practical sites. In the hands of the researchers who worked and associated themselves with it, colchicine was not a substance but a tool for the investigation of molecular phenomena. In other words, through the practices of the botanists and cell biologists, colchicine molecularized. By the end of this period, the now molecularized colchicine reentered therapeutic space, extending its repertoire to new parts of the body.

By studying an object over time, the biographical approach yields some useful insights (Kopytoff, 1986; Lubar and Kingery, 1993). One of these is that the object's identity at any time is the result of a process of negotiation and exchange of claims of ownership. The object's identity provides its classification and categorization. Focusing on the object and its identity is therefore tantamount to analyzing the process of negotiation and exchange and its social and cultural construction. Once this is seen dynamically, as has been attempted in this chapter, the biography of the object becomes a social and cultural history not just of the object but of all those human (and non-human) actors involved in the process. The colchicine molecule was transformed between 1930 and 1975 as the claims made on it, as those who spoke on its behalf and as its network partners and alliances changed substantially over this period. Colchicine can therefore be assigned the role of a historical probe as it winds its way temporally through sites of theory and practice (Pickering, 1995).

PROLEGOMENON: (RE)ENTER COLCHICINE: MOLECULE OF THE YEAR, 1937

Colchicine was isolated in 1820 from the meadow saffron, *Colchicum autumnale*, by Pelletier and Caventou in Paris as part of their grand interest in extracting and identifying the active principle of plant medicines in use in the European materia medica (Lesch, 1984). They identified the extract as veratrine and this remained the case until 1833 when Philippe Geiger proposed that the alkaloid of Pelletier and Caventou was not veratrine but something else which he called colchicine (Cook and Loudon, 1952, p. 264).

The meadow saffron and colchicine extract were used therapeutically principally for the treatment of gout though many other uses were indicated, including rheumatic disorders (Rodnan and Benedek, 1970; Wallace, 1973). Pharmacological studies and the elucidation of colchicine's chemical structure were carried out mostly by European researchers until the early 1930s. Most of the work concerned the pharmacology of toxicity, a topic that found favour as much in pathological studies as in forensic medicine (Eigsti and Dustin, 1955, pp. 155–158; Cook and Loudon, 1952, pp. 291–293). Confirmation came of colchicine's intense toxicity, an observation well-established in folk medicine.

These investigations were conducted on the whole organism. In the first decade of the twentieth century, however, a few researchers began to probe colchicine's action more deeply, particularly on blood cells. In 1908, W.E. Dixon and W. Malden published the results of their studies of colchicine in which they reported the effect of colchicine on bone-marrow function, and principally on the changes in white-blood cell count over the course of the treatment (Dixon and Malden, 1908). Their research showed an unequivocal steep rise in the count of white-blood cells over the first twelve hours following injection of colchicine, followed by a slower fall. Earlier, Dixon remarked that colchicine affected mitosis in blood cells and suggested that the effect "should not be regarded as specific to the leucocytes, but rather a type of action which goes on to a greater or less degree in other tissues of the body" (Dixon, 1906, p. 96). The "wandering cells of the blood" were a more convenient object for studying this phenomenon.

Dixon and Malden's work did not, apparently, receive critical attention, though their observation that colchicine increased the white-blood cell count became a standard item in medical textbooks (Eigsti and Dustin, 1955, p. 189). Moreover, hardly any further research was done on colchicine itself. The little that did appear concerned itself predominantly with earlier issues of toxicology and chemical structure (Eigsti, 1947, pp. 65–114).

Colchicine may be said to have been placed back on the laboratory shelf. In 1924, Windaus proposed a structural formula for the molecule, but even this did not stimulate much interest in colchicine (Windaus, 1924; Cook and Loudon, 1952, pp. 273–290; Dewar, 1945). Colchicine was not so much 'black-boxed' as lacked interest. Moreover, and this may be the key factor, colchicine was not central to the activities of any scientific social world; no groups identified with it, nor privileged its operations or mechanisms, nor rejoiced in its social and cultural achievements.

Interestingly and perhaps surprisingly, Dixon's statement did not cross over into other centres of rather intense research activities on the nature of cellular movements, most especially mitosis. Why Dixon's observations on mitosis did not align with similar observations in other laboratories is an intriguing problem in contingency but which cannot be pursued here (Fujimura, 1987).

One of the most intense research efforts where issues of mitosis and the effects on it by chemical and other agents (X-rays, radium and heat) were paramount, was in the investigation of malignant tumours (Dyer, 1949; Patterson, 1987, pp. 63–64). Several different approaches and agents of choice characterized this research activity but mitosis was the central concern (Ludford, 1936). Dyestuffs in particular were prominent for several reasons. First, Paul Ehrlich's resurrection of chemotherapy at the beginning of the twentieth century was founded on his insights about and experience with dyestuffs as research tools — indeed, he turned his attention to the problem of cancer after his success with the anti-syphilitic drug Salvarsan (Travis, 1989; Patterson, 1987, p. 64). Secondly, it was widely believed that some dyestuffs were biologically active (Lesch, 1995); and finally, and of substantial importance, dyestuffs were relatively easy to obtain and were, at the time, the only source of standardized, mass-produced, bioactive molecules (Travis, 1993; Marsh and Simpson, 1927).

The scene shifts to the pathology laboratory at the University of Brussels which concentrated much of its efforts on understanding tumour development at the cellular level (Dustin, 1946). Headed by Albert Pierre Dustin from 1929 to 1939, the work of the laboratory focused on what Dustin himself termed 'poisons caryoclasiques' or mitotic poisons. In one of the first papers published by Dustin as head of the laboratory, he suggested the use of mitotic poisons as tools of cytological analysis (Dustin, 1929). The poisons included several well-known dyestuffs and dimethyl-arsenic acid, or sodium cacodylate, arsenic compounds being a favourite candidate group for a chemotherapeutic approach to cancer (Dyer, 1949, pp. 42–45). Sodium cacodylate quickly became Dustin's chief chemical agent in tumour research (Ludford, 1936, p. 418).

One of Dustin's medical students, F. Lits, became interested in replicating Dixon and Malden's work with colchicine, after two German researchers, Beck and Levison, independently experimented with it on dogs and humans (Lits, 1934). His interest was broader than that of Dixon and Malden and included the effect of colchicine on lymphatic tissue, especially on the thymus, an organ that Dustin himself had investigated. Lits confirmed Dixon and Malden's observations on blood cells. More importantly, he concluded that colchicine caused an upsurge in mitosis in all the examined tissues of the laboratory mouse and that this happened with low concentrations; yet the effect was between 1000 and 2000 times stronger than with sodium cacodylate (Lits, 1934, p. 1422).

Immediately, and not surprisingly, sodium cacodylate was abandoned by Dustin and his co-workers and the laboratory homed in on colchicine. Dustin published his first colchicine paper very rapidly after Lits' own results were known (Dustin, 1934).

Judging from the publications of Dustin's laboratory in the few years immediately after 1934, it is clear that colchicine not only became the research group's special molecule but that the laboratory came to be identified with it. The work of the laboratory appears to have headed in two directions that ultimately met. On the one hand, Dustin concentrated on the effect of colchicine on cancerous tumours in animals (Dustin, 1934); some of his co-workers, particularly Laszlo Havas, on the other hand, pursued colchicine's effect on plant material, work that began in 1936 and appeared in publications (Eigsti and Dustin, 1955, p. 17; Eigsti, 1947, pp. 66–67). By 1937, the research group was able to establish that colchicine had a dramatic effect on mitosis in both plants and animals. In other words, they had established that the cellular impact of colchicine was general: its potential as a generalized research tool was immense.

Colchicine was not, however, exclusively the domain of the Brussels Pathology Laboratory. In 1935, for example, E.C. Amoroso, of the Royal Veterinary College in London, published a short notice in *Nature* providing evidence that colchicine inhibited tumour growth in experimental mice and dogs (Amoroso, 1935). In this article, Amoroso referred to earlier observations on colchicine's effects on tumours going back to 1927 as well as to continuing work at the Wellcome Bureau of Scientific Research. Amoroso's published work, in addition to his personal role in alerting attention to colchicine, appears to have been the stimulus behind a great number of experiments carried out at the laboratories of the Imperial Cancer Research Fund in London by Reginald J. Ludford (Ludford, 1936, p. 421). Ludford appears to have been a key figure in trying to understand the chemotherapeutic approach to cancer (Austoker, 1988). His own work during the 1920s was concerned primarily with understanding malignant cytology: in the early 1930s he began to turn his attention to the cytological effects of various well-known cellular poisons, including dyestuffs, urethane and sodium cacodylate (Ludford, 1929, 1933, 1935). Ludford was well-acquainted with the experiments and results of Dustin's laboratory though one cannot tell from the published results when this began. Certainly by 1936, Ludford was fully aware of the Brussels group's published work.

In the same year and according to various published versions of the events, the colchicine work of Dustin's laboratory came to the attention of G.M. Smith of the Yale University Medical School when he attended the Second International Cancer Congress in Brussels (Eigsti and Dustin, 1955, p. 19). For an endocrinology laboratory working on estrogenic hormones, colchicine was a perfect chemical research tool because of its direct effect on mitosis. Smith's experimental results on colchicine and the hormone theelin were reported in a short notice in 1936, shortly after his return to the United States: a longer, more detailed, report was published in the following year (Allen *et al.*, 1936, 1937). 1936 also saw the publication on the effect of colchicine on mitosis in the regenerating rat's liver (Brues, 1936).

These papers, together with those of Dustin and his co-workers, pushed colchicine as an analytical tool for research in physiology, in endocrinology and in the problem of cancer. By using colchicine, mitotic growth was made more visible and facilitated analytical procedures. As the understanding of mitosis was a problem common to some research workers in these areas, it is not surprising that colchicine itself came to circulate as a generalized tool of analysis.

At the end of 1936, however, colchicine was largely confined to experiments with animals and tissue cultures. This was soon to change, and dramatically so. As mentioned above, in 1936 Dustin's laboratory broadened its investigations to include plant material. At the same time, the endocrinological colchicine research at Yale came to the attention of geneticists at the Carnegie Institution at Cold Spring Harbor, out of which came two complementary and striking pieces of research (Eigsti and Dustin, 1955, p. 19). In both cases it was reported that colchicine produced polyploidy, a doubling of the number of chromosomes, in plants. The first published results to appear were by Albert Blakeslee who had spent many years working on mutations in the datura plant, and who had become interested in the artificial induction of mutation by X-rays, radium, heat and latterly chemicals (Blakeslee, 1937; Blakeslee and Avery, 1937a). Blakeslee's inclusion of colchicine was, as he himself acknowledged in print, entirely due to O.J. Eigsti, who, during his tenure of a position in the same department, had observed polyploidy in the cells of *Allium* roots (Blakeslee and Avery, 1937b, p. 395). Eigsti's research was reported a year after that of Blakeslee (Eigsti, 1938). Also from the Yale Medical School, colchicine was passed to D.F. Jones of the Connecticut Agricultural Research Station in New Haven and ultimately to Bernard Nebel of the New York State Agricultural Experiment Station who was made aware of some of colchicine's properties at the annual meeting of the American Association for the Advancement of Science in 1937 (Dermen, 1940, p. 603). Nebel's first results were reported late in 1937 and then in a more extensive form in the following year (Nebel, 1937; Nebel and Ruttle, 1938). As in the case of Blakeslee and Eigsti, it was colchicine's remarkable ability to induce polyploidy that was the subject of this research.

According to one of the observers of colchicine's emergence out of relative obscurity, it was the 1937 paper by Blakeslee and his co-worker Amos Avery that started the 'colchicine fad' (Wellensiek, 1939). The number of publications concerned with colchicine in 1938 and 1939 together were 10 times as many in number as in 1937 (Dermen, 1940, pp. 600–603). The colchicine work extended throughout the world, in laboratories throughout Europe, the United States, India and Japan, to mention but the most active sites (Eigsti, 1947). A mere fad it was not. There certainly was a flurry of research publications but more profound things were happening. A clue to what this was is contained in the final paragraph of a seminal paper by Blakeslee and Avery:

> The ability to induce chromosome doubling, therefore, is of importance to practical as well as to theoretical genetics. With increasing knowledge of the constitution of chromosomes and of methods whereby their structure and behavior may be altered, there arises an opportunity for the genetics engineer who will apply knowledge of chromosomes to building up to specifications forms of plants adapted to the surroundings in which they are to grow and suited to specific economic needs (Blakeslee and Avery, 1937b, p. 410).

In other words, colchicine, in the voice of Blakeslee and his colleagues, was a useful, commercially valuable molecule producing practical, marketable results in plant breeding, an activity of immense economic and political importance (Kloppenburg, 1988; Elliot, 1958; Palladino, 1990, 1996; Harwood, 1994). Plant breeding was big business and highly organized with substantial state, industrial and scientific interests. Papers from plant geneticists accounted for the bulk of the colchicine fad.

STRAND 1 — THERE IS NO CALM

Impassioned but measured, the biologist Laszlo Havas reeled off a letter from Brussels on 25 November 1939 to the editors of the *Journal of Heredity*. "As you may remember", he wrote, "I have written to you at the beginning of this year suggesting, among others, that the pioneer work on the effects of colchicine on plants, with which your JOURNAL seems to have credited American workers, was in fact initiated in Belgium: by the writer and his collaborators (1936), and in France: by Gauvadan and co-workers (1937)" (Havas, 1940).

Despite the calm tones, Havas was incensed. He wanted to set the record straight. As he told the editors, he came away from his recent trip to the Seventh International Genetical Congress, held in Edinburgh in late summer of the same year, feeling that his work and that of his colleagues in Europe was being ignored (Crew, 1939; Emerson, 1940). The colchicine section of the congress was extraordinarily attended, he remarked, but no mention was made of the European research on its biology, physiology and pathology. Instead, the audience heard papers which described colchicine as a chemical for inducing polyploidy (chromosome doubling), in plants. As Havas so lyrically put it: "And thus, in splendid isolation suspended betwixt earth and heaven, like the coffin of Mahomet, *polyploidy* was offered alone to the worship of the conference".

Setting the record straight was the chief, but not the only reason for writing to the editors. There was also the issue of national pride. "… colchicine is not a Miraculous Presence suddenly revealed, — between longitudes 65 and 125 west of Greenwich", Havas railed and continued: "It is the result of more than two decades of laborious and steadfast work consecrated by Professor A.P. Dustin and his collaborators in Brussels to the discriminating analysis of … effects of a long list of "nuclear" poisons in relation to pathological conditions in animals."

If there was anyone that Havas had in his sights between longitudes 65 and 125 west of Greenwich, it was the botanist Albert Francis Blakeslee then in the Department of Genetics, Carnegie Institution of Washington located in Cold Spring Harbor, New York. Judging from his publications, Blakeslee's interests were focused on the genetics of Datura, especially on variability resulting from the plant having extra chromosomes (Blakeslee, 1939, p. 171). It is not that Blakeslee, in print at least, was taking full credit for the 'discovery' of colchicine: indeed, in an important and widely cited paper, he acknowledged his place in the colchicine story, alongside that of other researchers (Blakeslee, 1939, p. 163). Interestingly, Blakeslee's short coverage did not include the name of Havas. Nevertheless, it was not Blakeslee but others who attributed primacy in the use of colchicine in botanical investigations to him. Statements such as "After Blakeslee and Avery had made the sensational discovery of the induction of chromosome doubling by colchicine treatment" were typical in the attribution notices appearing in the immediate aftermath of Blakeslee's colchicine researches (Levan, 1938, p. 471).

Blakeslee's omission of Havas may or may not have been a cause of the latter's irritation. But the issue of priority disturbed him deeply even though he played this down in his letter to the editors of the *Journal of Heredity*. "… the work of your countrymen", he wrote, "to commence with that of Professor Blakeslee has my fullest admiration and the question of priority becomes secondary in comparison with the intrinsic value of the results which they have obtained. It would be also ungrateful of my part not to acknowledge what my own work owes to their researches." (Havas, 1940, p. 117). After the end of the war, Havas returned to the priority issue in an article comprehensively listing the colchicine literature up to 1947, on the occasion of the tenth anniversary of the use of colchicine in plant genetics, and prepared by the botanist O.J. Eigsti. Havas' contribution to Eigsti's article was short but abruptly to the point. "Owing to the persistence of the erroneous attribution of the discovery of the effects of colchicine in plants the following list, showing the chronological order of the dates of the *first* publications, on the different aspects of the problem, may not be superfluous" (Eigsti, 1947, p. 66). This was followed by a list of five papers, beginning with Havas' for 27 February 1937 in *Nature* and ending with Blakeslee's for 13 September 1937 in *Comptes rendus de l'Académie des Sciences de Paris*.

Others got in on the act to set the record straight. The first was the botanist Haig Dermen of the Bureau of Plant Industry at the U.S. Horticultural Station in Beltsville, Maryland. In an article surveying colchicine polyploidy, Dermen unequivocally linked the researches on colchicine by Blakeslee and others to the Brussels laboratory of Albert Pierre Dustin, particulary the work of F.J. Lits and Havas (Dermen, 1940, pp. 601–603). In the same year, Dermen and his coworker, Nellie Brown, also published an article on the effect of colchicine on plant tumours and Havas' name appeared as the first word of the text (Dermen and Brown, 1940, p. 197). Two years later J.M. Krythe and S.J. Wellensiek published their survey of colchicine with their version of events supporting Dermen's own (Krythe and Wellensiek, 1942). If this were not enough, when O.J. Eigsti and Pierre Dustin published their story in the first book-length treatment of colchicine that appeared in 1955, they discussed Blakeslee's contributions in the context of research connections across the Atlantic and highlighted the importance of direct contact between European and American researchers (Eigsti and Dustin, 1955, pp. 16–21).

A storm in a teacup? What was at stake in this part of the colchicine story?

ENTR'ACTE

Colchicine was fundamentally transformed as a result of the work done by Dustin, Havas, Lits, Blakeslee, Eigsti and many others. That much is clear from the account given above. The transformation, however, had multiple components. In the first place, the experimental results of these researchers brought colchicine out of the obscurity it had fallen into during the first few decades of the 20th century. This period of obscurity helped to eclipse the specifically medico-legal interest in colchicine which dominated the 19th century since its chemical isolation (Eigsti and Dustin, 1949). Secondly, the results established colchicine as a powerful biochemical probe for the understanding of mitosis and therefore cellular growth; and as a remarkable chemical method for inducing mutations in plants, a process with substantial theoretical and practical applications. But most importantly for the purposes of the theme of this volume, colchicine was molecularized by this work. The identity of colchicine fundamentally altered from that of a substance with effects on the organism to that of a molecule with specific cellular and sub-cellular actions. Colchicine's rapid translation from a pathology laboratory in Brussels to a plant genetics laboratory in Cold Spring Harbor, New York, was only made possible by decontextualizing it; by thinking of colchicine as a biochemical research tool for investigating cellular and sub-cellular processes. Though the actions of colchicine were not at this stage charcaterized in terms of interactions between it and specific molecules in the cell, the possession of a molecular tool was important in thinking molecularly of biological processes.

The purpose of this short interlude in the narrative is not only to clarify the meaning of molecularization as it is used here but also to make a statement on method to which it is related. Here, I wish to make explicit the particular approach I am taking to the unravelling of the history of colchicine as a research tool and as a therapy, the two main ways in which colchicine has been used since at least the 1930s; and to highlight the implications that the colchicine story has for our understanding of the circulation of standard research materials and their impact on the construction and organization of scientific knowledge.

Turning now to the first of these. It is my intention in this chapter to analyse colchicine in biographical terms, that is, to follow colchicine, as Bruno Latour has advised on many occasions, through society, through its resonances and sites of intersection over a specific period of its life (Latour, 1988). The biographical approach to colchicine should make colchicine's twists and turns, its movements within and between sites of consumption and sites of knowledge production especially visible (Marcus, 1995). By sites of consumption, as the title of the paper suggests, I mean the plants, cells and human bodies that were subjected to colchicine's molecular mode of action: sites of knowledge production refers specifically to the laboratories where colchicine became a standard research tool.

By following colchicine through the resonances and intersections, no *a priori* judgment is being made about its temporal pathways and its identity. Indeed, this approach allows colchicine to be seen in its full array of multiple identities (Graham, 1994; Smith, 1994): as an inducer of polyploidy; an arrester of cell division; as a chemotherapeutic cancer agent; and as a reliever of pain and inflammation, in many differing and unrelated contexts.

Igor Kopytoff insists on the usefulness of a biography of objects. To him, not only can the life history of an object be biographized, but through this process it can reveal more than itself.

> … a person's social identities are not only numerous but often conflicting, and there is no clear hierarchy of loyalties that makes one identity dominant over the others … the drama of personal biographies has become more and more the drama of identities … The drama, in brief, lies in the uncertainty of identity … The biography of things in complex societies reveals a similar pattern … an eventful biography of a thing becomes the story of the various singularizations of it, of classifications and reclassifications in an uncertain world of categories whose importance shifts with every minor change in context. As with persons, the drama here lies in the uncertainties of valuation and of identity (Kopytoff, 1986, pp. 89–90).

Lest it be misunderstood, this approach does not amount to the reification of the object. Quite the opposite. It highlights the dual and interactive agency of human and non-human actors and thus resonates with the insights of Michel Callon and Bruno Latour and their insistence on free association between humans and non-humans (Callon, 1995; Latour, 1987).

The biographical approach also throws light on issues raised in discussions about the standardization and circulation of research materials, an area of increasing interest in the sociology and history of science, technology and medicine (Clarke and Fujimura, 1992; Fujimura, 1992; Oudshoorn, 1990). Studies on these issues have generally argued that standardization and the circulation of standard research materials are crucial in mapping locally generated knowledge onto generalized and stable sites (Knorr-Cetina, 1981; Fujimura, 1988; Fujimura, 1992). Put another way, standardization and circulation play a vital role in bringing disparate and separate sites of knowledge production and social worlds together to present a unified field. Standardization and circulation can therefore be understood as disciplinary tools, in the Foucauldian sense. The biographical approach emphasizes contingency; the chance meeting at sites of social and cognitive action of human actors who respond to a common but contingent language (Rorty, 1989); the unpredictable strategies and practices of humans and non-humans. The biographical approach makes no claims about the relationship between circulating research materials and social worlds; research materials can be thoroughly decontextualized in their migration from one research site to another, bringing with them no previous case history to affect their incorporation in a new social space (Fujimura, 1991). In these migrations, new identities of the research material are shaped.

This excursion into colchicine's biography is, in fact, a study of just such a phenomenon. Colchicine's multiple identities appear as individual strands winding their way through time and space. Now and then, there are sharp moments of resonance and intersection but without long-lasting effects. What colchicine is at any time is the result of how and by whom it is being handled. Categories and classifications are unstable, unpredictable and mutable.

STRAND 2 — PLANTS: POLYPLOIDY IN EXPERIMENT AND PRACTICE

In their monograph on colchicine, O.J. Eigsti and Pierre Dustin Jr credited the latter author's father with the identification of colchicine as a research tool. "Colchicine", he wrote, "enables the detection of the otherwise invisible state of preparedness to mitosis" (Eigsti and Dustin, 1955, p. 216). Mitosis was regarded as the key to understanding growth and the chief advantage of colchicine over other techniques for observing and understanding mitosis was that it slowed down the process of cellular change. With this, quantitative studies of mitotic processes could proceed more effectively as could the identification of the most active sites of growth (Eigsti and Dustin, 1955, p. 214).

The utility of colchicine in exploring problems of growth was substantially increased by the appearance of the 1936 and 1937 papers by the team of Allen, Gardner and Smith at Yale University Medical School. This research as well as much of it that followed can be directly related to the increasing availability in this very period of purified and standardized sex hormones (Oudshoorn, 1990; Oudshoorn, 1994). The congruence of two standard research materials more than adequately explains the surge of research activity in endocrinology (Eigsti and Dustin, 1955, pp. 249–254).

For endocrinologists, colchicine was a molecule that performed an enabling function in the service of observation, particularly in studies of the growth effects of hormones: in this sense, the molecule *prepared* the object under investigation for observation and analysis. For plant geneticists, however, the importance of the colchicine molecule was of another order of magnitude.

As mentioned earlier, in his major contribution to the application of colchicine in plant genetics, Blakeslee was quick to point out both the theoretical and practical implications of this new technique. The various techniques for inducing mutations in plants before the discovery of the colchicine method were generally disappointing (Imperial Bureau of Plant Genetics, 1936; Bates, 1939; Dermen, 1940, pp. 603–607; Randolph, 1941). In a comparative study of the heat method and the colchicine method carried out in 1940, the researchers reported that polyploidy was induced in one out of every two cotton plants when treated with colchicine: in heat-treated plants the ratio was a meagre 1:600 (Beasley, 1940). Moreover, it would appear that colchicine helped dispense with large and cumbersome machinery that had been used to induce polyploidy (Eigsti and Dustin, 1955, p. 274).

While he was visiting the United States in 1937, Albert Levan, the Swedish geneticist, was shown experimental results of colchicine and polyploidy by Bernard Nebel of the New York State Experimental Research Station. Upon his return to the cytogenetic laboratories of the Swedish Seed Association in Svalöf, Levan embarked on a large number of experiments concerned with induced polyploidy and colchicine (Levan, 1948). His first published account of the colchicine method clearly and unambiguously supported Blakeslee and Avery's results and expectations. He also provided a succinct account of colchicine's unique biochemical effectiveness:

> A great many facts go to prove that colchicine will constitute the agent long sought for, which, without detrimental secondary effects and with full certainty, induces polyploidy. From a practical point of view colchicine has many advantages over the methods hitherto prevalent for increasing the chromosome number. Of importance is, for instance, that the colchicine action is specific and total. No other disturbances besides the inactivation of the spindle apparatus are observed … The completeness of the action of colchicine is remarkable. All the mitoses taking place within the exposed tissue turn into c-mitoses and bring about chromosome doubling (Levan, 1938, pp. 484–5).

This much explains its experimental functions, but Levan goes on to emphasize the remarkable nature of colchicine's impact on the cell after it has arrested its mitosis. Here is the key to colchicine's practical uses:

> Of importance is also the reversibility of the inactivation process of the spindle. After the conclusion of the exposure the spindle apparatus recovers and is able to function in the normal manner. Cells with increased chromosome numbers are thus brought back to normal mitoses after their period of rest (Levan, 1938, p. 485).

Polyploids occur naturally. Plant breeders had known for a long time that polyploid plants, especially crop plants, were often larger and more vigorous than the diploids (Elliott, 1958, p. 146). According to G.H. Bates of the Farm Institute in Stafford, England, Mendelian techniques of producing new and improved varieties of crop plants had become exhausted by the 1930s. In his words: "A point has been reached where the hybridizer can only wait for some new mutation to turn up" (Bates, 1939, p. 315).

Blakeslee literally ran with the colchicine method. Two years after his breakthrough paper with Avery, he was able to report that he and his co-workers had been able to induce polyploidy in 65 different kinds of plants (Blakeslee, 1939, p. 163). Obviously, this kind of result boosted his confidence in colchicine. Indeed, his confidence spilled over into a genetically engineered view of the future of plants. In a passage he described how in evolutionary time, fertile hybrids with double the number of chromosomes appeared spontaneously and these had been valued and preserved for their superior qualities by our prehistoric human ancestors. These plant varieties included wheat, oats, tobacco and cotton. "Now" he added, referring directly to colchicine-induced polyploidy, "we … no longer have to wait ages for the *chance hybridization* (my emphasis) between species and the later spontaneous doubling of their chromosomes in order to secure such superior varieties. *We can now make them up to order* (my emphasis)" (Blakeslee, 1939, p. 170).

Made up to order may not have been how plant researchers saw their work, but it is the case that a huge number of plants were given the colchicine treatment following Blakeslee's example. By the time they wrote their overview of the subject in 1942, Krythe and Wellensiek of the Institute for Plant Improvement in Wageningen, Holland, were able to report that polyploidy had been induced by colchicine in 243 species (Krythe and Wellensiek, 1942). After reviewing the enormous literature generated from 1937, they were also able to state, however, that results of artificial polyploidy had been largely over-confident and over-optimistic. They concluded that the value of newly induced polyploids was small and, in practical terms, restricted to ornamental plants, where large flowers, for example, have an appeal: new polyploids were best considered as material for breeding (Krythe and Wellensiek, 1942, p. 37). Even Albert Levan, who it may be recalled was one of the first great enthusiasts of colchicine, revised his initial exuberant evaluation, and put it quite simply thus: 'the mere production of new polyploids must therefore be regarded only as the first step in a laborious breeding work' (Levan, 1942, p. 245).

Artificial polyploidy appears to have had a chequered experience in the decades following World War II. One particularly pessimistic opinion was voiced by G. Ledyard Stebbins in 1956 and, according to the author of one overview of the subject published many years later, Stebbins was, if anything, too optimistic (Stebbins, 1956; Dewey, 1980, p. 448). Even so, research activity on induced polyploidy remained brisk throughout this period, though as an analysis of research output during the 1970s showed, there were substantial national differences: the Soviet Union and India together accounted for over half of the world's interest in this subject (Dewey, 1980, p. 449). As far as the techniques for inducement are concerned, however, colchicine held its place as the preferred method (Hagberg and Åkerberg, 1961, p. 40).

One small, but interesting aside. The investigation of polyploidy was not confined to plants. Other life forms, such as bacteria, fungi, yeast and algae, were subjected to induced polyploidy, with varying results (Eigsti and Dustin, 1955, pp. 120–125). More startlingly, experiments were also conducted on inducing polyploidy by colchicine on animals, including frogs, rabbits and pigs (Eigsti and Dustin, 1955, pp. 58–59, 380–383; Pincus and Waddington, 1939). These too were largely failures. Obviously, the animal experiments on colchicine-induced polyploidy were conducted after 1937, one of the first results being published in 1939. Yet, somehow, the American press seized on Blakeslee and Avery's paper and extrapolated its plant results to that of other life forms. The editor of the *Journal of Heredity*, which had published Blakeslee and Avery's results, felt compelled to rebut the statements made in the press. This is what he wrote:

> The tidal wave of weirdly misleading publicity distributed by the Hearst newspapers on November 21, has doubtless aroused hopes and fears that have no foundation whatever. Colchicine is not a "growth elixir", and evidence is lacking as to its effects on animal chromosomes, so that doctors may still make their calls safe from attack by giant rats, and ladies are in no danger of being gobbled up by caterpillars suddenly gone carnivorous (Editor, 1937, p. 411).

Still, he could not deny the implications of what Blakeslee and Avery reported. He continued thus:

> It offers no Magical Royal Road to the production of new varieties of plants and animals. In spite of the cold water which must be thrown on over-enthusiastic popularizations, the effect which colchicine has on doubling the chromosomes in the cells makes it unquestionably one of the most important genetic discoveries of the year (*Ibid.*, 1937, p. 411).

Colchicine became a molecular black box among plant geneticists very quickly. Within no more than a couple years after 1937 it had become a standard research tool and material. As a highly effective biochemical tool, colchicine had earned a substantial reputation, but there seemed to be no need to look into the molecule any further.

STRAND 3 — CELLS: MICROTUBULES AND THE TUBULIN-BINDING SITE

As we have seen, 1937 was a turning point in colchicine's biography. Plant geneticists and plant breeders firmly enrolled it in their social world. The practice of plant breeding produced its own knowledge about colchicine; when and how much to use, which plants to use it with, how experimental polyploids could be used in comprehensive plant breeding programs, etc. How colchicine acted on the cell with the resultant doubling of chromosomes was of little interest in this area of practice. In 1947, Pierre Dustin, in reviewing work done on mitotic poisons over the past decade, reported that, though colchicine's action was undoubtedly mitotic arrest at metaphase, there was still little understanding (or agreement) of its precise mode of action: what was going on in the cell during colchicine's presence? Where was colchicine's site of action? (Dustin, 1947). Dustin himself proposed investigating the possibility that colchicine inhibited enzymes involved in mitosis and that the nucleus might be a rewarding place to look for a site of action (Dustin, 1947, p. 797).

Judging from the published work of the 1950s, while the problems of mitosis and other issues in cell development continued to interest researchers, colchicine faded from the spotlight. Controversies about colchicine's mode of action were not so much settled as attracted little attention. Colchicine's chief endowment to the biology of the cell was the designation c-mitosis (colchicine mitosis) a term used to describe the type of mitotic inhibition produced by molecules such as colchicine (Wilson and Bryan, 1974, p. 39). Moreover, the surge in interest in cancer chemotherapy in the late 1940s and 1950s, which might have reintroduced colchicine into the mainstream of cell biology, given its early association with experimental tumour research, actually marginalized it further largely because the clinical experiences were very discouraging (Goodman, 1995).

In addition, colchicine as a biochemical probe of the internal structure of the cell was being superseded by other less obtrusive and generalized techniques. Seen through the gaze of the 1950s, these techniques, especially electron microscopy, held sway and helped establish cell biology as an independent academic discipline (Bechtel, 1990; Bechtel, 1995) It is difficult to find a place for colchicine in the controversies and discoveries made by the electron microscope during the 1950s. Yet, by a strange turn of events, colchicine appeared a decade later right in the middle of a ferment of knowledge making: its biochemical moment had come once again.

For all its theoretical promise in the 1930s and 1940s, the electron microscope posed serious problems when applied to biological specimens. There were many obstacles that turned on the question of whether the techniques themselves were producing artefacts: there was concern that the electronic beam might damage the biological material; there was the difficulty of obtaining thin enough slices of tissue without causing damage; and finally, there was the issue of the fixation method (Bechtel, 1990, pp. 561–562). These problems notwithstanding, experimental results flooded the literature. It was not until the end of the 1950s that the technical and cognitive difficulties were worked out in interesting and complex ways.

Some of the electron microscopic results seemed to contradict earlier and well-regarded conclusions. For example, one electron microscope study on the cell's spindle concluded that there were no fibres (Rozsa and Wyckoff, 1951). This opposed other results which, using an improved light microscope, postulated the existence of these spindle fibres (Inoué, 1952). This controversy was not resolved until the mitotic apparatus itself was isolated and stabilized at about the same time (Mazia and Dan, 1952).

Electron microscopy studies of mitosis flourished in the late 1950s. Spindle fibres were observed, though the fixation method used in this decade destroyed some of the more ubiquitous fibres (Wilson and Bryan, 1974, p. 22). The precise structure of these fibres was not clearly articulated until 1963 when they were given the name microtubules (Slautterback, 1963; Olmsted and Borisy, 1973). By that time the fixative had been changed and electron microscopy now revealed the true ubiquity of these intracellular structures: the microtubules were the central element of the spindle fibre (Sabatini *et al.*, 1963; Porter, 1966, pp. 317–319).

Looking back over the years of research into the structure and organization of the cell made visible by the electron microscope, Daniel Mazia, whose team isolated the mitotic apparatus, commented on the matter-of-fact nature of the discovery of microtubules. "Once we knew we were looking for microtubules, they knew we were looking for them and there they were. By the early 60s we had many of the answers about the comportment of spindle fibres in mitosis that had been sought for decades" (Mazia, 1975, p. 10). One does not have to agree with Mazia's rather lyrical view of discovery: the main point of it is, however, that the revelation of the existence of microtubules, while not resulting from any single radical change in instrumentation or theory, nevertheless had a profound impact on cell biology.

Reenter colchicine. By the early 1960s, Eigsti and Dustin's exhaustive monograph on and bibliographical guide to colchicine in virtually all of it aspects had become out-of-print. As the biologist Lynn Margulis put it: "many are unaware of the enormity of the classic literature assembled by these investigators on the biological effects of colchicine and its analogs in the premicrotubule days" (Margulis, 1973, p. 344). In effect, colchicine had been forgotten and neglected: its use as experimental method had become so routine, so unquestioning that its presence in the laboratory was taken for granted. This is not that unusual in science: many other cases of neglect have been documented (Sacks, 1995).

There was, however, one exception. In their early electron microscopic studies of the cell's ultrastructure, Etienne de Harven and Wilhelm Bernhard observed the existence of spindle filaments and described what would later be called microtubules (De Harven and Bernhard, 1956, p. 394). These spindle filaments were, they argued, tubular in shape and the centrioles, from which the filaments extended, were also made of a tubular construction. Of particular note is that they were able to confirm earlier studies on the mitotic effects of colchicine, but now at a sub-light microscopic level: they strongly urged researchers to pursue colchicine's mode of action at this level and specifically for the study of cilia and flagella (De Harven and Bernhard, 1956, pp. 395–397). To put it another way: de Harven and Bernhard were recommending that colchicine might yet prove to be a most valuable biochemical companion to the electron microscope.

These were certainly lone voices in cell biology during the 1950s. De Harven, in particular, had published a few papers on colchicine during the 1950s and on several occasions had teamed up with Pierre Dustin: he conducted some of his research in Pierre Dustin's laboratory in Brussels together with Wilhelm Bernhard (Dustin, 1963; Dustin, 1980, p. 60).

There is little evidence in the published record that de Harven and Bernhard's colchicine recommendations were acted upon directly: the biological, as opposed to the chemical, research output on colchicine in the period until the early 1960s simply continued the concerns raised in Eigsti and Dustin's monograph: readers of a review of colchicine published in 1960 looking for biological research information were referred back to this work without further comment (Wildman, 1960). Unfortunately, it is not possible to account entirely for colchicine's dramatic reappearance in 1965. There are some clues, however small, to its pathway and its human supporters in the years leading up to 1965. One of these was Stephen Malawista at the Yale University Medical School. His concern was to try to understand the relationship between colchicine's effect on mitosis which, of course, was well-established and demonstrated at the microscopic and submicroscopic level by Sahnya Inoué in 1952, and its therapeutic effects in acute gouty arthritis (Malawista, 1965; Inoué, 1952). His suggestion that colchicine had a specific cellular target with which it bound, leading to the destruction of "organized, labile, fibrillar systems concerned with structure and movement" certainly provided sufficient ground for a molecular understanding of colchicine's mechanism of action: his further suggestion that colchicine's function as a biochemical tool of analysis could be extended beyond its limited use in studying the mitotic apparatus to the broader problem of the biology of the cell was particularly powerful (Malawista, 1965, p. 381).

That was one strand. The other was provided by Edwin Taylor's biophysics laboratory at the University of Chicago which had been interested in issues relating to the formation of the spindle and the effects on it by chemical agents (Taylor, 1959). At some time in the early 1960s the research group turned their attention to colchicine and this became their investigative molecule. Taylor's work with colchicine until 1963 culminated in a 1965 article in which he described his work with radioactive colchicine and concluded that colchicine entered the cell and bound to a structural component of the spindle: its effect on mitosis was direct (not through the action of an enzyme, for example) and he suggested that colchicine may bind to a microtubule element, perhaps a protein, of the spindle (Taylor, 1965, p. 159). Taylor continued this line of enquiry for the next few years and in 1967 published work he had conducted with Gary Borisy. In two important articles they claimed that the colchicine binding site was a protein, a subunit of the microtubule: colchicine's binding to this subunit prevented it from assembling the microtubules and this was colchicine's mechanism of action (Borisy and Taylor, 1967, p. 547). By 1968, this microtubule subunit was confirmed as being a protein, isolated and named tubulin (Shelanski and Taylor, 1967; Wilson and Friedken, 1967; Weisenberg *et al.*, 1968; Mohri, 1968).

A powerful link had been constructed between colchicine and microtubules at the molecular level — the binding of colchicine to a very common cellular protein. Colchicine entered a new phase as a diagnostic tool: colchicine guided researchers through the cellular system, in static and dynamic configurations (Wilson *et al.*, 1974). As Daniel Mazia stated in the first microtubule conference held in 1974:

> ... the way to our present knowledge ... was through the invocation of much older cytological knowledge of the action of colchicine ... the very fact of colchicine binding was all-important; it provided a biochemical handle on some component of the cell that is crucial for mitosis; biochemical experience tells us we can not purify anything without such a handle (Mazia, 1975, pp. 10–11).

That deluge of publications and the research activity it reflected, witnessed in the aftermath of the discovery of polyploidy by colchicine, was repeated with the establishment of this biochemical handle. Between 1966 and 1980, for example, publications reporting on colchicine and microtubules increased from an average of 10 per year at the beginning to over 100 per year at the end (MEDLINE 1966–1980).

By the time that colchicine's identity had become associated with cellular processes at the molecular level, the cell itself had become a site of investigation, drawing into it researchers, techniques, materials and questions from a variety of disciplines and approaches. In contrast to its earlier experience in the 1930s when colchicine was enrolled into and absorbed by plant geneticists as their singular molecular tool, in the 1960s and 1970s, colchicine adopted an identity, thanks to its cellular actions, that could articulate with a wide variety of social actors: biologists, pharmacologists, cancer specialists, arthritis and rheumatism researchers. The colchicine-binding domain on tubulin became a kind of industrial standard, a cellular space that could be shared with and negotiated by heterogeneous actors: for example, it acted as a gateway for investigations into possible anti-cancer drugs whose mechanism of action was, in some manner, anti-mitotic (Creasey, 1974; Wilson, 1975a; Wilson, 1975b; Schiff and Horwitz, 1981; Wilson and Jordan, 1994).

STRAND 4 — BODIES: GOUT, CANCER AND MORE

We have already seen in the previous strand of colchicine's biography that the cellular investigation of gout provided an important launch pad for colchicine's explosive emergence out of the biological wilderness. In this, the final strand, we need to return to colchicine as a therapeutic molecule and to follow its vicissitudes as a chemical agent for the body as body.

Nineteenth-century researchers, as pointed out earlier, were preoccupied with colchicine's toxicity. This would continue to be a concern in the twentieth century but in a different guise. The work of Dustin and his colleagues in Brussels during the 1930s opened up the possibilities for understanding toxicity in a different way. One of the insights was, of course, that arrested mitosis due to the presence of colchicine might define toxicity. Unfortunately, the problem was not so easy to solve because, in the body, the issue of toxicity depended on how colchicine was administered and for what reasons (Eigsti and Dustin, 1955, pp. 175–196).

Once colchicine had been isolated from the colchicum plant, it appeared in nineteenth-century pharmacopoeias as the treatment for gout, taking the place of the crude plant preparation (Rodnan and Benedek, 1970, pp. 145–152; Wallace, 1973). For the first half of the twentieth century, colchicine was the preferred therapy though there was little idea why it worked nor much interest in the question itself. It was considered an efficacious substance. Its molecular structure and the biological and botanical laboratory investigations went unnoticed by clinicians. Gout itself had become unproblematic, not in the sense that its diagnosis and treatment had become a matter of routine. Gout had disappeared as a disease. Abraham Cohen, Chief of Arthritis Clinics at Philadelphia General and Jefferson Hospitals, said as much in a paper he read to the College of Physicians in Philadelphia in 1936 (Cohen, 1936). According to Cohen, gout was "conspicuously absent from the literature" and, in the cases he reviewed, he discovered that most of the patients had been diagnosed with arthritis — incorrectly according to him (Cohen, 1936, p. 488).

Gout as the forgotten disease was a theme to which others turned. W.H. Herrick and T. Lloyd Tyson, both clinicians at the College of Physicians and Surgeons of Columbia University, for example, wrote a scathing attack on their fellow practitioners for ignoring gout. Interestingly, they argued that the recent newly awakened interest in infectious and degenerative forms of arthritis diverted attention away from those of chemical origin, namely gout (Herrick and Tyson, 1936, p. 487). They too prescribed colchicine. Ten years later, gout, the forgotten disease, was brought to the attention of the medical profession in an article by three clinicians in Boston. They too were incensed about the misdiagnosis of acute gout as arthritis and reproved the profession for causing needless suffering from a disease that was well-known and for which there was a simple and very effective therapeutic remedy (McCracken, Owen and Pratt, 1946).

In the introduction to a symposium on gout whose proceedings were published in 1957, John Talbott referred in his opening paragraph to the reasons why specific maladies should be given attention or not, as the case might be (Talbott, 1957). Among his list of factors, he singled out serendipity, fashion and coincidence in addition to the availability of drugs and investigative tools. Talbott stated that gout had come back into fashion in the last two decades though, irritatingly, he gave no further details and no explanation. Leaving aside the issue of fashion, it is probably true to say that gout reemerged as a disease partly because of the spectacular effects of steroid preparations in the treatment of rheumatoid arthritis made at the Mayo clinic in 1949 and reported by Philip Hench and his co-workers (Marks, 1992). Hench had been closely involved in gout treatment at the Mayo clinic since the 1920s and it is not surprising therefore that cortisone and ACTH (pituitary adrenocorticotropic hormone) should have been tried on gout soon after the success with rheumatoid arthritis (Hartung, 1957, p. 206; Friedlander, 1951).

Pessimism over the understanding of gout was the predominant mood of researchers in the 1950s. Leon Sokoloff opened a review of the pathology of gout as it was understood in 1957 with a very down-beat tone (Sokoloff, 1957). His review could not offer much in the way of furthering the understanding of the pathogenesis of acute gouty arthritis. He was, however, able to support Philip Hench's assertion that gout was not caused by uric acid, as was commonly thought, since colchicine suppressed the condition while not affecting uric acid (Sokoloff, 1957). The treatment of gout also showed little advance: colchicine was "still the best 'home' remedy"; combination treatments with colchicine and ACTH had some success (Bartels, 1957).

By the early 1960s, gout and colchicine were partnered, but the nature of the relationship between the two was unknown. As Stanley Wallace of the State University of New York College of Medicine stated: "the mechanism of action of colchicine in acute gouty arthritis is as poorly understood as the pathogenesis of the attack itself" (Wallace, 1961, p. 446). Wallace had had a long acquaintance with colchicine and was one of the few researchers who had devoted considerable attention to chemical approaches to the disease. Judging from his publications, he was also well acquainted with the colchicine literature that had emerged from Dustin's laboratory in the 1930s and all those who were motivated by it. Interestingly, however, in his review of colchicine's clinical pharmacology, he pointed out that, as far as he could tell, there was no direct relationship between its by then well-known anti-mitotic effects and its anti-gout activity (Wallace, 1961, p. 446).

In the early 1960s, colchicine was firmly absorbed into the practical and theoretical work of plant geneticists and, to a lesser extent, organic chemists interested in its molecular structure and problems of synthesis, and cell biologists studying mitosis (Wildman, 1960; Fell and Ramsden, 1967). We have the situation, therefore, that colchicine's multiple identity in botany, chemistry, biology and rheumatology, was, in a sense, frozen. As far as gout was concerned, there was little to expect from colchicine studies.

As it turned out, however, rheumatology's interest in the mechanism of action of colchicine derived not so much from insights beyond the rheumatology laboratory and clinic as from within. Many of the details were worked out in laboratories in Philadelphia and the National Institute of Arthritis and Metabolic Diseases. What emerged was both surprising and enlightening: surprising in that sodium urate crystals were implicated in the painful gouty attack — they had been banished from the explanatory model as we have already noted; and enlightening in that the acute attack was now understood as an inflammation reaction provoked by the presence of the crystals in the tissue; and that the entire gouty episode was complicated and consisted of several developmental cycles (Seegmiller and Howell, 1962; Seegmiller *et al.*, 1962; Seegmiller, 1965; Malawista, 1968).

Attention on colchicine focused rapidly on its anti-inflammatory agency. Synovial fluids of gouty patients contained concentrations of polymorphonuclear leucocytes (PMN), a class of cells whose metabolic activity was implicated in a vicious circle of chemical activity, including crystallization, leading to acute gouty attacks (Malawista, 1968, p. 191). Colchicine, it was argued, targeted PMN, broke the vicious circle and, therefore, relieved the attack. Colchicine's mechanism of action could now be revealed thanks to these insights and discoveries by cell biologists of microtubules and the evidence of colchicine's inhibition of the microtubule function. Malawista and his co-worker Klaus Bensch pulled these strands together in 1967 when they reported that PMN contained microtubules which, after treatment with colchicine, disappeared (Malawista and Bensch, 1967). Colchicine, as they argued, had a common mechanism for its anti-inflammatory and anti-mitotic effects. Malawista concluded his exploration of colchicine's dual effects by stating that, "it seems likely that colchicine, beyond its unique and prestigious career in cytogenetic studies and in gout, will continue to be a biophysical tool of considerable power" (Malawista, 1968, p. 195). Interestingly and obviously unknown to Malawista, E.C. Amoroso had in 1935 drawn a common link between the pain in acute attacks of gout, cell mitosis and tumour growth (Amoroso, 1935).

One can be forgiven for thinking that aligning rheumatology to cell biology through the common vector of a molecule and its mechanism of action would close the argument about gout and colchicine. That was not the case and the major attack on Malawista's representation of colchicine came, in fact, from Stanley Wallace who, as we have seen, had identified with colchicine in the 1950s. At the Second Conference on Gout and Purine Metabolism, Malawista and Wallace occupied distant positions, not on colchicine's effectiveness, but on its cellular target. Malawista stuck to PMN whereas Wallace, drawing on evidence from an analogue of colchicine which was as effective as colchicine in treatment but did not inhibit mitosis, suggested another candidate cell (Malawista, 1975; Wallace, 1975). As late as 1990, a review of colchicine could report that its mechanism of action was still not fully understood and that the conflict raised between Malawista and Wallace had not been resolved (Levy *et al.*, 1991, p. 198).

Stephen Goldfinger, who had been involved in research at the National Institute of Arthritis and Metabolic Diseases with co-workers of Stephen Malawista, reported in 1972 that he had successfully administered colchicine prophylactically for the relief of the disabling arthritic-like attacks of familial Mediterranean fever (Goldfinger, 1972). His rationale for trying colchicine was based on the observation that the effusion of patients with familial Mediterranean fever contained PMN, and on his knowledge of Malawista's insights into the mechanism of colchicine. Interestingly, the major review of familial Mediterranean fever which had been published in 1967 dismissed colchicine as being of no value for the episodic attack form of the disease (Sohar *et al.*, 1967). Goldfinger's recommendation was to administer colchicine as a chronic, not an emergency, therapy. The medical team that had earlier dismissed colchicine swiftly undertook a controlled trial of colchicine using Goldfinger's recommendation, and this proved highly successful (Zemer *et al.*, 1974). Goldfinger's own double-blind trial was reported at the same time and also showed colchicine to be highly effective (Dinarello *et al.*, 1974).

Since 1974, many other conditions and diseases have been shown to respond in varying degrees to colchicine. They include cirrhosis of the liver, various skin disorders such as psoriasis and scleroderma, malaria and eye problems (Levy *et al.*, 1991, p. 201).

Colchicine's identity as a medical therapy has been much more problematic than as a biochemical probe. Its toxicity has always been of concern and continues to be so (Levy *et al.*, 1991, pp. 200–207; Roberts *et al.*, 1987). This brings us to one of the ironies in colchicine's biography. The anti-mitotic effects of colchicine were of great interest to cancer researchers during the 1930s and 1940s. Many of those who were involved in the studies on colchicine in the early 1930s were, in fact, primarily interested in cancer, particularly in trying to understand the difference between normal and malignant cells, between normal and malignant growth (Eigsti and Dustin, 1955). Studies on colchicine's effect on tumours were not conclusive, though most research agreed that because colchicine's mitotic effects were temporary, any therapy using colchicine would have to be continuous and likely lead to near-toxic dosages (Levine, 1951). Reginald Ludford who, it will be recalled, was engaged in colchicine research as part of his general interest in cancer chemotherapy, concluded that tumour regression observed after the administration of colchicine, was due not to its anti-mitotic effects but to its side effects of causing haemorrhaging (Ludford, 1945; 1948).

There is little doubt that colchicine's toxicity has been an impediment to its incorporation into cancer chemotherapeutic regimes. Whether this is the whole story is an open question and a full answer would go well beyond the concerns of this paper. It is, however, likely that an answer to the question would involve a deep analysis of the history of cancer chemotherapy and its relation to other therapeutic approaches (Bud, 1978; Goodman, 1995). It would also need an analysis of alternative antit-mitotic, microtubule-sensitive drugs such as the vinca alkaloids and, more recently, taxol especially in terms of the social and economic context, the role of pharmaceutical companies and patents (Dustin, 1963, pp. 460–461; Rowinsky and Donehower, 1991; Suffness, 1995).

CONCLUSION

Following colchicine from 1930 to 1975 has brought out a number of issues which it may be useful to review.

The first is that this multi-sited excursion with colchicine has confirmed a much-noted phenomenon, that the meaning of an object is unstable, contextually specific and is a result of how it is used. But I would go further than this and argue that meaning is not just a question of use. What also matters to an object's meaning is who speaks for it, who else wants to speak for it and how it is spoken about. The only way to get at this is to follow the object into all the heterogeneous sites it enters and to listen to the voices emanating from these sites.

We saw earlier in this chapter how Laszlo Havas reacted to being sidelined in reports of experimental studies on plants using colchicine. This was not just a matter of setting the record straight. No, it was about who should speak for colchicine — Europeans in a pathology laboratory in Brussels or Americans in a genetics laboratory on Long Island. And more, whoever got to speak for colchicine, established him/herself as an obligatory passage point (to borrow from Bruno Latour) and, therefore, gave it their own special identifiable meaning. This was what Havas feared and rightly so. What would have happened had this episode *not* taken place in the Europe of 1939 is anyone's guess. And that is just the point. The war made it possible for Havas' voice to be silenced, contrary to that of Blakeslee. Havas was a participant to the process of molecularizing colchicine, of employing it experimentally to study biological processes. Blakeslee was not. He only saw colchicine's commercial possibilities.

This brings me to the second point. There has been a strong tendency in the literature on the sociology of science, technology and medicine to view the standardization of research materials, instruments and the like involved in practice in integrative terms; as enabling the mapping of local practices onto a larger canvas (Clarke and Fujimura, 1992; Fujimura, 1992). While this way of thinking about practice has proved very useful, there is at least some evidence that there are more tensions in the process of mapping than the literature would lead one to expect (Hogle, 1995). The case of colchicine offers an additional complication; namely, following the research material over time, as it moves from place to place, from one social and cognitive arena to another, shows that there is nothing inevitable, nor predictable, about the mapping process.

Finally, there is the question of molecularization, the overall theme of this book. The case of colchicine provides an insight into this process though perhaps in an unexpected way. Seen from colchicine's point-of-view, molecularization was not an autonomous, continuous process, with a clear unbroken history. It had to be worked on, sustained and argued for, by associations of human and non-human actors, in laboratories, in journals, in conferences and meetings, in different countries. Molecularization, like colchicine's meaning, was unstable, in the way suggested by Foucault's genealogy (Foucault, 1991). The process of molecularizing colchicine resulted from practices associated specifically with the pathology laboratory in Brussels during the 1930s and their insistence that colchicine was a biochemical probe of cellular processes. When Blakeslee and others interested in the practice of plant breeding began to speak on colchicine's behalf, they inadvertently interrupted the molecularization process. For them, chromosome doubling was the objective; the plant was the object and colchicine was the chemical *tout court*. Molecularization began anew in the 1960s when cell biologists and rheumatologists returned colchicine to the cell and spoke for it molecularly, in the laboratory and in the clinic. But, as I have argued, there was nothing inevitable in the resumption of the process. It was contingent.

ACKNOWLEDGEMENTS

An earlier and shorter version of this chapter was given at the Science Museum in London. Mahmoud Ezzamel, Jonathan Harwood, Eric Hirsch, Peter McHugh, Dallas Sealy and the editors of this book have all read earlier drafts. I am very grateful for their valuable suggestions and criticisms. I would also like to thank Les Wilson for our 'colchicine chat'.

REFERENCES

E. Allen, M. Smith and W.U. Gardner, "Accentuation of the Growth Effect of Theelin on General Tissues by Arrest of Mitosis with Colchicine" *Anatomical Record* (1936), 67 supplement 1: 1.

E. Allen, M. Smith and W.U. Gardner, "Accentuation of the Growth Effect of Theelin on Genital Tissues of the Ovariectomized Mouse by Arrest of Mitosis with Colchicine" *American Journal of Anatomy* (1937), 61: 321–342.

E.C. Amoroso, "Colchicine and Tumour Growth" *Nature* (1935), 135: 266–267.

J. Austoker, *A History of the Imperial Cancer Research Fund 1902–1986* (Oxford: Oxford University Press, 1988).

E.C. Bartles, "Treatment of Gout" *Metabolism: Clinical and Experimental* (1957), 6: 297–307.

G.H. Bates, "Polyploidy Induced by Colchicine and its Economic Possibilities" *Nature* (1939), 144: 315–316.

J.O. Beasley, "The Production of Polyploids in *Gossypium*" *Journal of Heredity* (1940), 31: 39–48.

W. Bechtel, "Scientific Evidence: Creating and Evaluating Experimental Instruments and Research Techniques" in A. Fine, M. Forbes and L. Wessels (eds), *PSA 1990* (East Lansing, Michigan: Philosophy of Science Association), pp. 559–572.

W. Bechtel, "Defining Cell Biology: Explicit Activities Directed at Defining the Discipline" paper presented at History of Science Society Conference, Minneapolis/ St. Paul, 1995.

A.F. Blakeslee, "Dédoublement du nombre de chromosomes chez les plantes par traitement chimique" *Comptes rendus de l'Academie des Sciences de Paris* (1937), 205: 476–479.

A.F. Blakeslee, "The Present and Potential Service of Chemistry to Plant Breeding" *American Journal of Botany* (1939), 26: 163–172.

A.F. Blakeslee and A.G. Avery, "Methods of Inducing Chromosome Doubling in Plants by Treatment with Colchicine" *Science* (1937a), 86: 408.

A.F. Blakeslee, "Methods of Inducing Doubling of Chromosomes in Plants By Treatment with Colchicine" *Journal of Heredity* (1937b), 28: 393–411.

G.G. Borisy and E.W. Taylor, "The Mechanism of Action of Colchicine. I and II" *Journal of Cell Biology* (1967), 34: 525–548.

A. Brues, "The Effect of Colchicine on Regenerating Liver" *Journal of Physiology* (1936), 86: 63–64.

R.F. Bud, "Strategy in American Cancer Research after World War II: A Case Study" *Social Studies of Science* (1978), 8: 425–459.

M. Callon, "Four Models for the Dynamics of Science" in S. Jasanoff, G.E. Markle, J.C. Petersen and T. Pinch (eds), *Handbook of Science and Technology Studies* (London: Sage Publications, 1995), pp. 29–63.

A.E. Clarke and J.H. Fujimura (eds), *The Right Tools for the Job: At Work in Twentieth-Century Life Sciences* (Princeton, NJ: Princeton University Press, 1992).

A. Cohen, "Gout" *American Journal of the Medical Sciences* (1936), 192: 488–493.

J. W. Cook and J.D. Loudon "Colchicine" in R.H.F. Manske and H.L. Holmes (eds), *The Alkaloids Vol. II* (New York: Academic Press, 1952), pp. 261–329.

W.A. Creasey, "Vinca Alkaloids and Colchicine" in A.C. Sartorelli and D.G. Johns (eds), *Antineoplastic and Immunosuppressive Agents* (New York: Springer-Verlag, 1974), pp. 670–694.

F.A.E. Crew, "The Seventh International Congress of Genetics" *Nature* (1939), 144: 496.

E. De Harven and W. Bernhard, "Etude au microscope electronique de l'ultrastructure du centriole chez les vertébrés" *Zeitschrift für Zellforschung* (1956), 45: 378–398.

H. Dermen, "Colchicine Polyploidy and Technique" *The Botanical Review* (1940), 6: 599–635.

H. Dermen and N.A. Brown, "Cytological Basis of Killing Plant Tumors by Colchicine" *Journal of Heredity* (1940), 31: 197–199.

M.J.S. Dewar, "Structure of Colchicine" *Nature* (1945), 155: 141–142.

D.R. Dewey, "Some Applications and Misapplications of Induced Polyploidy to Plant Breeding" in W.H. Lewis (ed.), *Polyploidy: Biological Relevance* (New York: Plenum Press, 1980), pp. 445–470.

C.A. Dinarello, S.M. Wolff, S.E. Goldfinger, D.C. Dale and D.W. Alling, "Colchicine Therapy for Familial Mediterranean Fever: A Double-Blind Trial" *New England Journal of Medicine* (1974), 291: 934–937.

W. Dixon, *A Manual of Pharmacology* (London: Arnold, 1906).

W. Dixon and W. Malden, "Colchicine with Special Reference to Its Mode of Action and Effect on Bone-Marrow" *Journal of Physiology* (1908), 37: 50–76.

A.P. Dustin, "Les poisons caryoclasiques, moyens d'analyse cytophysiologique" *Archives d'Anatomie Microscopique* (1929), 25: 37–48.

A.P. Dustin, "Contribution à l'étude de l'action des poisons caryoclasiques sur les tumeurs animales. II. Action de la colchicine sur le sarcome greffé, type Crocker, de la souris" *Bulletin de l'Académie Royale de Médecine Belgique* (1934), 14: 487–502.

P. Dustin, Jr, "L'activité du Laboratoire d'Anatomie Pathologique de la Faculté de Médecine de l'Université Libre de Bruxelles, sous la direction du Professeur Albert-Pierre Dustin, de 1929 à 1939" *Archives de Médecine Belgique* (1946), 1: 157–167.

P. Dustin, Jr, "Some New Aspects of Mitotic Poisoning" *Nature* (1947), 159: 794–797.

P. Dustin, Jr, "New Aspects of the Pharmacology of Antimitotic Agents" *Pharmacological Reviews* (1963) 15: 449–480.

P. Dustin, Jr, "Microtubules" *Scientific American* (1980), 243: 59–68.

H.M. Dyer, *An Index of Tumor Chemotherapy* (Washington, DC: Federal Security Agency, Public Health Service, 1949).

Editor, "Colchicine and Double Diploids" *Journal of Heredity* (1937), 28: 411–412.

J.T. Edsall, "Proteins as Macromolecules: An Essay on the Development of the Macromolecule Concept and Some of Its Vicissitudes" *Archives of Biochemistry and Biophysics* (1962), Supplement 1: 12–20.

O.J. Eigsti, "A Cytological Study of Colchicine Effects in the Induction of Polyploidy in Plants" *Proceedings of the National Academy of Sciences* (1938), 24: 56–63.

O.J. Eigsti, "Colchicine Bibliography" *Lloydia* (1947), 10: 65–114.

O.J. Eigsti and P. Dustin, Jr, "Colchicine Bibliography III" *Lloydia* (1949), 12: 185–207.

O.J. Eigsti, *Colchicine — In Agriculture, Medicine, Biology and Chemistry* (Ames, Iowa: The Iowa State College Press, 1955).

F.C. Elliott, *Plant Breeding and Cytogenetics* (New York: McGraw Hill, 1958).

R.A. Emerson, "Impressions of the Seventh International Congress of Genetics" *Journal of Heredity* (1940), 31: 129–130.

K.R. Fell and D. Ramsden, "Colchicum: A Review of Colchicums and the Sources, Chemistry, Biogenesis and Assay of Colchicine and its Congeners" *Lloydia* (1967), 30: 123–140.

M. Foucault, "Nietzsche, Genealogy, History" in P. Rabinow (ed.), *The Foucault Reader* (Harmondsworth: Penguin Books, 1991), pp. 76–100.

R.D. Friedlander, "Cortisone as an Adjunct in the Therapy of Acute Gout" *Journal of the American Medical Association* (1951), 145: 11–14.

J.H. Fujimura, "Constructing 'Do-able' Problems in Cancer Research: Articulating Alignment" *Social Studies of Science* (1987), 17: 257–293.

J.H. Fujimura, "The Molecular Biological Bandwagon in Cancer Research: Where Social Worlds Meet" *Social Problems* (1988), 35: 261–283.

J.H. Fujimura, "On Methods, Ontologies, and Representation in the Sociology of Science: Where Do We Stand?" in D.R. Maines (ed.), *Social Organization and Social Processes: Essays in Honor of Anselm Strauss* (New York: Aldine de Gruyter, 1991), pp. 207–248.

J.H. Fujimura, "Crafting Science: Standardized Packages, Boundary Objects, and 'Translation'" in A. Pickering (ed.), *Science as Practice and Culture* (Chicago: University of Chicago Press, 1992), pp. 168–211.

J.L. Fyfe, "The Action and Use of Colchicine in the Production of Polyploid Plants" (Cambridge: Imperial Bureau of Plant Breeding and Genetics, School of Agriculture, 1939).

S.E. Goldfinger, "Colchicine for Familial Mediterranean Fever" *New England Journal of Medicine* (1972), 287: 1302.

J. Goodman, "Who Speaks for the Body? Cancer, Cancer Chemotherapy and the Cell, 1945–1970", unpublished paper (1995).

J. Goodman, "Can it Ever be Pure Science: Pharmaceuticals, the Pharmaceutical Industry and Biomedical Research in the Twentieth Century" in J.-P. Gaudillière, I. Löwy and D. Pestre (eds), *The Invisible Industrialist and the Construction of Scientific Knowledge* (London: Macmillan, in press, 1998).

L.D. Graham, "Critical Biography Without Subjects and Objects: An Encounter with Dr. Lillian Moller Gilbreth" *The Sociological Quarterly* (1994), 35: 621–634.

A. Hagberg and E. Åkerberg, *Mutations and Polyploidy in Plant Breeding* (London: Heinemann, 1962).

E.F. Hartung, "Historical Considerations" *Metabolism:Clinical and Experimental* (1957), 6: 196–208.

J. Harwood, "Did Theory Transform Practice? Mendelism and Plant-Breeding in Germany, 1880–1920", paper presented at workshop 'Ideas, Instruments and Innovation', California Institute of Technology (1994).

L.J. Havas, "A Colchicine Chronology" *Journal of Heredity* (1940), 31: 115–117.

W.W. Herrick and T. Lloyd Tyson, "Gout — A Forgotten Disease" *American Journal of the Medical Sciences* (1936), 192: 483–488.

L.F. Hogle, "Standardization across Non-standard Domains: The Case of Organ Procurement" *Science, Technology, & Human Values* (1995), 20: 482–500.

Imperial Bureau of Plant Genetics, *The Experimental Production of Haploids and Polyploids* (Cambridge: School of Agriculture, 1936).

S. Inoué, "The Effect of Colchicine on the Microscopic and Submicroscopic Structure of the Mitotic Spindle" *Experimental Cell Research* (1952), Supplement 2: 305–318.

L.E. Kay, "Laboratory Technology and Biological Knowledge: The Tiselius Electrophoresis Apparatus, 1930–1945" *History and Philosophy of the Life Sciences* (1988) 10: 51–72.

J.R. Kloppenburg, Jr., *First the Seed: The Political Economy of Plant Biotechnology, 1492–2000* (Cambridge: Cambridge University Press, 1988).

K. Knorr-Cetina, *The Manufacture of Knowledge: An Essay on the Constructivist and Contextual Nature of Science* (Oxford: Pergamon Press, 1981).

I. Kopytoff, "The Cultural Biography of Things: Commoditization as Process" in A. Appadurai (ed.), *The Social Life of Things: Commodities in Cultural Perspective* (Cambridge: Cambridge University Press, 1986), pp. 64–91.

J.M. Krythe and S.J. Wellensiek, "Five Years of Colchicine Research" *Bibliographica Genetica* (1942) 14: 1–132.

G.C. Kunkle, "Technology in the Seamless Web:'Success' and 'Failure' in the History of the Electron Microscope" *Technology and Culture* (1995), 36: 80–103.

B. Latour, *Science in Action* (Cambridge, MA: Harvard University Press, 1987).

B. Latour, *The Pasteurization of France* (Cambridge, MA: Harvard University Press, 1988).

J.E. Lesch, *Science and Medicine in France: The Emergence of Experimental Physiology, 1790–1855* (Cambridge, MA: Harvard University Press, 1984).

J.E. Lesch, "Chemistry and Biomedicine in an Industrial Setting: The Invention of the Sulfa Drugs" in S.H. Mauskopf (ed.) *Chemical Sciences in the Modern World* (Philadelphia: University of Pennsylvania Press, 1995), pp. 158–215.

A. Levan, "The Effect of Colchicine on Root Mitoses in *Allium*" *Hereditas* (1938), 24: 471–486.

A. Levan, "Plant Breeding by the Induction of Polyploidy and Some Results in Clover" *Hereditas* (1942), 28: 245–246.

A. Levan, "The Cyto-genetics Department, 1931–1947" in A. Åkerman (ed.) *Svalöf, 1886–1946* (Lund: Carl Bloms Boktryckeri AB, 1948), pp. 304–323.

M. Levine, "The Action of Colchicine on Cell Division in Human Cancer, Animal, and Plant Tissues" *Annals of the New York Academy of Sciences* (1951), 51: 1365–1408.

M. Levy, M. Spino and S.E. Read, "Colchicine: A State-of-the-Art Review" *Pharmacotherapy* (1991), 11: 196–211.

F. Lits, "Contribution à l'étude des réactions cellulaires provoquées par la colchicine" *Comptes Rendus de la Société de Biologie de Paris* (1934), 115: 1421–1423.

S. Lubar and W. David Kingery (eds), *History from Things: Essays on Material Culture* (Washington, DC: The Smithsonian Institution Press, 1993).

R.J. Ludford, "The Vital Staining of Normal and Malignant Cells — II. The Staining of Malignant Tumours with Trypan Blue" *Proceedings of the Royal Society of London* (1929), 104/B: 493–511.

R.J. Ludford, "The Cytological Action of Methylene Blue" *Archiv für experimentelle Zellforschung* (1935), 17: 339–358.

R.J. Ludford, "The Action of Toxic Substances upon the Division of Normal and Malignant Cells *in vitro* and *in vivo*" *Archiv für experimentelle Zellforschung* (1936), 18: 411–441.

R.J. Ludford, "Colchicine in the Experimental Chemotherapy of Cancer" *Journal of the National Cancer Institute* (1945), 6: 89–101.

R.J. Ludford, "Factors Determining the Action of Colchicine on Tumour Growth" *British Journal of Cancer* (1948), 2: 75–86.

S.E. Malawista, "On the Action of Colchicine: The Melanocyte Model" *Journal of Experimental Medicine* (1965a), 122: 361–384.

S.E. Malawista, "The Action of Colchicine in Acute Gout" *Arthritis and Rheumatism* (1965b), 8: 752–756.

S.E. Malawista, "Colchicine: A Common Mechanism for its Anti-Inflammatory and Anti-Mitotic Effects" *Arthritis and Rheumatism* (1968), 11: 191–197.

S.E. Malawista, "The Action of Colchicine in Acute Gouty Arthritis" *Arthritis and Rheumatism* (1975), 18: 835–846.

S.E. Malawista and K.G. Bensch, "Human Polymorphonuclear Leukocytes: Demonstration of Microtubules and Effect of Colchicine" *Science* (1967), 156: 521–522.

G.E. Marcus, "Ethnography In/Of the World System: The Emergence of Multi-Sited Ethnography" *Annual Review of Anthropology* (1995), 24: 95–117.

L. Margulis, "Colchicine-Sensitive Microtubules" *International Review of Cytology* (1973), 34: 333–361.

H.M. Marks, "Cortisone, 1949: A Year in the Political Life of a Drug" *Bulletin of the History of Medicine* (1992), 66: 419–439.

M.C. Marsh and B.T. Simpson, "Chemotherapeutic Attempts with Coal-Tar Derivatives on Spontaneous Mouse Tumors" *Journal of Cancer Research* (1927), 11: 417–435.

D. Mazia, "Microtubule Research in Perspective" *Annals of the New York Academy of Sciences* (1975), 253: 7–13.

D. Mazia and K. Dan, "The Isolation and Biochemical Characterization of the Mitotic Apparatus of Dividing Cells" *Proceedings of the National Academy of Sciences* (1952), 38: 826–838.

J.P. McCracken and J.H. Pratt, "Gout: Still a Forgotten Disease" *Journal of the American Medical Association* (1946), 131: 367–372.

H. Mohri, "Amino-Acid Composition of 'Tubulin' Constituting Microtubules of Sperm Flagella" *Nature* (1968), 217: 1053–1054.

N. Morgan, "The Strategy of Biological Research Programmes: Reassessing the 'Dark Age' of Biochemistry, 1910–1930" *Annals of Science* (1990), 47: 139–150.

B.R. Nebel, "Mechanism of Polyploidy through Colchicine" *Nature* (1937), 140: 1101.

B.R. Nebel and M.L. Ruttle, "The Cytological and Genetical Significance of Colchicine" *Journal of Heredity* (1938), 29: 3–9.

J.B. Olmsted and G.G. Borisy, "Microtubules" *Annual Review of Biochemistry* (1973), 42: 507–540.

N. Oudshoorn, "On the Making of Sex Hormones: Research Materials and the Production of Knowledge" *Social Studies of Science* (1990), 20: 5–33.

N. Oudshoorn, *Beyond the Natural Body: An Archeology of Sex Hormones* (London: Routledge, 1994).

P. Palladino, "The Political Economy of Applied Research: Plant Breeding in Britain, 1910–1940" *Minerva* (1990), 28: 446–468.

P. Palladino, "Science, Technology and the Economy: Plant Breeding in Great Britain, 1920–1970" *Economic History Review* (1996), 49: 116–136.

J.T. Patterson, *The Dread Disease: Cancer and Modern American Culture* (Cambridge, MA: Harvard University Press, 1987).

A. Pickering, *The Mangle of Practice: Time, Agency, and Science* (Chicago: University of Chicago Press, 1995).

G. Pincus and C.H. Waddington, "The Effects of Mitosis-Inhibiting Treatments on Normally Fertilized Pre-Cleavage Rabbit Eggs" *Journal of Heredity* (1939), 30: 515–518.

K.R. Porter, "Cytoplasmic Microtubules and Their Functions" in G.E.W. Wolstenholme and M. O'Connor (eds), *Principles of Biomolecular Organization* (London: J. & A. Churchill Ltd, 1966), pp. 308–345.

L.F. Randolph, "An Evaluation of Induced Polyploidy as a Method of Breeding Crop Plants" *The American Naturalist* (1941), 75: 347–363.

W.N. Roberts, M.H. Liang and S.H. Stern, "Colchicine in Acute Gout" *Journal of the American Medical Association* (1987), 257: 1920–1922.

G.P. Rodnan and T.G. Benedek, "The Early History of Antirheumatic Drugs" *Arthritis and Rheumatism* (1970), 13: 145–165.

R. Rorty, *Contingency, Irony, and Solidarity* (Cambridge: Cambridge University Press, 1989).

E.K. Rowinsky and R.C. Donehower, "The Clinical Pharmacology and Use of Antimicrotubule Agents in Cancer Chemotherapeutics" *Pharmacology and Therapeutics* (1991), 52: 35–84.

G. Rozsa and R.W.G. Wyckoff,"The Electron Microscopy of Onion Root Tip Cells" *Experimental Cell Research* (1951), 2: 630–641.

D. Sabatini, K. Bensch and K. Barnett, "Cytochemistry and Electron Microscopy: The Preservation of Cellular Ultrastructure and Enzymatic Activity by Aldehyde Fixation" *Journal of Cell Biology* (1963), 17: 19–58.

O. Sacks, "Scotoma: Forgetting and Neglect in Science" in R.B. Silvers (ed.), *Hidden Histories of Science* (New York: NYREV Inc., 1995), pp. 141–187.

P.B. Schiff and S.B. Horwitz, "Tubulin: A Target for Chemotherapeutic Agents" in A.C. Sartorelli, J.S. Lazo and J.R. Bertino (eds), *Molecular Actions and Targets for Cancer Chemotherapeutic Agents* (New York: Academic Press, 1981), pp. 483–507.

J.E. Seegmiller, "The Acute Attack of Gout Arthritis" *Arthritis and Rheumatism* (1965), 8: 714–725.

J.E. Seegmiller and R.R. Howell, "The Old and New Concepts of Acute Gouty Arthritis" *Arthritis and Rheumatism* (1962), 5: 616–623.

J.E. Seegmiller, R.R. Howell and S.E. Malawista, "The Inflammatory Reaction to Sodium Urate" *Journal of the American Medical Association* (1962), 180: 469–475.

M.L. Shelanski and E.W. Taylor, "Isolation of a Protein Subunit from Microtubules" *Journal of Cell Biology* (1967), 34: 549–554.

D.B. Slautterback, "Cytoplasmic Microtubules I. Hydra" *Journal of Cell Biology* (1963), 18: 367–388.

L.M. Smith, "Biographical Method" in N.K. Denzin and Y.S. Lincoln (eds) *Handbook of Qualitative Research* (Thousand Oaks, CA: Sage, 1994), pp. 286–305.

E. Sohar *et al.*, "Familial Mediterranean Fever: A Survey of 470 Cases and Review of the Literature" *American Journal of Medicine* (1967), 43: 227–253.

L. Sokoloff, "The Pathology of Gout" *Metabolism: Clinical and Experimental* (1957), 6: 230–243.

G.L. Stebbins, "Artificial Polyploidy as a Tool in Plant Breeding" in *Genetics in Plant Breeding. Brookhaven Symposia in Biology*, Vol. 9 (Upton, NY: Brookhaven National Laboratory, 1956), pp. 37–52.

M. Suffness (ed.), *Taxol: Science and Applications* (Boca Raton, FL: CRC Press, 1995).

J.H. Talbott, "Introduction to Symposium on Gout" *Metabolism: Clinical and Experimental* (1957), 6: 193–195.

E.W. Taylor, "Dynamics of Spindle Formation and its Inhibition by Chemicals" *Journal of Biophysical and Biochemical Cytology* (1959), 6: 193–196.

E.W. Taylor, "The Mechanism of Action of Colchicine Inhibition of Mitosis I" *Journal of Cell Biology* (1965), 25: 145–160.

A.S. Travis, "Science as Receptor of Technology: Paul Ehrlich and the Synthetic Dyestuffs Industry" *Science in Context* (1989), 3: 383–403.

A.S. Travis, *The Rainbow Makers: The Origins of the Synthetic Dyestuffs Industry in Western Europe* (Bethlehem, PA: Lehigh University Press, 1993).

S.L. Wallace, "Colchicine: Clinical Pharmacology in Acute Gouty Arthritis" *American Journal of Medicine* (1961), 30: 439–448.

S.L. Wallace, "Colchicum: The Panacea" *Bulletin of the New York Academy of Medicine* (1973), 49: 130–135.

S.L. Wallace, "Colchicine and New Antiinflammatory Drugs for the Treatment of Acute Gout" *Arthritis and Rheumatism* (1975), 18: 847–851.

R.C. Weisenberg, G.G. Borisy and E.W. Taylor, "The Colchicine-Binding Protein of Mammalian Brain and Its Relation to Microtubules" *Biochemistry* (1968), 7: 4466–4479.

S.J. Wellensiek, "The Newest Fad, Colchicine, and Its Origins" *Chronica Botanica* (1939), 5: 15–17.

W.C. Wildman, "Colchicine and Related Compounds" in R.H.F. Manske (ed.), *The Alkaloids* Vol. 6 (New York: Academic Press, 1960), pp. 247–288.

L. Wilson (ed.), "Symposium: Pharmacological and Biochemical Properties of Microtubule Proteins" *Federation Proceedings* (1974), 33: 151–174.

L. Wilson, "Microtubules as Drug Receptors: Pharmacological Properties of Microtubule Protein" *Annals of the New York Academy of Sciences* (1975a), 253: 213–231.

L. Wilson, "Action of Drugs on Microtubules" *Life Sciences* (1975b), 17: 303–310.

L. Wilson and J. Bryan, "Biochemical and Pharmacological Properties of Microtubules" *Advances in Cell and Molecular Biology* (1974), 3: 21–72.

L. Wilson and M. Friedkin, "The Biochemical Events of Mitosis II. The *in vivo* and *in vitro* Binding of Colchicine in Grasshopper Embryos and Its Possible Relation to Mitosis" *Biochemistry* (1967), 6: 3126–3135.

L. Wilson and M.A. Jordan, "Pharmacological Probes of Microtubule Function" in J.S. Hyams and C.W. Lloyd (eds) *Microtubules* (New York: Wiley-Liss, 1994), pp. 59–83.

A. Windaus, "Untersuchungen über die Konstitution des Colchicins" *Liebig's Annalen der Chemie* (1924), 439: 59–75.

J.B. Wyngaarden, "Gout" in R. Levine and R. Luft (eds) *Advances in Metabolic Disorders*, Vol. 2 (New York: Academic Press, 1965), pp. 1–78.

D. Zemer, M. Revach, M. Pras, B. Modan, S. Schor, E. Sohar and J. Gafni, "A Controlled Trial of Colchicine in Preventing Attacks of Familial Mediterranean Fever" *New England Journal of Medicine* (1974) 291: 932–934.

CHAPTER 2

Chemistry in the Clinic: The Research Career of Donald Dexter Van Slyke

Olga Amsterdamska

While the interactions between laboratory research and the practice of medicine have been an abiding interest for both historians of science and of medicine, the development of routine laboratory techniques used for diagnostic and monitoring purposes has received relatively little attention (but see Keele, 1963; Reiser, 1978; Büttner, 1983). Historians of science tend to focus on the research laboratory and often take the clinical applications of biomedical research pretty much for granted; historians of medicine who focus on the medical practices, pay relatively little attention to the processes by which the techniques used by the physicians are developed in the laboratories and clinics. And yet, the development of laboratory — for example chemical — methods and techniques which are to be used in the clinical practice is a complex process shaped not only by the available biomedical knowledge and the current medical practice, but also by the changing institutional and intellectual relations among research laboratories, clinical laboratories, and the bedside.

In the early twentieth century, when clinical laboratories charged with the routine performance of various diagnostic tests were being established in hospitals, and preclinical biomedical disciplines were gaining greater institutional autonomy, the development of new diagnostic methods and their incorporation into medical practice involved transformations in the manner in which medical problems were understood and handled both in the laboratory and the clinic, new modes of collaboration between laboratory researchers and clinicians, and, as Joel Howell (1995) has shown, a variety of organizational changes in hospitals.

In his analysis of the emergence of biochemistry as an independent discipline in the U.S., Robert E. Kohler (1982) emphasized the institutional conditions which allowed biochemistry in the U.S. to develop as a field that was, on the one hand, independent of physiology and chemistry, on the other hand, firmly embedded within a clinical context of a medical school or a hospital laboratory. As a result of both the growing disciplinary autonomy vis-à-vis other biomedical fields and the service role of biochemists whose legitimacy depended on association with medical practice and medical research, American pre-World War II biochemistry is then said to have been distinctively oriented towards clinical applications and problems. Given his focus on the development of biochemistry as a biological discipline, Kohler, however, does not treat this "national style" of American prewar biochemistry on its own terms: He does not ask how the clinical chemists who were so prominent in American biochemistry in the prewar era conducted their research, how they conceptualized medical problems or how biochemical techniques, practices, and styles of thinking came to be incorporated into medical practice. If we wish to understand the process of incorporation of (bio)chemical knowledge and methods into medicine, we should thus supplement the institutional analysis of the establishment of biochemistry in medical settings with a more fine-grained study of the biochemical research practices.

As a rule, however, when the development of biochemical research conducted during the period under consideration here had been investigated by other historians, the interdependence of biochemical and medical knowledge has often been ignored or marginalized (e.g., Fruton, 1990; Edsall, 1972 and 1985), and interest tended to focus instead on the contributions made by the prewar — especially European — biochemists to the development of basic biochemical knowledge of life processes.[1] Thus while Kohler has been criticized for playing down the development of knowledge in favor of emphasizing institutional strategies (e.g., Holmes, 1983; Wilson, 1983; Servos, 1983), other historians tended to regard biochemistry's involvement in medicine as an inconsequential — institutionally advantageous, but intellectually irrelevant — characteristic of the field whose main commitment was purely scientific.[2] If, however, we follow Kohler and accept his characterization of American biochemical research during the first half of the twentieth century as "clinically oriented," then any attempt to understand its development must examine the nature and characteristics of this medical orientation.[3]

Making chemical research relevant to medicine is, after all, neither a self-evident nor an unequivocal process, even when the commitment to unite clinical observation and laboratory experiments is strong: a clinical problem (usually already prestructured by the current scientific knowledge and the existent medical practices) can become a subject of laboratory research only after it has been transformed into a problem interpretable within the current theoretical vocabulary of a given biomedical field and addressable with the tools and techniques available in the laboratory (Amsterdamska, 1993). Such translations of medical problems into research problems take many different forms, but the choice of which translation to adopt shapes the interactions between research and the practice of medicine. Moreover, medical concerns affect not only the initial formulations of problems, but also the development of specific research strategies and directions, the evaluation of research, and the actual investigative practices of laboratory or hospital scientists.

This chapter focuses on the research career of one of the most eminent of the prewar American clinical biochemists, Donald Dexter Van Slyke, to examine how specific medical problems were made subject to chemical investigation, how biochemical and clinical research were brought to bear on each other, and what kind of understanding of disease was being developed when clinical chemistry was being incorporated into medical practice. In other words, my goal here is to examine one specific form of reductionism as a research practice both in the laboratory and in the clinic.

THE SETTING

When the Hospital of the Rockefeller Institute was established in 1910, its founders intended it to be an embodiment of scientific medicine and both a model for research hospitals elsewhere and a training ground for the new breed of clinical researchers (Corner, 1965). Given its privileged funding situation, the fact that it admitted only patients suffering from selected diseases which were chosen for investigation, and the priority given to research, the Rockefeller Hospital most certainly cannot be considered as a typical example of other hospitals at the time — even those associated with medical schools elsewhere. It did, however, provide a post-graduate training ground for many future professors in the most important medical schools in the U.S., and the manner in which research was organized was supposed not only to promote the advance of medical knowledge but also to serve as a model to be emulated elsewhere.

The hospital was to be organized so as to foster a close integration of the bedside study of a selected number of diseases with experimental investigations conducted in the laboratories:

> The laboratories of the hospital were to be designed, not merely to carry out routine diagnosis, but also to promote experimental studies disclosing the real nature of the diseases being observed in patients in the hospital wards. It was believed that by continued observational and experimental studies, fruitful results would sometimes be secured.[4]

To further this integration of research conducted in the laboratory and clinical investigation conducted at the bedside, Rufus Cole, the Hospital's first director, rejected the option of making the Hospital an annex to the Institute's Laboratories which had been established some eight years earlier. Research conducted at the Hospital was to be independent of the work of the Institute's Laboratories, and the Hospital was provided with its own well-equipped research facilities which were suited for research and not only for routine diagnostic work. Although cooperations between scientists working in the Laboratories and those working at the Hospital sometimes developed and were not discouraged, the Hospital was not intended as a testing ground for the findings of the Laboratory scientists nor a source of "material" for their investigations, but rather as an independent research unit. Doctors working at the Hospital were not permitted to engage in consultancy or private practice and were expected to devote their entire time to their duties at the Hospital. Cole also insisted that every Hospital physician, whether senior or junior, be engaged in research and that no resident be involved exclusively in the care of patients. Ideally, the staff was to consist of people who had clinical as well as "fundamental" scientific training. In the early years, it was apparently not always easy to find graduates of medical schools who possessed the necessary scientific qualifications and the Hospital employed not only medical doctors but also scientists trained in other fields, particularly chemists.[5] Cooperation between the clinicians and laboratory scientists (whether those were trained in medicine or not) was strongly encouraged, and while the laboratory scientists could conduct research which did not stem directly from the clinic, there seems to have been an expectation that they would be directly involved in clinically relevant research as well.

The founders of the Institute and the Hospital had little doubt that medical progress could be secured only through the integration of "fundamental" biological, chemical, and physical studies into research directed to the solution of medical problems. The strongly reductionist notion of science to which they also subscribed led them to put a particular emphasis on (bio)chemistry. At the beginning of the twentieth century, the founders of the Rockefeller Institute were certainly not unique (in the U.S. or Europe) in subscribing to this belief as an article of faith, but at the time they were in a relatively privileged situation of being able to organize research and provide resources necessary for the realization of this ideal. But how did the integration of chemical and clinical research actually take place? What kinds of biochemical research were conducted at the Rockefeller Hospital and how was this research related to the clinical practices?

Donald Dexter Van Slyke was one of those scientists whose employment in the Rockefeller Hospital was supposed to facilitate the integration of laboratory and clinical research. When he came to the Hospital, he had no medical degree and no experience in medicine; both his training and his previous experience were in organic chemistry. A son of an agricultural chemist, he came to the Rockefeller Institute immediately upon receiving his Ph.D. in organic chemistry from the University of Michigan where he worked with Moses Gomberg. His first position at the Institute was as a research assistant to Phoebus Levene where Van Slyke worked on proteins and amino acids and in 1911 he went to Berlin to work with Emil Fischer, "the sun and god" of organic chemistry (Olch, 1971, p. 8).[6] When, in 1913, Simon Flexner asked Van Slyke to move to the position of the chief chemist at the Hospital, he was well aware that Van Slyke would have to make some gradual adjustments.[7] The expectation was that he would develop a department of clinical chemistry while continuing with his "basic chemistry."[8] In a report of the Director of the Hospital to the Board of Scientific Directors of the Institute, this expectation was explicitly formulated:

> For the present Dr. Van Slyke will continue his work on protein metabolism, but will from time to time, it is hoped, become interested in problems more directly bearing on those under investigation in the hospital. He has already been of great assistance in making suggestions and rendering aid to the work now in progress.[9]

Van Slyke took his new responsibilities seriously: he invested some time in studying medicine and physiology, toured laboratories and clinics with chemical laboratories in Britain, France and Germany,[10] and within a few years he was publishing articles on topics in clinical chemistry. Although he continued with protein studies, he also began devoting attention to diabetes, or, more specifically, to the development of methods which could be useful in the early detection of diabetic acidosis. During the following 34 years, Van Slyke collaborated with a large number of chemists and clinicians as he worked on such problems as acid-base balance and its pathology, respiratory diseases, nephritis, metabolism of proteins and amino acids, and gas and electrolyte equilibria in blood. A large number of his contributions were methodological in character as Van Slyke devised and repeatedly improved quantitative tests for a variety of chemical substances whose deficiency or excess in the organism could be used as indicators of diseases. Here, I will focus only on a few of Van Slyke's research lines — his work on acidosis and blood chemistry and on nephritis — in order to examine the manner in which chemical, physiological and clinical research were integrated in his and his associates' work. I have selected these research interests of Van Slyke not only because they occupied a rather prominent place in his career, but also because they allow us to ask how Van Slyke maintained continuity in his research and afford us an opportunity to examine the variety of ways in which Van Slyke's collaboration with clinicians, mediated through molecules and techniques, was an essential aspect of his research.

ACIDOSIS

The problem of early diagnosis and monitoring of diabetic acidosis was the first medical problem which Van Slyke addressed after he had moved to the hospital. The problem was apparently suggested to him by Edgar Stillman, a clinician working in the Hospital (Corner, 1964, pp. 274–275).[11]

Diabetes was among the diseases selected for special attention in the Hospital. Frederick Allen, a clinician and physiologist in charge of the diabetic patients, was developing what came to be known as "the starvation diet" for diabetic patients. Following experiments on dogs, Allen attempted to keep diabetes under control in his patients by drastically reducing not only the intake of carbohydrates, but of all caloric foods. Although Allen's diet and the strict regimen it imposed on patients (Feudtner 1995) seemed to produce some beneficial results, Allen's patients occasionally quite unexpectedly went into the diabetic coma. The problem Stillman suggested to Van Slyke was how to monitor best the development of acidosis in patients before it was so advanced as to produce coma. Interestingly, the plans to extend Allen's work in more chemical directions were aboard *before* the decision was made to focus on acidosis. In October 1914, the Director of the Hospital reported:

> Dr. Allen … has used mainly biological or physiological methods. It is important that, at the same time, chemical and more extensive metabolic studies should be undertaken and it is therefore planned that, during the coming year, a part of the work of the chemical laboratory will be devoted to this subject.[12]

The link between diabetes, defective metabolism and the circulation of acetoacetic and ß-hydroxybutyric acid was established in the late 19th century by a series of physiological, chemical and clinical investigations of Frederick Walter, Eugen Hallervorden, E. Stadelmann, Oskar Minkowski, August Magnus-Levy, and others. In 1906, Bernhard Naunyn referred to the circulation and excretion of these acids with the term acidosis and within the next few years the term was generalized to other pathologies in the acid-base equilibrium. Acidosis was then linked to various forms of nephritis (Straub and Schlayer, 1912; Sellards, 1914; Henderson and Palmer, 1913), cholera (Sellards, 1910), "food intoxication" and cyclic vomiting in children (Edsall, 1903; Howland and Marriott, 1916), and a series of other disorders. Both the definition of the condition and its relation to other symptoms such as the presence of specific acids in blood or urine, the actual pH of the blood, or abnormal alveolar CO_2 tension, remained subjects of dispute. At the same time, the chemistry and physiology of blood gases were a subject of extensive research among British physiologists such as John Scott Haldane and Joseph Barcroft and Danish physiological chemists, such as Christian Bohr, August Krogh, and K.A. Hasselbach.

Beginning in 1909, Lawrence Henderson, working at Harvard, published a series of theoretical and experimental studies on the regulation of neutrality in the body (for example, Henderson, 1908 and 1913; Henderson and Palmer, 1913). Henderson placed his discussion of the regulation of the body's reaction in the context of the organism's general ability to maintain a dynamic equilibrium. Discussing the regulation of the acid-base equilibrium, he was more concerned with the chemical mechanisms involved in what he saw as a fundamental principle of the organization of life, than with any pathologies or clinical problems with which the disturbance of this equilibrium might be associated.

> Neutrality is quite as definite, quite as fundamental and quite as important a characteristic of an organism as its temperature, or osmotic pressure, or in fact anything else that we know. And it turns out to possess … remarkable characteristic of stability … Within wide limits of amount any acid or base may be poured into an organism, and this constancy will protect all enzymatic processes, the function of respirations and the whole distribution of material throughout the body (Henderson, 1913, p. 395).

Henderson analyzed the chemical and physiological mechanisms used to regulate the acid-base balance, suggested how acidosis develops in a variety of pathological conditions, and on the grounds of theoretical biochemical analysis, argued that the pathological condition should be identified with the diminution of the bicarbonates of the blood rather than either circulation of free acids in the blood or the blood's pH.[13]

When Van Slyke took up the problem of developing a clinically useful diagnostic method, he adopted Henderson's definition[14] as well as his argument that the alkali reserve in the blood is a physiological constant.[15] Accordingly, Van Slyke hoped that "acidosis" could be given as "definite a meaning as 'fever' or 'tachycardia'" (Van Slyke and Cullen, 1917, p. 292). Note that although Van Slyke seems to be simply taking over Henderson's argument, he is in fact changing a comparison: Van Slyke compares acidosis (a pathological state) to other pathological conditions (e.g., tachycardia); Henderson compares normal physiological states to each other (neutrality of blood reaction to temperature and to osmotic pressure).

Thus, during the first decade of this century, acidosis ceased to be a pathological condition specific to diabetes, and became a pathological condition whose causes could be variable and which could appear as a symptom in a variety of conditions, only one of which was diabetes. Once acidosis became equated with bicarbonate deficit, it became, according to Van Slyke, an accurately "definable and determinate phenomenon" whose severity could be measured by determination of the blood bicarbonate. At the same time, however, it was analytically disconnected from its physiological and pathological causes: the diminution of alkali reserve could be just as much a result of the accumulation of acids in the blood as of their slower excretion through the kidneys or lungs. Acidosis was still "a symptom of disturbed function" (Van Slyke and Cullen, 1917, p. 292), but what exact function was disturbed was, according to the new definition, immaterial.[16] Van Slyke's work on acidosis was to carry this purely quantitative and chemical definition of the condition even further. At the same time, the redefinition of acidosis in chemical terms became the basis for the establishment of the method by which it could be measured.

In devising the method for the measurement of blood bicarbonate Van Slyke resorted to an analytical procedure by which the amount of bicarbonate in blood could be derived from its carbon dioxide capacity, and this capacity, in turn, estimated by use of a gasometric apparatus (see Figure 2.1) which he developed while still working in Levene's lab. The purely chemical arguments used to justify the technique were, however, inseparable from Van Slyke's attention to the clinical conditions in which the method was to be shown useful.

To begin with, the need for a quick and sensitive method was linked closely to Allen's "starvation treatment:"

> One point in which the blood and alveolar air tests promise to be of aid in the Allen Treatment of diabetes is in controlling the condition of fasting patients. Some develop no acidosis, or very little; but in other cases acidosis can develop rapidly, becoming marked after a single day's fast. A control by quick and sensitive methods for detecting the presence and degree of acidosis seems important, for as yet there is no way of telling which cases will develop acidosis in the treatment and which will not.[17]

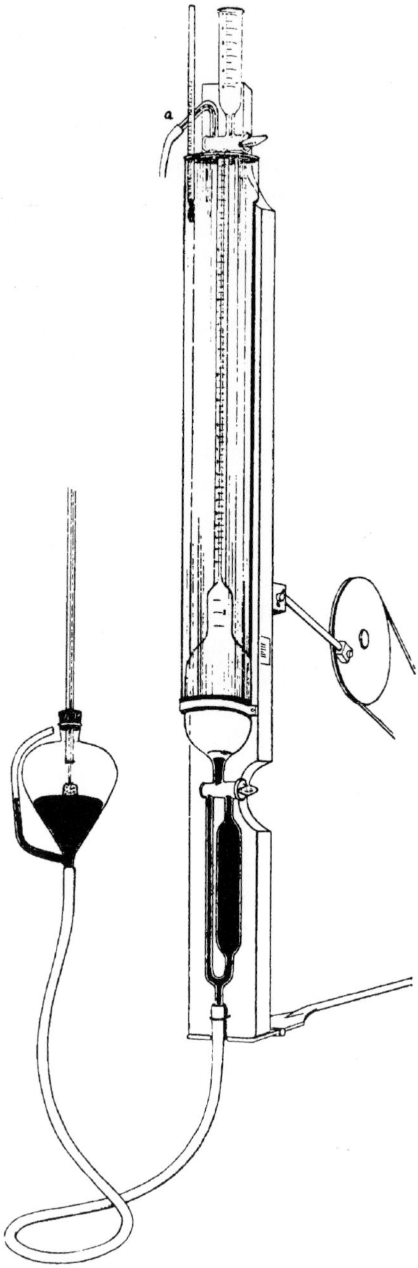

Figure 2.1. Van Slyke volumetric apparatus with water jacket and shaking device.

Thus, the development of a test was closely linked to the local clinical practice which imposed specific requirements. For example, the methods had to be rapid and simple enough to allow for repeated close monitoring of the condition of fasting patients. Van Slyke repeatedly emphasized the simplicity of the apparatus he developed and the ease and speed with which it could be used to perform routine measurements. Already in his first article in the series entitled "Studies of Acidosis" which was primarily devoted to the theoretical — largely biochemical — justification of the method he was proposing, Van Slyke argued not only that his method is appropriate for the measurement of the clinically interesting value and that it is particularly sensitive, but also that "The simplicity of the technique and the few minutes required for the determination make it available, not only for physiological experiments, but also for clinical routine" (Van Slyke and Cullen, 1917, p. 344).[18] Simplicity of the technique was also used as an argument for the superiority of the new technique over certain alternative methods developed by others. Thus "the failure of Walter's method for detecting acidosis by determination of venous CO_2 to attain general clinical use even in hospitals must be attributed chiefly to the lack of a sufficiently simple technique" (p. 297); while the Haldane procedure for measuring the concentration of carbon dioxide in alveolar air is described as appropriate for "normal individuals" but: "Under pathological conditions, or under the influence of drugs, of decreased atmospheric oxygen tension, of anxiety or of excitement, the sensitiveness of the respiratory control may vary, so that the alveolar carbon dioxide is not under all conditions even an approximate measure of the bicarbonate reserve" (p. 299).

The development of the method's scientific rationale was inextricably linked to the development and justification of its practical utility in the clinic. By the same token, dealing with the exigencies of the method's clinical use was part and parcel of Van Slyke's research from its very inception. For example, although on purely physiological grounds it was the level of alkali in arterial blood that constituted the quantity that should be measured, Van Slyke spent considerable time and effort establishing experimentally (on dogs and later also on patients) that estimations made from venous blood (which could be easily drawn from patients) were sufficiently accurate and reliable to be used instead of the estimations from arterial blood which were not practicable in the clinic (Stadie and Van Slyke, 1920). Similarly, he reported on the effects (or lack of such effects) of various methods of drawing blood under different conditions, conducted experiments to determine the length of time that a sample could stand before being tested without the results being compromised, and attempted to show that estimations could be made not only on whole blood but also on blood plasma (this was preferable in the clinic since it extended the time which could elapse between the drawing of the blood and running of the test without loss of accuracy) (Van Slyke and Cullen, 1917; Van Slyke, 1922).

Although one of Van Slyke's most abiding concerns was the possibility of providing quantitative data for the clinic, his very notion of precision was shaped by clinical considerations. The method had to be accurate enough to distinguish clearly between pathological and normal amounts of bicarbonate in the plasma, but as long as the errors associated with the performance of the test were considered too small to be clinically significant, they could be disregarded in favor of ease or simplicity of the procedure.[19] Accuracy was understood primarily in terms of its clinical significance: a test had to be precise enough to allow for a clear distinction to be drawn between normal and pathological ranges; beyond that accuracy was only desirable if it did not compromise the speed or simplicity of the determination.

It is difficult to imagine that all of these considerations would have been taken into account and all the experiments conducted were it not for the fact that Van Slyke's gasometric apparatus for the measurement of acidosis was being developed in a clinical context in active cooperation with practicing physicians. In his recollections, Van Slyke links his manner of working to Cole's idea "that the laboratory and the clinic would work as one organism. As the Hospital started out, most of the young men there that were working on patients did their own laboratory work" (Olch, 1971, p. 30). In his own work, Van Slyke recalls:

> I always had a couple of chemists working with me that were Ph.D. chemists, but I also had usually four or five M.D.s working with me, but they also worked in the laboratory. All of them. (...) Yes, they had patients, they were in charge of patients, but the idea was to have so few patients per doctor that the doctor could also do laboratory work (Olch, 1971, p. 31).

The clinicians provided Van Slyke with much more than a clinical problem to solve, their practices and concerns shaped many of the aspects of the technique being developed.

The fact that "providing clinically useful diagnostic methods" was not just a matter of applying biochemical knowledge to clinical problems supplied by the clinicians is also evident when further steps in the development of the test are considered. In order to establish the value of the newly proposed method and to determine the normal and pathological ranges of the CO_2 of blood, Van Slyke and his clinical collaborators: (1) compared the newly developed test to other methods or other indicators of related values (eg., combining power for CO_2 was compared to acid excretion in the urine) (Van Slyke, 1917; Fitz and Van Slyke, 1917); (2) compared the results of tests in the same individuals under different conditions (e.g., before and after eating) (Van Slyke *et al.*, 1917); (3) compared the values obtained from normal and pathological cases from the use of Van Slyke's test and of other related tests; and (4) compared the plasma bicarbonate levels established with the use of the new test to other indicators of the clinical severity of acidosis (Stillman *et al.*, 1917). Each of these comparisons was predicated, at one and the same time, on some physiological or chemical reasoning so that expected results could be formulated and then compared with the results of the new method and each such comparison involved active cooperation with clinicians without whom the measurements could not be taken, clinical cases could not be compared, and normal values could not be established. All the comparisons required also active participation of the chemists: in many cases the methods used for comparative purposes were themselves newly invented or adjusted for the purpose at hand, (quantitative) laws linking the values of the various tests were theoretically derived and articulated, and the various measurements which were used were by no means a matter of clinical routine.

The close collaboration between biochemists and physicians on which Van Slyke's studies of acidosis depended is quite striking if we examine the tables, graphs and illustrations used to present data in the various articles. Tables containing the results of calculations based on theoretically derived quantitative chemical formulas (for example, Van Slyke and Cullen, 1917, p. 316) and graphs presenting the results of measurements on experimentally manipulated plasma and blood (for example, tables in Van Slyke and Cullen, 1917) were presented along with measurements from individual subjects (for example, in Van Slyke, Stillman, and Cullen, 1917) and charts accompanied by detailed descriptions of clinical case histories of individual diabetic patients which included measurements of the degree of acidosis monitored over extended periods of time (for example, in Van Slyke, Cullen, and Fitz, 1917, see Figure 2.2). The combination of information presented in these tables and graphs allowed Van Slyke and his collaborators to establish the reliability and validity of the proposed method, to specify the range of normal and abnormal values for alkali reserve of blood, and to lay a link between these values, other indicators of acidosis and the associated chemical changes, and the progress of the disease. Or, we might say, that by collaborating with the clinicians, linking their concerns and their methods of studying disease with the biochemical methods and determinations, Van Slyke and his co-workers *simultaneously constructed both the biochemical and the clinical meaning of the indicator they developed.*

The interaction between the chemical and clinical arguments, methods and procedures which Van Slyke used in the development of his volumetric gas apparatus for the measurement of the severity of diabetic acidosis provided Van Slyke with a variety of opportunities to extend his research in a number of both clinical and physiological directions.

It is of course difficult to draw a clear distinction between research which led to the establishment of a method for clinical diagnosis of diabetic acidosis and physiological research in which the various chemical processes accompanying diabetes were investigated. In his attempts to validate the method, Van Slyke examined its correlation with various other chemical signs of diabetes (e.g., acid and ammonia excretion or the amount of hydroxybutyric acid in the blood), and the measurement of these other substances in the blood and urine constituted a contribution to the understanding of diabetes as a physiological and a chemical process while at the same time serving to strengthen Van Slyke's claim about the plasma bicarbonate method. Van Slyke, however, did not really take up the possibility of linking his new method to an investigation of the physiology or

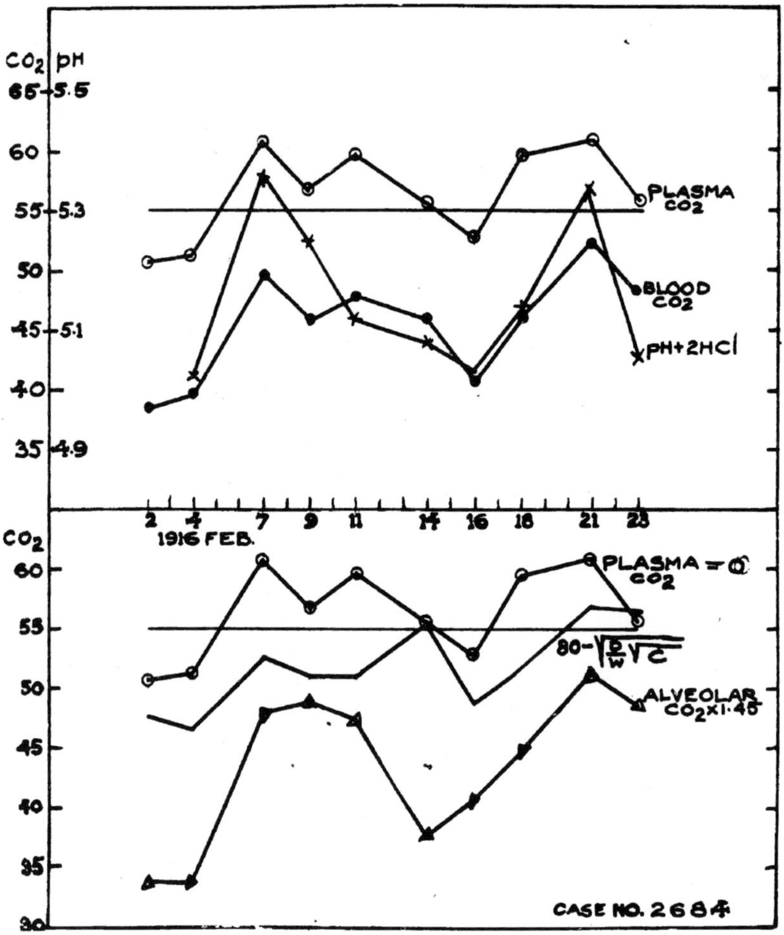

Figure 2.2. Monitoring acidosis in a diabetic patient. The curves trace acidosis in terms of acid excretion and plasma bicarbonate in "Case 2684, female, age 43," suffering from extreme diabetes.

pathology of diabetes. Although to some extent he had to rely on physiological arguments in order to devise his method in the first place and used such arguments also in establishing the relevant comparisons, Van Slyke did not attempt to contribute to the understanding of the physiological mechanisms underlying diabetic acidosis. At no point was his research structured by physiological concerns or considerations. Van Slyke did not ask about the causes of acidosis, about the significance of his findings for the understanding of the disease, or about mechanisms underlying variations in the degree of acidosis in various diabetic patients. Chemical and physiological arguments legitimized the test, but

Van Slyke did not use the test to investigate the causes or physiological mechanisms of diabetic — or any other — form of acidosis, nor did he turn his attention to the examination of other chemical or physiological changes in diabetes. In fact when in 1918 Frederick Allen resigned from the Rockefeller Hospital, he complained about being unable to develop an adequately broad program of studying diabetes at the Institute:

> The conception of diabetes as a disorder of total metabolism rather than of carbohydrate combustion alone, which is the underlying basis of this treatment, presents a broader problem and requires more extensive experimental support than has apparently been appreciated in this institute.[20]

Neither the Scientific Reports nor the correspondence found in the Archives provide direct leads for explaining Allen's disparaging comments. But Allen's quitting, as well as the fact that at no point of his work on acidosis did Van Slyke publish an article together with Allen (though Van Slyke did publish together with clinicians, and Allen with other chemists), suggests that even if their work would intersect in the clinic, it did not intersect much in the laboratory: Allen went back to study physiology using dogs; Van Slyke went back to the chemical laboratory.

Secondly, the apparatus developed for the measurement of CO_2 could be further improved and standardized or alternatively adapted for the measurement of other, possibly clinically interesting, substances. During his career Van Slyke published a very large number of methods articles in which he detailed improvements to the original apparatus: its transformation into a micrometric apparatus in which determinations could be made on smaller quantities of material, the changes required to make it into a manometric apparatus with which much greater accuracy could be claimed, and so on. The original apparatus was also adapted to the measurement of a great many substances other than CO_2, among them, oxygen, carbohydrates, fats, proteins, amino acids, chlorides, phosphate, and calcium (see Figure 2.3). Insofar as the substances measured were interesting because of their association with some physiological or pathological process, the collaboration with clinicians was again an essential part of research. A number of improvements and extensions were also introduced in attempts to extend more purely biochemical investigations whose connection to the clinic was less immediate.

Thirdly, Van Slyke — again in collaboration with the clinicians — attempted to demonstrate the usefulness of his gasometric method in monitoring or studying diseases other than diabetes, for example in the study of toxemias of pregnancy (eclampsia) or respiratory diseases such as pneumonia and tuberculosis. (There were relatively few such attempts: perhaps because once the CO_2 method became a standard test, these types of studies were basically clinical in character).

Fourthly, Van Slyke's redefinition of acidosis in terms of the reaction of the blood and the chemical buffering processes (largely in terms of the balance of

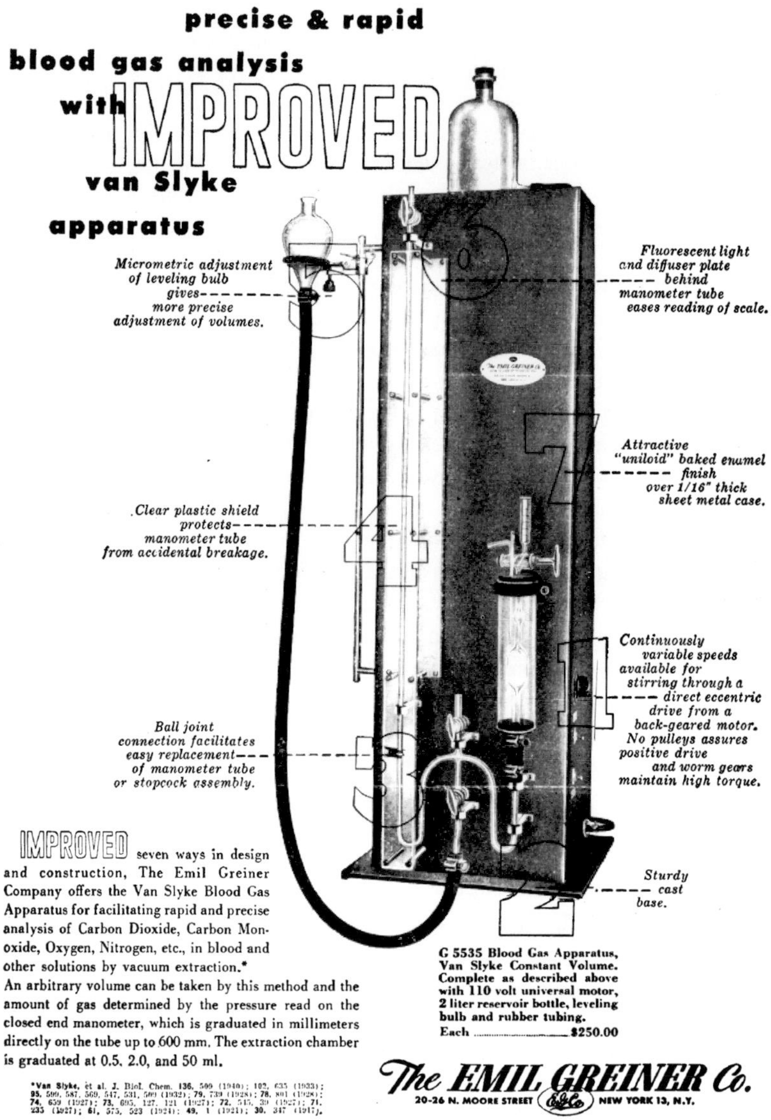

Figure 2.3. Modified Van Slyke apparatus sold by The Emil Greiner Company, *ca.* 1948–1952.

bicarbonate in relation to carbonic acid) required to keep the blood pH within a narrow range, led him to consider other possible disturbances in the reaction of the blood and in the acid-base balance. In this manner, Van Slyke, using theoretical chemical arguments, established "normal values" for pH and bicarbonate reserve. He also elaborated a classification of nine *theoretically* possible states of the acid-base balance of the blood, eight of which were expected to be pathological. Van Slyke's classification relied basically on theoretical arguments and no experiments were reported in the article in which the extension and classification were first discussed. Van Slyke was aware that the classification he was proposing was theoretical and based on chemical possibilities, but he emphasized that its support is also clinical and physiological, and argued that even though when he began his studies of acidosis he had considered only two of the pathological possibilities, it was now "known that the other six abnormal possibilities can be produced experimentally, and that at least some of them occur clinically" (Van Slyke, 1921, p. 153). Both in the text of the article and in its tables, Van Slyke correlated the various potential chemical abnormalities with known pathological states and diseases (such as hysteria, encephalitis, vomiting, emphysema, diabetes, nephritis or diarrhoea), but his starting points were not the various clinical conditions and differences between them, but the various (bio)chemical possibilities. Pathologies or abnormalities were conceived of primarily not as disturbances of specific physiological functions, but rather as various chemical states of the organism whose causes could vary. The same pathological condition could result in different chemical states and a specific chemical state was associated with a variety of different pathologies. Moreover, "different means, depending on the nature of the primary case, may be required in each cause to restore and maintain normality" (p. 164). The new classification of abnormalities in blood reaction constituted in many respects a new chemical representation of disease, and while Van Slyke made attempts to correlate this representation with clinical phenomena, he paid very little attention to its correlation with physiological or pathological mechanisms. Van Slyke's portrayal of the normal and abnormal reactions of the blood can also serve, in my view, as the marker of the specific kind of reductionism involved in Van Slyke's research: *it is not just that physiological or pathological states are redescribed in chemical terms, or that physiological phenomena are given a chemical interpretation, but rather that the chemical theoretical representation itself became autonomous, structuring classifications, ideas of disease, and the course of further research.*

The theoretical discussion served to establish reaction of the blood as a subject of clinical interest extending well beyond the problem of acidosis. At the same time, it led from Van Slyke's studies of acidosis to biochemical investigations of gas and electrolyte equilibria in blood. In these studies the possibilities offered by the new apparatus were extended and combined with questions about the chemical mechanisms "by which blood is enabled to carry its allotted load of carbon dioxide, and nevertheless maintain its hydrogen ion concentration at such a low and remarkably constant level" (Van Slyke, 1921a, p. 153). The questions raised in this series of investigations were more distant from both clinical practice and clinical problems,[21] and although a few of Van Slyke's co-authors in the series of articles on gas and electrolyte equilibria were clinicians, the investigations resulted from a cooperation with Lawrence Henderson at Harvard begun in 1920.[22] Van Slyke and Henderson together with their collaborators engaged here in studies of blood as a physico-chemical system. Through a series of theoretical arguments and quantitative analyses and a variety of complex measurements, they attempted a purely physico-chemical description of electrolytic equilibria, buffering, and osmotic pressure of the different phases of blood.

L.J. Henderson summarized the collaboration in a series of lectures at Yale University which were published in 1928 in a monograph entitled *Blood: A Study in General Physiology*. The controversy which followed the publication of Henderson's book, in which Van Slyke predictably intervened on his collaborator's side, brought to the fore the distinctiveness of L.J. Henderson's as well as Van Slyke's approach to biochemistry: their treatment of the subject purely as a problem in physical chemistry and their disregard for physiological processes. The critiques of the Yale physiological chemist, Yandell Henderson and of John Scott Haldane disputed the possibility of reducing the treatment of blood as a system to such purely chemical language:

> If the discussion of the regulation of the composition of the blood is taken for what it seems to mean, it presents a conception of equilibrium which is sound physical chemistry, as it deals with stationary states, but is essentially erroneous as a description of physiological functions, since the characteristic equilibria of life are dynamic. The energy equilibria of the organism are as much in flux as the chemical equilibria of metabolism. (Y. Henderson, 1928, p. 39)

Similarly, Haldane alleged:

> L.J. Henderson treats the constancy of reaction in the living body as if it depended on the physico-chemical properties of blood. In actual fact this constancy depends during health on the coordinated activity of the kidneys and respiratory organs, in accordance with Bernard's principle; and in various individual parts of the body the constancy depends on the coordinated or regulated influence of the circulation. Not all the buffering in the world would keep the reaction constant otherwise, though the buffering greatly smooths the regulation … (Haldane, 1929, p. 453)

Van Slyke rejected both critiques,[23] but there can be little doubt that both Henderson and Van Slyke tended to treat blood as an isolated system, studied and described it purely in terms of physical chemistry, and paid relatively little attention to the physiological processes involved.[24] Their critics — either in the name of "organicism," like Yandell Henderson[25], or in the name of Claude Bernard's "milieu interieur," like J.S. Haldane — rejected this form of reductionism. I have discussed this controversy elsewhere (Amsterdamska, 1995); for the purposes of the discussion here, however, it is illuminating to mention only the treatment of the pathologies of blood in Henderson's book: Henderson refers there to Van Slyke's classification of acidoses and alkaloses discussed above, and lists anemia as a condition "in which the concentration of hemoglobin in the blood is considerably diminished below what is regarded as its normal value" (L.J. Henderson, 1928, p. 264). The rest of the chapter consists of a discussion of joint variations in the concentration of the various chemical components of the blood and their agreements and presentation of numerous charts (nomograms) for various patients at various stages of disease. Henderson does not attempt to account for the various disturbances in blood reaction or composition or function in terms of pathologies of other organs or in terms of physiological processes: anemia, diabetes, nephritis, myxedema, etc., are presented exclusively with the help of charts listing the chemical conditions of blood, serum and cells in venous and arterial blood (water content, pH, chloride, bicarbonate, oxygen, hemoglobin concentration, and so on). Dismissing "verbal analysis of … pathological states" as unprofitable and extolling the elegance of mathematical language, Henderson regarded the presentation of the pathologies in the nomographic charts as entirely sufficient:

> all that has been said in this chapter and the last concerning the blood in pernicious anaemia, in nephritis, and in the normal man during work is either explicitly presented in the nomograms and tables or else easily deducible from them. The verbal analysis is here carried through chiefly to illustrate the meaning and use of these nomograms and tables …
>
> […] But we should reject one of the most precious advantages of mathematics … if we were to translate into English that which can be well said only in the nomographic idiom. Therefore, the results of a few other pathological and clinical studies are presented in the following pages exclusively in the form of nomograms and tables (*Ibid.*, p. 297).

In effect, Henderson, who unlike Van Slyke, did not work in the clinic and did not seem to be particularly interested in making the results of his work clinically relevant,[26] carried the chemical representation of disease even further than Van Slyke. But there are telling similarities between the two in the attempt to make quantitative, chemical language self sufficient, and in any case, independent of physiological or anatomical language.

As the enumeration makes clear, Van Slyke's studies of acidosis provided him with a number of various possibilities to maintain continuity in his research. Some of these possibilities relied again on very close cooperation with clinicians, though biochemical extensions which were not obviously clinically motivated could also be taken up. Both the methods developed by Van Slyke and the theoretical arguments on the basis of which the test for acidosis was developed provided Van Slyke with starting points for new research paths. In so far as they all went back to the initial method, these new research directions were clearly extensions and elaborations of previous research, but apart from sharing a common descent, they bore relatively little relationship to one another.

NEPHRITIS

Just as Van Slyke's initial interest in acidosis followed an interest of clinical researchers at the Hospital, so also studies of nephritis (or Bright's disease) had been conducted in the Hospital before Van Slyke became directly involved. As early as 1915, Franklin McLean studied the renal function in diabetes and a few patients suffering from nephritis were being treated at the hospital. Prior to World War I, nephritis, and particularly edema that was one of its most serious symptoms, was also studied by Walter Palmer and Reginald Fitz. Van Slyke appears to have been marginally involved in these prewar investigations of nephritis but a decision to focus on the disease was made after World War I, when, following the departure of Allen, diabetes ceased to be a focus of attention in the hospital. The first of the many clinicians who worked with Van Slyke on the investigations of nephritis was Edgar Stillman who had already worked together with Van Slyke on acidosis. It appears that a plan for the study of nephritis was drawn up jointly with Stillman.

Preferences of the clinicians, however, were not the only reason Van Slyke became interested in nephritis. As he remembered it many years later, he took up "the renal problem because there was a chance for quantitative chemistry there — measurement of functions and the metabolic things that the kidney did" (Olch, 1971, p. 25). Van Slyke continued to explore these opportunities for quantitative chemistry throughout the extended period when research on nephritis and kidney function was conducted at the Rockefeller Hospital. But whereas Van Slyke's studies of acidosis and other aspects of the acid-base balance were linked by a common method (which meant that eventually the fact that the research began with the study of diabetes receded from view), the long term investigations of nephritis were held together by their object, the disease. Van Slyke did in fact develop his own measure of renal function and repeatedly improved on the methods and instruments which could be used to determine the range of chemical substances whose presence in blood or urine was affected by renal problems, but the method (or rather, methods) used in this study were not as central to the research program as a whole as they were in the study of acidosis. The stated aim of the investigations undertaken in the early 1920s together with Stillman and Austin was to develop a clinically useful classification of nephritis by collecting an extensive and systematic set of clinical data on patients suffering from the disorder.[27] Studies of specific problems such as links between edema and blood protein content or the apparent disturbances in fat metabolism were also envisaged.

Initially, however, Van Slyke's research did focus on the development and improvement of methods for measuring various aspects of the functioning of the kidneys. The method for assessing the functional capacity of the kidney through the measurement of the rate blood urea clearance had been developed a few years earlier by two French researchers, L. Ambard and A. Weill. Van Slyke and his clinical colleagues, however, "experienced trouble" when they attempted to use it: "The relationship between concentration of urea in the urine and rate of urea excretion assumed by Ambard holds so loosely that it is often difficult to ascertain any relation at all."[28] Van Slyke, Stillman, and Austin were not the first ones to find the results of applying Ambard's formula subject to inexplicable variations[29] and their attempt to reduce them involved a revision of Ambard's formula by taking into account the volume of urine (and weight of the patient) in their calculations. In their first article on the subject, Austin, Stillman and Van Slyke (1921) supported the theoretical arguments for the necessity of modifying the existing method by presenting repeated measurements of the relevant values in both animals and healthy human subjects. Although Van Slyke's avowed aim was clinical application in diagnosis and in clinical studies of the course of nephritic disorders, the measure described in the article was believed to require further investigation in clinical research:

> The work presented in this paper has been undertaken with the frank intention of obtaining knowledge of normal urea secretion in order that pathological deviations therefrom might be recognized and studied with additional accuracy. Before application of the results to routine diagnosis can appear justified, however, it is necessary to compare the relationships of urea excretion to the governing factors as established in normal individuals with the relationships found in a sufficient number of patients under the most complete and prolonged clinical observation. The results of such comparison may be expected to indicate the possible variety of interpretations, diagnostic and prognostic, that may attach to observed abnormalities in the excretion. Until this has been accomplished, the functional index must be regarded as an object of investigation, rather than as an aid in the clinic. (*Ibid*., p. 110)

As the extended quote makes clear, research on the method for estimating blood urea clearance not only had an explicit clinical goal, but the considerations which led to the various modifications of the techniques and equations were also clinical in character. The problem of the variability of the measure in normal subjects was a foremost consideration, since it was this variability which rendered the specification of normal and pathological ranges difficult. Accordingly, Van Slyke's modifications were designed so as to render the measure as stable and as standardized as possible by taking into account the various factors on which variations in the rate of urea excretion depended. Clinical studies of rates of urea excretion in patients suffering from different types of nephritis and in healthy subjects were thus an integral part of Van Slyke's research (Möller, McIntosh, and Van Slyke, 1928a and 1928b).

It is instructive to compare Van Slyke's understanding of the rate of urea clearance with that of Thomas Addis, a clinician and researcher working at the Stanford University Medical School. For Addis, the utility of the measurement of kidney function depended primarily on how well such measurement could serve as an indicator of morphological damage to the kidneys: functional tests were meaningful to a clinician if they could be interpreted as measures of the extent of the renal lesion (Addis *et al.*, 1918; Addis, 1928; see also Peitzman, 1986 and 1989). For Addis biochemical measurements, along with qualitative and quantitative assessments of casts, blood, and urine were signs of renal lesions and parts of a "qualitative and particular" "picture he [a physician] is making in his mind as to the disease and the effect of the disease on this particular patient" (Addis, 1948, p. 4). Addis was profoundly concerned with providing clinicians with methods for "making visible a part of his patient that would otherwise be invisible" (Addis, 1948, p. 3) and since "in a living patient we can only measure what kidneys do," his goal was to find out if "a fixed relation exists between the amount of renal tissue and some quantitative measure of renal function" (Addis, 1928, p. 623).

Van Slyke was far less concerned than Addis with the precise interpretation of the variations in renal function in (visual) morphological terms. Blood urea clearance was for him not an *indirect indicator* of a lesion, but *a direct measure* of how well or badly the kidneys as "chemical machines" were performing their work. The test was useful in clinical practice and in clinical research not when it could be given a clear interpretation in terms of a structural lesion or in terms of a specific physiological malfunction, but when it provided quantitative results which could be correlated in a stable and reliable manner with other clinical signs or symptoms of disease, allowing for a clear differentiation of pathological and normal ranges and permitting the making of a clinical prognosis. In the article describing the measurement of the rate of urea clearance, Austin, Stillman, and Van Slyke (1921, p. 111) explicitly asserted that their methods "afford a basis for detecting abnormalities in urea excretion, but not for interpreting the significance of such abnormalities." The ultimate value of the method, however, was not its precise meaning but the fact that it was "more sensitive than previously used indicators of deficit in renal function," and in all cases of severely diminished clearance,

> the clinical history, hematuria, albuminuria, and other data have indicated the presence of renal disease, and when the condition has progressed there have been eventual retardation of phthalein excretion, increase in blood urea, and creatinine. But compared with the blood urea clearance these signs are comparatively tardy warnings of decreased function.[30]

Improvement of diagnostic and prognostic methods was an intermittent feature of Van Slyke's research on nephritis. Such improvements led to modifications both in the equations used to calculate the relevant values and in the techniques used to conduct the determinations. The most important of such procedural changes was probably the modification of the gasometric apparatus (earlier used in the studies on acidosis) which was now adopted for the measurement of CO_2

formed by the decomposition of urea with urease (Van Slyke, 1927). Considerations of speed and simplicity and accuracy in routine clinical use were again cited as the major advantages of the gasometric measurement.

These attempts to validate and improve the proposed methods were, however, not as prominent in Van Slyke's collaborative studies on nephritis as the systematic efforts to describe the various chemical changes which accompany it. Between 1921 and 1948,[31] Van Slyke and his clinical collaborators attempted to chart the clinical course of different types of nephritis in terms of a large variety of chemical changes with which the disease and its progress were associated. Beginning with an extensive study of changes in quantity and ratio of plasma proteins (Linder, Lundsgaard and Van Slyke, 1924; Linder, Lundsgaard, Stillman and Van Slyke, 1924), Van Slyke worked together with clinicians in describing fat and carbohydrate metabolism in nephritic patients (Hiller *et al.*, 1924; Linder *et al.*, 1925), their excretion of proteins, glucose, ammonia and acid, phthalein, blood creatinine, plasma electrolytes, etc. (for example, Van Slyke *et al.*, 1925; Hawkins *et al.*, 1929; Hiller *et al.*, 1927). In contrast to the work on acidosis the results of which were published mainly in biochemical journals and particularly the *Journal of Biological Chemistry*, most of the studies on nephritis were published in clinical journals such as the *Journal of Clinical Investigations* or *Medicine*. Virtually all of the articles reported on correlations between the clinical histories of patients suffering from different forms of the disease and the various chemical changes accompanying the disease and its progress. The most extensive such study, "Observation on the Course of Different Types of Bright's Disease," published in *Medicine* in 1930 (Van Slyke *et al.*, 1930)[32] presented data on 67 clinical cases of Bright's Disease and attempted to correlate both different forms of the disease (according to Addis' classification) and their progress with measures of blood urea clearance, plasma protein content, proteinuria, hematuria, anemia, blood pressure, and urine sediment (see Figure 2.4). In Van Slyke's research, nephritis became a collection of quantitative chemical indicators and changes which could be measured by biochemical methods and procedures. The clinical study of that disease on which he embarked together with his clinical colleagues was designed to chart these chemical changes. Again, the collaboration between the chemists and the clinicians was a *sine qua non* of this type of research: it was clinical research conducted by (sometimes innovative) chemical means and as such would be impossible without a close collaboration between the chemists and the clinicians.

Van Slyke's studies of nephritis eventually led him to engage not only in chemical and clinical investigations but also in experimental physiological research. Working together with C. Rhoads, the Hospital pathologist who developed a technique of explanting kidneys of dogs, Van Slyke attempted to determine experimentally "the factors which are responsible for the relative constancy of urea clearance" and for its change (Van Slyke *et al.*, 1934, p. 337).[33] Using his measurement techniques for urea clearance, Van Slyke also addressed the question of the adequacy of two competing theories of the mechanisms of urine

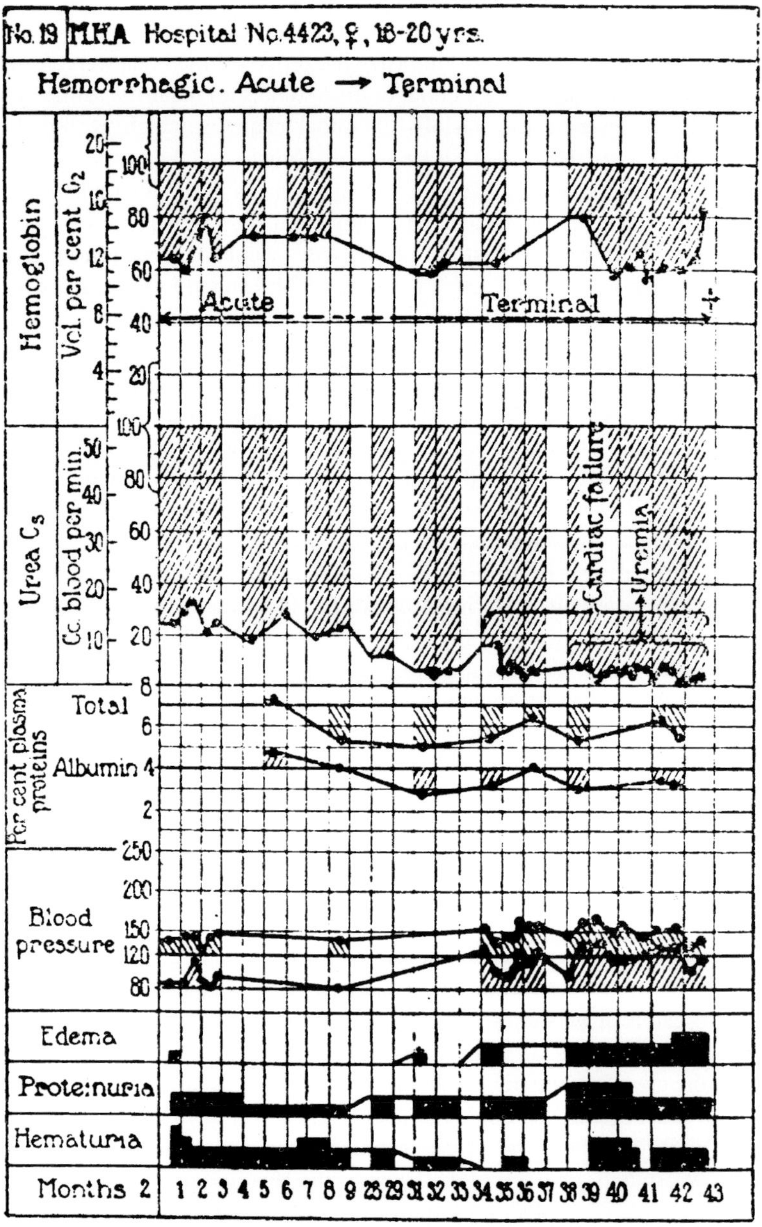

Figure 2.4. Charting the course of nephritis in chemical parameters.

formation.[34] These more physiologically oriented studies fitted well with other physiological research on kidneys conducted at that time, which, as Steven Peitzman (1986) has described it, was dominated by the development of chemical techniques and assays for functional measurements. As long as kidneys were regarded primarily as "chemical machines," Van Slyke saw a direct continuity between the clinical and the physiological investigations: both focused on chemical determinations of the constituents of blood and urine as the means of measuring function.

Van Slyke's collaboration with clinicians and with physiologists depended on this shared chemical definition of kidney function and on the idea that chemical representations of nephritis should be primary in both clinical practice and clinical research. When these conditions were not met, the collaboration was not always successful. In 1928, Addis[35] was invited to spend a year at the Hospital to take part in the studies of nephritis "in association with Dr. Van Slyke."[36] No articles co-authored with Van Slyke resulted from the visit, and neither in the scientific reports nor in his published work does Van Slyke ever mention collaboration with Addis (even though he does use his classification of different types of nephritis). When in 1930, Cole attempted to get Addis back to the Hospital, Van Slyke insisted that this would be possible only if they were to work completely independently of one another:

> Addis is a delightful person and a man of ideas. He would not at all fill the place that Hannon has had, however. We could take Addis only as an entirely independent member, with his own staff and laboratory. He would require also his own clinical interns if he were interested in observing patients, which he was not to any great extent when he was here a year ago. He is full of his own excellent ideas, which grow and change from day to day as do the ideas of any thinking man. The only way to have him would be to have him utterly free, and unfettered, to me or anyone else. This became entirely apparent when he was here. If you or Dr. Flexner could see your way to add a completely new and independent division to the hospital for Addis, no one could be more delighted than I. But it would be utterly impossible to expect him to give up his own progressing line of thought and experimental work in order to adapt himself to the line along which my work has been developing. It would be splendid if we were both working in the same place, but independently. There would be a continual and profitable exchange of ideas, but there could be no constraint to cooperation in work.[37]

Van Slyke never explicitly stated why he considered Addis' "line of thought" to be incompatible with his own. Addis' own published references to Van Slyke's research are noncommittal. In the book on glomerular nephritis addressed to clinicians in which Addis summarizes his own research on the disease and its clinical management, however, Addis does express explicit misgivings about letting the chemists "do the work" for the clinicians:

> The ordinary clinical methods may reveal that something is wrong, but what and how much is left obscure. Faced with this situation the modern clinician, like Dr. Bright,

> turns to the chemist for help, and the records of his patients are embellished with the results of determinations of the quantities of various materials in the urine and blood. However, these figures have not become a part of the clinicians armamentarium. They do not give him what he wants. [...]
>
> We have waited long enough in the hope that chemists will do the work for us. They never will, for to do so would be a prostitution of all they are (Addis, 1948, p. 3–5).

Addis was objecting here (and repeatedly through his book) precisely to the kind of division of labor between chemists and clinicians on which Van Slyke's entire research program depended: the application and development of ever more complex tests and methods of analysis provided by the chemists and tested and elaborated on by clinicians. And, as Addis also realized, this kind of collaboration depended on the primacy of a biochemical conception of disease which was to replace the idea that a clinician comprehends the chemical signs of disease "not in the form of number but as a part of the picture he is making in his mind as to the disease and the effect of the disease in this particular patient" (Addis, 1948, p. 4).

TRANSLATIONS AND PRACTICES

When he translated clinical problems into biochemical problems, Van Slyke relied on the available stock of physiological and chemical knowledge. In this manner, the problem of accurately diagnosing the severity of diabetic acidosis became first, the problem of measuring the acid-base balance in an organism, then of measuring the alkali reserve of the blood, and eventually of estimating the CO_2 capacity of blood plasma. Once such a translation was made and incorporated into a diagnostic method and an instrument, Van Slyke repeatedly returned to the clinic where the procedure had to be standardized and made clinically meaningful. Simultaneously, he generally proceeded also to extend his (bio)chemical theorizing without linking it further to physiology. Chemistry of the blood would then serve as the basis for further research *and* as a means of classification for pathological conditions. Disturbances in the acid-base balance, defined as a chemical problem, gradually came to dominate the manner in which the various clinical problems were to be understood, while the apparatus and the techniques developed in collaboration with the clinicians made the chemical translations clinically appropriate.

Similarly, the study of nephritis which had begun as an attempt to develop or improve appropriate diagnostic techniques, gradually became restructured into an investigation of primarily chemical phenomena to be investigated in a clinical context. Physiological research was undertaken only when the entire functioning of the kidney was translated into chemical terms. Clinical definitions and classifications remained central as Van Slyke charted the clinical course of the disease in terms of a variety of chemical changes in the organism.

Theoretically, Van Slyke's biochemical and physiological translations could have also provided the basis for the formulation of questions about physiological function and could have sustained a development of his program in the direction of chemical physiology. But Van Slyke rarely attempted to elaborate on these intermediate, physiological contexts. Once they were "built into" the diagnostic procedures and the apparatus for the quantitative determination of clinically relevant chemical substances, they disappeared as relevant contexts. Collaboration with physiologists (Allen), for whom biochemical techniques were only means to another end, was difficult. In a sense, Van Slyke was continually testing the appropriateness of his biochemical translations by relating them to the clinical contexts: establishing that the CO_2 capacity is indeed a good measure of acidosis, or that changes in the protein content of plasma do indeed accompany changes in the severity of nephritis. In other words, Van Slyke's translations were generally not used to take him away from the clinic and into the laboratory for any length of time: these translations were continually checked, articulated, elaborated, and modified in terms of how well they fitted into the clinic. And, when the clinic was abandoned for the laboratory, research usually became purely chemical both in terms of the methods used and in terms of the interpretations provided. This marginalization of physiology both in the laboratory and in the clinic can be understood as a direct outcome of the local arrangements in which the chemists were expected to collaborate directly with the clinicians, provide solutions to clinical problems, and participate in clinical research: questions about physiological causes and mechanisms receded into the background as the number of chemical parameters used to diagnose, monitor, or investigate disease became ever more prominent.[38]

In describing two research lines in which Donald Van Slyke was engaged for much of his career at the Rockefeller Institute Hospital, I have emphasized the close integration between the clinic and the laboratory, and thus also between the clinician and the biochemist, which the type of research conducted by Van Slyke required. The collaboration between the physicians and the chemists is visible in almost every aspect of Van Slyke's research: in how research problems were formulated, in the manner in which the many methods and the apparatus for determination of chemical substances were designed, tested and evaluated, in the kinds of data that were collected and presented, and in the very conduct of investigations which required the participation of both chemists and physicians.

Analyzing the emergence of biochemistry in the United States, Robert Kohler pays particular attention to the first generation of American clinical chemists who, by establishing a new relationship between clinical medicine and chemistry, laid the foundations for the emergence of biochemistry as a distinct and autonomous biomedical discipline. According to Kohler, the success of clinical biochemists depended on the establishment of a new partnership between clinicians and the biomedical researchers:

> Each partner accommodated his professional role and ideology to the needs and expectations of the other. Professional biochemists trained clinical investigators and developed ever more sophisticated diagnostic procedures. Hospital physicians provided

> biochemists with abundant experimental materials and attractive research problems. Clinicians enjoyed the prestige of academic connections and an aura of "objective" and "disinterested" science. Biochemists, physiologists and others enjoyed larger resources and the prestige in participating in the progress of medicine. Clinicians learned to live with independent specialists; biochemists learned to adapt clinical imperatives to their disciplinary goals (Kohler, 1982, p. 217).

In his description of the clinical biochemists, Kohler specifies the institutional conditions which made the partnership between chemists and physicians possible and emphasizes the institutional benefits that accrued to both groups as "clinical problems provided opportunities to extend and improve analytical procedures; in turn, new techniques led to discoveries in the physiology of disease" (*Ibid.*, p. 245). But while Kohler sees the relations between the biochemists and the clinicians in terms of exchange relationships, he pays little attention to the fact that this partnership also depended on collaboration in research and involved mutual accommodations between distinct research practices and distinct conceptions of health and disease. Van Slyke was surely one of the most prominent representatives of this first generation of biochemists, and the examination of his research makes it amply apparent that it would be misleading to treat his relationship with the clinicians as only a marriage of convenience, a matter of professional ideology or a strategy of discipline formation. It was also an integral component of the research process itself: problems encountered by the physicians were not taken up by the biochemists without undergoing substantial transformations, the development of methods was attuned to the practical exigencies of their future clinical use, and their validity, significance and usefulness were established with extensive participation of biochemists in clinical research. The cooperation between the chemists and the physicians involved the articulation and the mutual adjustment of a number of chemical and clinical evidential contexts and the theoretical and experimental elaboration of mutually constraining chains of translations. The partnership was thus grounded not only in the political economy of their disciplines, but also in the research process itself and in the intellectual cooperation on which success depended. In Van Slyke's case, this partnership was also built on the basis of a conception of disease as primarily a collection of chemical disturbances which could be quantitatively measured using the instruments biochemists developed for this purpose and on the biochemists' conscious attempts to conform to the constraints of clinical practice in designing diagnostic instruments and techniques and their willingness to participate actively in clinical research. But the collaboration also depended on the willingness of the clinicians to accept the biochemical mode of understanding disease and the presence of chemists at the bedside — when such an understanding of pathology was not shared (as in the case of Addis), collaboration did not develop.

The development of such collaborative forms of biomedical research was greatly facilitated by the organizational arrangements and ideological commitments at the Rockefeller Institute Hospital. The faith of the Hospital's founders and directors that laboratory research aimed at the chemical understanding of disease will bring improvements in clinical medicine found its organizational expression in their attempts to staff the Hospital with (bio)chemists as well as physicians, their insistence that all clinicians be involved in laboratory and clinical research on one of the diseases treated in the Hospital, and the provision of laboratory facilities close to the bedside.

In many ways, Van Slyke's research was a continuation of the old practices of clinical chemistry: he focused on analyses of blood and urine and on diagnostic techniques. It might be surprising to see how well these traditional practices fitted into what was proclaimed the most advanced medical research institute. Van Slyke's lab in the Hospital might not have been intended for "routine diagnostic determinations," but it certainly devoted much of its effort to the development of "routine diagnostic determinations" which at the Rockefeller Hospital mediated the relations between laboratory and clinical research. Van Slyke is occasionally described as a quintessential "methods man" and his name seems to be permanently associated with his gasometric apparatus which was widely used in clinical and chemical laboratories, and bore his name.[39] His research, however, not only brought new methods and instruments to the bedside practice and to the laboratory, but also promoted a way of treating disease as a collection of chemical phenomena and brought with it a way of classifying and understanding pathologies in terms of quantitative chemistry. Several obituaries of Van Slyke referred to him in their titles as a "modern iatro-chemist" (for example, Hastings, 1964). Both the traditionalism of Van Slyke's approach and his way of understanding disease make this a uniquely appropriate label.

ACKNOWLEDGEMENTS

I would like to thank Anne Löhnberg for her help in gathering the material used in this paper, Patricia Gossell for providing me with trade catalogues and advertisements of the Van Slyke apparatus which she found at the Smithsonian, and the archivists at the Rockefeller Archives and the American Philosophical Society whose help made it possible to complete research for this article during my exceedingly brief visits to Terrytown and Philadelphia. I am also grateful to the editors of this volume, the participants in the workshop on the "Molecularization of Biology and Medicine," and members of the "medical group" in the Department of Science and Technology Dynamics who have offered valuable comments, criticisms, and suggestions on the various drafts of this paper.

NOTES

1 Institutional and practical contexts of research tend to disappear when historians attempt to trace the development of biochemical knowledge in specific areas (for example, intermediary metabolism, structure of large molecules, or bioenergetics). In prewar biochemistry, research on any given problem

was likely to proceed in a variety of different institutional settings and intellectual contexts; by treating such work as a part of a single tradition, both the researchers and the historians tend to decontextualize it. For example, Edsall's (1972 and 1985) reconstruction of the studies of the blood reaction brings together research conducted in clinical contexts (conducted, for example, to gain a better understanding of diabetes) with research conducted as part of organic chemistry or as part of physiology. Holmes (1992) is explicit about this problem in his discussion of intermediary metabolism as an "investigative stream" which involved work conducted in a variety of intellectual and institutional contexts, but he does not attempt to examine how the "potential relevance of the chemistry of life for a better understanding of disease" (p. 57) shaped the work of the scientists whose contributions he studied. For example, Holmes mentions research directed to the understanding of diabetes only in terms of its contribution to the pursuit of knowledge of intermediary metabolism.

2 Thus, Fruton (1992, p. 185, see also 1983) says that he deplores Kohler's assumption that "biochemistry is, and has been, 'a biomedical discipline'". In his detailed analysis of the career of Hans Krebs, Holmes (1991) explicitly asserts that he sees no connection between Krebs' activities as a physician (and more generally the various institutional settings in which he worked) and his biochemical investigations of intermediary metabolism. At most, Krebs' clinical responsibilities are seen as an interference because they took Krebs' time away from the laboratory. While such a complete separation of clinical and scientific work seems to have characterized Krebs' work, it appears untenable when one examines the activities of many American biochemists working during the same period.

3 Curiously, it might be that both Kohler and his critics downplay the interest of the medically oriented research because they share a common interest in the later development of biochemistry as an independent (i.e., not subordinate to medicine, physiology, or organic chemistry) discipline, a chemical science of life, a "basic biological science," and a precursor or competitor of molecular biology.

4 Simon Flexner's "Report to the Board of Scientific Directors of the Rockefeller Institute," *Scientific Reports of the Rockefeller Institute for Medical Research*, Box 5 (1930), p. 421 (Rockefeller University Archives, Rockefeller Archive Center, Tarrytown, N.Y.). Hereafter abbreviated as *Scientific Reports*.

5 In a retrospective assessment of the first years of the Hospital, S. Flexner recalled: "One of the greatest difficulties originally related to finding scientific staff to carry on the work of a research hospital. To a large extent the staff had to be created" (*The Developments of the First Twenty Years of the Rockefeller Institute and an Outlook on Future Growth*, Report presented at the meeting of the Board of Scientific Directors, *Scientific Reports*, Box 4 (April 17, 1926), p. 67.)

6 For more biographical information on Van Slyke see Hastings (1964, 1972, and 1976), Archibald (1971), Sendroy (1971).

7 Finding the staff with expected "scientific attitude" was clearly difficult. In a 1914 report, Cole referred to these difficulties when announcing Van Slyke's appointment: "An important change in the organization of the Hospital staff has taken place during the past year by the appointment of Dr. Donald D. Van Slyke to take charge of the chemical laboratory. The organization of this department of medical work has offered the most serious difficulties, and it has been very difficult to obtain a man of proper training and scientific inclination to act as director of this laboratory" ("Report of the Director of the Hospital to the Corporation, 1913–14," *Scientific Reports*, Box 2, Vol. 3 (16 October 1914), p. 387).

8 In a letter offering Van Slyke a position at the Hospital, Flexner carefully spelled out the conditions for Van Slyke's transfer, citing a letter from R. Cole: "Dr. Van Slyke will for the present carry on his work on protein metabolism as he has already outlined it … This will give him time and opportunity to become acquainted with pathological problems which might profitably be studied on patients in the Hospital … While I hope that Dr. Van Slyke will become interested in clinical problems and in the work now being carried on in the Hospital, I understand that he is free to do so or not as he sees fit, and if he so desires during the period it is now arranged for … he is at liberty to occupy his entire time on work such as he would have carried on in the Institute laboratories." (Simon Flexner to D.D. Van Slyke, November 6, 1913, *S. Flexner Papers*, Van Slyke file #1, American Philosophical Society, Philadelphia). See also Corner (1964, 274–75).

9 "Report of the Director of the Hospital to the Board of Scientific Directors," *Scientific Reports*, Box 2, Vol. 3 (10 January, 1914), p. 290.

10 D.D. Van Slyke to S. Flexner, June 22, 1914; Aug 11, 1914; S. Flexner to D.D. Van Slyke, July 22, 1914; August 15, 1914. *S. Flexner Papers*, Van Slyke file #1, American Philosophical Society, Philadelphia. D.D. Van Slyke to R. Cole August 16, 1914, *R.I. Cole Papers*, Van Slyke file #1, American Philosophical Society, Philadelphia.

11 Van Slyke remembers that he teamed up with Stillman when the latter had problems with the determinations of oxybutyric acid in the urine of diabetic patients and that they worked together on diabetes from early on, but he does not mention Stillman in his recollections of how he came to work on diabetes (see Olch, 1971, pp. 71–72 and 23–24).

12 "Report of the Director of the Hospital to the Corporation, 1913–14," *Scientific Reports*, Box 2, Vol. 3 (October 16, 1914), p. 398.

13 It should be noted here, that this was precisely the period when methods for the determination of reaction in terms of hydrogen ion concentration (pH) were elaborated by S.P.L. Sørensen (1909) and the measurements of the reaction of complicated organic substances such as blood and urine were being worked out.

14 "We define acidosis as a condition in which the concentration of bicarbonate in the blood is reduced below the normal level. The definition appears a necessary preliminary because of present confusion in the literature, different authors regarding acidosis differently as 'acid intoxication,' as a condition in which acetone bodies are formed, or as an actual increase in the hydrogen ion concentration of the blood" (Van Slyke and Cullen, 1917, p. 291).

15 Van Slyke also remembered the importance of Henderson's work for the early studies of acidosis at the Rockefeller: "the paper that really got us steered on our work was a paper by Henderson on the acid-base balance of the blood, that was published in German" (Olch, 1971, p. 43). The paper being referred to is probably Henderson (1909).

16 For a discussion of the critical attitude of the British physiologists J.S. Haldane and C.G. Douglas towards this purely chemical and one-dimensional understanding of acidosis, see Sturdy (1987, pp. 335–350). Sturdy argues that Haldane's opposition was based both on his physiological understanding of the various acidoses and on the specific conception of medical practice which he embraced.

17 "Report of the Directors of the Hospital to the Board of Scientific Directors," *Scientific Reports*, Box 3, Vol. 4 (June 12, 1915), p. 113.

18 Similarly, in the internal reports, Van Slyke emphasized that the simplicity of the method will make it popular in the clinic: "This method is extremely simple and has already come into wide use and will in the future probably prove to be one of the standard laboratory methods." ("Report of the Director of the Hospital to the Corporation, 1916–17," *Scientific Reports*, Box 3, Vol. 5 (October 1917), p. 272).

19 For example, comparing two methods for the estimation of the same substance, Van Slyke, Stillman and Cullen wrote: "In some types of experimental work the theoretical perfection of the titration will doubtless give it an advantage, but for ordinary purposes, especially for clinical diagnosis, the choice between the two methods appears entirely one of convenience. The carbon dioxide capacity has the advantage that the extraction and measurement of the gas require slightly less time than even the few minutes necessary for titration, and require no accurate standard solutions; while the titration has the advantage that it requires no special apparatus. [...] For clinical and most experimental purposes, however, it appears that the two methods give so nearly identical results that they may be used interchangeably" (Van Slyke *et al.*, 1919, p. 177). Similarly, in discussing the method used for the drawing of blood, Van Slyke considers the effects of the exposure of the sample to air, and concludes that the error thus introduced is clinically irrelevant: "The carbon dioxide loss [...] measurably lowers the bicarbonate content of the plasma. As a matter of experience, however, the gas exchange which occurs in the interval ... during which the blood is falling is only enough to bring the carbonic acid content of the venous blood to that of the arterial" (Van Slyke and Cullen, 1917, p. 306).

20 "Report to the Board of Scientific Directors," *Scientific Reports*, Box 3, Vol. 6 (April 20, 1918), pp. 228–229.

21 According to Van Slyke, however, "the blood gas work all developed out of the diabetes" (Olch, 1971, p. 27).

22 Edsall (1985) places this collaboration in the context of research on the physical chemistry of blood. On the role of physical chemistry in the development of other chemical fields, including biochemistry, in the U.S., see Servos (1990).

23 "I have read Haldane's discussion of L.J. Henderson's book in *Science*, and can understand Haldane's remarks only on the assumption that he has not read Henderson's book and is unfamiliar with the work on which it was based" (D.D. Van Slyke to S. Flexner, May 2, 1929, *Flexner Papers*, Van Slyke file #6, American Philosophical Society, Philadelphia, PA). Van Slyke's public response to Yandell Henderson was published as "The 1928 Silliman Lectures" (Van Slyke, 1929). It focuses on allegations that the book is metaphysical and that Henderson did not acknowledge the work of other physiologists and biochemists.

24 L.J. Henderson is often remembered for his "organicism", his treatment of both social organization and organization of life as complex systems striving for equilibrium. In this context, the fact that he was criticized for his reductionism appears somewhat ironic. We should remember, however, that the problem of "integration," "homeostasis" or "equilibrium" might be posed on a variety of levels, and that Haldane's and Yandell Henderson's criticisms were addressed mainly to Henderson's treatment of such equilibria at a purely physico-chemical level. On Henderson's "organicism" see articles by Cross and Walbury (1987), Heyl (1968) and Parascandola (1971).

25 Yandell Henderson described himself as "'organicist' or 'emergent evolutionist'" and Lawrence Henderson as a 'mechanist' (Yandell Henderson to L.J. Henderson, February 7, 1929, *Lawrence Henderson Papers*, Box 7, Folder 7-2, Harvard Business School, Baker Library, Boston, MA).

26 Although he was trained in medicine, he never practiced it, and was in constant difficulties at the Harvard Medical School as being insufficiently interested in medicine.

27 "It is our intention to perform on a considerable number of patients physical examinations, blood analyses, and functional tests, according to a routine which will be established chiefly on the basis of the data already available from Stillman's work, to follow the clinical histories of the cases and repeat these examinations at intervals, and to use the data thus gained in obtaining eventually as logical a classification of nephritis as may be possible, and of treatment according to classification. It is intended ... to attack one a time certain specific problems that arise in connection with the study" ("Report of the Director of the Hospital to the Board of Scientific Directors," *Scientific Reports*, Box 3, Vol. (15 April 1922), p. 167).

28 "Report of the Director of the Hospital to the Corporation" *Scientific Reports*, Box 3, Vol. (15 October 1920), p. 263.

29 In 1916, for example, Thomas Addis and C.K. Watanabe (1916, p. 209) came to the conclusion that "other factors than urea concentration are important in determining the rate of urea excretion by the normal kidney."

30 "Report of the Director of the Hospital to the Board of Scientific Directors," *Scientific Reports*, Box 5, Vol. (20 April 1922), p. 285.

31 Nephritis was throughout this time not the only research line in Van Slyke's Hospital laboratories, though it was perhaps the most extended of his research lines. During the war, Van Slyke also studied the effect of shock on kidneys. In 1948 Van Slyke retired from the Rockefeller Hospital and moved to Brookhaven Laboratories.

32 It has to be noted that with the exception of Van Slyke all of the authors were physicians.

33 The physiological investigations were all conducted in the mid-thirties and resulted in some half a dozen articles published in the *Journal of Physiology*.

34 The filtration-reabsorption theory proposed by A.R. Cushny, A.N. Richards and Homer Smith, and the direct excretion theory of F. Volhard, and R.R. Bensley and W.B. Steen.

35 While at the Rockefeller in 1928, Addis studied the effect of nutrition on size of kidneys in rats. He was a clinical researcher who refused to come to the Rockefeller permanently because he did not want to abandon the patients he had followed for a decade at Stanford (T. Addis to R. Cole, 19 January, 1930, *S. Flexner Papers*, Addis file, Philadelphia Philosophical Society, Philadelphia, PA), and yet, when Van Slyke came to describe him, he claimed that Addis was not interested in working in the clinic. Considering Addis' attention to the special needs of clinicians in his writings (see Addis, 1948) and his insistence on the distinctiveness of clinical (as opposed to chemical or, more generally scientific) approaches to the understanding of disease, Van Slyke's claim of lack of interest appears odd.

36 S. Flexner to T. Addis, February 16, 1928, *S. Flexner Papers*, Addis file, American Philosophical Society, Philadelphia, PA. The draft of the letter inviting Addis to the Hospital bears a handwritten annotation "approved by Dr. Van Slyke."

37 D.D. Van Slyke to R. Cole, July 9, 1930, *R. Cole Papers*, Van Slyke file #3, American Philosophical Society, Philadelphia, PA.

38 It is possible that the relations between biochemistry and physiology also had a broader professional significance. As Kohler argues, one of the major institutional accomplishments of biochemists in the U.S. was precisely the extent to which the American institution-builders managed to establish their field as a discipline which was independent of both organic chemistry and physiology. In a medical setting, the dependence on physiology would have obviously been the most natural, and it is possible that Van Slyke's

strategy of building up direct links between chemistry and the clinic, and thus also of circumventing physiology, was a part of a broader professional trend. But here again, the professional agenda has an intellectual correlate.

39 The gasometric apparatus was widely produced and sold by a number of companies, and appears to have been also a part of the standard equipment of clinical chemical laboratories as well as academic biochemical labs (Amsterdamska and Löhnberg, forthcoming). The extent to which the chemical conception of disease travelled together with the apparatus, however, is far more difficult to determine and deserves a separate investigation.

REFERENCES

T. Addis, "The Renal Lesion in Bright's Disease" *American Journal of the Medical Sciences* (1928), 176: 617–637.

T. Addis, *Glomerular Nephritis: Diagnosis and Treatment* (New York: Macmillan, 1948).

T. Addis, G.D. Barnett and A.E. Shevky, "The Regulation of Renal Activity" *The American Journal of Physiology* (1918), 46: 1–10.

T. Addis and C.K. Watanabe, "The Rate of Urea Excretion. A Criticism of Ambard and Weill's Laws of Urea Excretion" *Journal of Biological Chemistry* (1916), 24: 203–220.

O. Amsterdamska, "From Pneumonia to DNA: The Research Career of Oswald T. Avery" *Historical Studies in the Physical and Biological Sciences* (1993), 24: 1–40.

O. Amsterdamska, "Blood Feuds: Chemistry, Physiology, and Medicine in the Controversy between L.J. Henderson and Yandell Henderson". Paper presented at the Conference of the International Society for the History, Philosophy, and Social Studies of Biology, Leuven, Belgium 19–23 July, 1995.

O. Amsterdamska and Anne Löhnberg, "Van Slyke's Gasometric Apparatus" in R. Bud, S. Johnston and D. Warner (eds.), *Instruments of Science. A Historical Encyclopedia* (New York: Garland Publishers, forthcoming).

R. Archibald, "Donald Dexter Van Slyke" *Yearbook of the American Philosophical Society* (1971): 194–200.

J.H. Austin, E. Stillman and D.D. Van Slyke, "Factors Governing the Excretion Rate of Urea" *Journal of Biological Chemistry* (1921), 46: 91–112.

J. Büttner (ed.), *History of Clinical Chemistry* (Berlin: Walter de Gruyter, 1983).

G.W. Corner, *History of the Rockefeller Institute. 1901–1953* (New York: Rockefeller Institute Press, 1964).

S.J. Cross and W.R. Walbury, "Walter Cannon, L.J. Henderson and the Organic Analogy" *Osiris* (1987), 2: 165–192.

D.L. Edsall, "Preliminary Communication Concerning the Nature and Treatment of Recurrent Vomiting in Children" *American Journal of the Medical Sciences* (1903), 125: 629.

J.D. Edsall, "Blood and Haemoglobin: The Evolution of Knowledge of Functional Adaptation in a Biochemical System. Part I: The Adaptation of Chemical Structure to Function in Haemoglobin" *Journal of the History of Biology* (1972), 5: 205–257.

J.D. Edsall, "Carbon Dioxide Transport in Blood: Equilibrium between Red Cells and Plasma. The Work of D.D. Van Slyke and L.J. Henderson, 1920–28" *History and Philosophy of the Life Sciences* (1985), 7: 105–120.

R. Fitz and D.D. Van Slyke, "Studies of Acidosis. IV. The Relationship Between Alkaline Reserve and Acid Excretion" *Journal of Biological Chemistry* (1917), 30: 389–400.

C. Feudtner, "The Want of Control: Ideas, Innovations, and Ideals in the Modern Management of Diabetes Mellitus" *Bulletin of the History of Medicine* (1995), 69: 66–90.

J.S. Fruton, Review of R.E. Kohler, *From Medical Chemistry to Biochemistry, American Scientist* (1983), 71: 93–94.

J.S. Fruton, *Contrasts in Scientific Style: Research Groups in the Chemical and Biochemical Sciences* (Philadelphia: American Philosophical Society, 1990).

J.S. Fruton, *A Skeptical Biochemist* (Cambridge: Harvard University Press, 1992).

J.S. Haldane, "Claude Bernard's Conception of the Internal Environment" *Science* (April 26, 1929), 69: 453–454.

A. Baird Hastings, "Donald Dexter Van Slyke. The 20th Century Iatro-Chemist" *Federation Proceedings* (1964), 23: 586–591.

A. Baird Hastings, "Donald Dexter Van Slyke, 1983–1971" *Journal of Biological Chemistry* (1972), 247: 1635–1640.

A. Baird Hastings, "Donald Dexter Van Slyke, March 29, 1883–May 4, 1971" *Biographical Memoirs, National Academy of Sciences* (1976), 48: 309–360.

J.A. Hawkins, E.M. McKay and D.D. Van Slyke, "Glucose Excretion in Bright's Disease" *The Journal of Clinical Investigations* (1929), 8: 107–121.

L.J. Henderson, "The Theory of Neutrality Regulation in the Animal Organism" *American Journal of Physiology* (1908), 21: 427–448.

L.J. Henderson, "Das Gleichgewicht zwischen Basen und Säuren im tierischen Organismus" *Ergebnisse der Physiologie* (1909), 8: 224–325.

L.J. Henderson, "The Regulation of Neutrality in the Animal Body" *Science* (1913), 37: 389–395.

L.J. Henderson, *Blood: A Study in General Physiology* (New Haven: Yale University Press, 1928).

L.J. Henderson and W.W. Palmer, "Intensity of Urinary Acidity in Normal and Pathological Conditions" *Journal of Biological Chemistry* (1913), 13: 393.

Y. Henderson, "Is this Science or Metaphysics?" *Science* (January 11, 1929), 69: 39–40.

B.S. Heyl, "The Harvard 'Pareto Circle'" *Journal of the History of the Behavioral Sciences* (1968), 4: 316–334.

A. Hiller, G.C. Linder, C. Lundsgaard, and D.D. Van Slyke, "Fat Metabolism in Nephritis" *Journal of Experimental Medicine* (1924), 39: 931–955.

A. Hiller, J.F. McIntosh, and D.D. Van Slyke, "The Excretion of Albumin and Globulin in Nephritis" *Journal of Clinical Investigations* (1927), 3: 235–251.

F.L. Holmes, "Intermediary Metabolism in the Early Twentieth Century" in W. Bechtel (ed.), *Integrating Scientific Disciplines* (Dordrecht: Martinus Nijhoff, 1986), pp. 59–76.

F.L. Holmes, *Hans Krebs: The Formation of a Scientific Life* (Oxford: Oxford University Press, 1991).

F.L. Holmes, *Between Biology and Medicine: The Formation of Intermediary Metabolism* (Berkeley: Office of History of Science and Technology, 1992).

J. Howell, *Technology in the Hospital: Transforming Patient Care in the Early Twentieth Century* (Baltimore: Johns Hopkins University Press, 1995).

J. Howland and W. McKim Marriott, "A Discussion of Acidosis with Special Reference to That Occurring in Diseases of Children" *Bulletin of the Johns Hopkins Hospital* (1916), 27: 63–68.

K.D. Keele, *The Evolution of Clinical Methods in Medicine* (London: Pitman Medical Publishing, 1963).

R.E. Kohler, *From Medical Chemistry to Biochemistry: The Making of a Biomedical Discipline* (Cambridge: Cambridge University Press, 1982).

G.C. Linder, A. Hiller, and D.D. Van Slyke, "Carbohydrate Metabolism in Nephritis" *Journal of Clinical Investigations* (1925), 1: 247–272.

G.C. Linder, C. Lundsgaard and D.D. Van Slyke, "The Concentration of the Plasma Proteins in Nephritis" *Journal of Experimental Medicine* (1924), 39: 887–920.

G.C. Linder, C. Lundsgaard, D.D. Van Slyke and E. Stillman, "Changes in the Volume of Plasma and Absolute Amount of Plasma Proteins in Nephritis" *Journal of Experimental Medicine* (1924), 39: 921–929.

E. Möller, J.F. McIntosh and D.D. Van Slyke, "Studies of Urea Excretion. II. Relationship between Urine Volume and the Rate of Urea Excretion by Normal Adults" *Journal of Clinical Investigations* (1928a), 6: 427.

E. Möller, J.F. McIntosh and D.D. Van Slyke, "Studies of Urea Excretion. IV. Relationship Between Urine Volume and Rate of Urea Excretion by Patients with Bright's Disease" *Journal of Clinical Investigations* (1928b), 6: 485–504.

P. Olch, *Donald Dexter Van Slyke. An Oral Memoir.* Transcript of interviews (Bethesda, Maryland: National Library of Medicine, 1971).

J. Parascandola, "Organismic and Holistic Concepts in the Thought of L.J. Henderson" *Journal of the History of Biology* (1971), 4: 63–113.

S.J. Peitzman, "Nephrology in the United States from Osler to the Artificial Kidney" *Annals of Internal Medicine* (1986), 105: 937–946.

S.J. Peitzman. "From Dropsy to Bright's Disease to End-stage Renal Disease" *The Milbank Quarterly* (1989), 67: 16–32.

J. Reiser, *Medicine and the Reign of Technology* (Cambridge: Cambridge University Press, 1978).

A.W. Sellards, "Tolerance for Alkalies in Asiatic Cholera" *Philippine Journal of Science* (1910), 5.

A.W. Sellards, "The Essential Features of Acidosis and Their Occurrence in Chronic Renal Disease" *Johns Hopkins Hospital Bulletin* (May 1914), 25: 141–153.

J. Sendroy Jr., "In Appreciation: Donald Dexter van Slyke" *Clinical Chemistry* (1971), 17: 670–673.

J.W. Servos, Review of R.E. Kohler, *From Medical Chemistry to Biochemistry, Isis* (1983), 74: 273–275.

J.W. Servos, *Physical Chemistry from Ostwald to Pauling: The Making of a Science in America* (Princeton: Princeton University Press, 1990).

S.P.L. Sørensen, "Enzymstudien II. Über die Messung und Bedeutung der Wasserstoffkonzentration bei enzymatischen Prozessen" *Biochemische Zeitschrift* (1909), 21: 131–200.

W.C. Stadie and D.D. Van Slyke, "Studies of Acidosis. XV. Carbon Dioxide Content and Capacity in Arterial and Venous Blood Plasma" *Journal of Biological Chemistry* (1920), 41: 191–194.

E. Stillman, D.D. Van Slyke, G. Cullen and R. Fitz, "Studies of Acidosis. VI. The Blood, Urine and Alveolar Air in Diabetic Acidosis" *Journal of Biological Chemistry* (1917), 30: 405–456.

H. Straub and D. Schlayer, "Die Urämie eine Säurevergiftung" *München medizinische Wochenschrift* (1912), 59: 569.

S.W. Sturdy, *A Co-ordinated Whole: The Life and Work of John Scott Haldane*. Ph.D. Dissertation, University of Edinburgh (1987).

D.D. Van Slyke, "Studies of Acidosis. VII. The Determination of ß-hydroxybutyric Acid, Aceto-Acetic Acid, and Acetone in Urine" *Journal Biological Chemistry* (1917), 32: 455–494.

D.D. Van Slyke, "The Carbon Dioxide Carriers of the Blood" *Physiological Reviews* (1921a), 1: 141–176.

D.D. Van Slyke, "Studies of Acidosis. XVII. The Normal and Abnormal Variations in the Acid-Base Balance of the Blood" *Journal of Biological Chemistry* (1921b), 48: 153–176.

D.D. Van Slyke, "Studies of Acidosis. XVIII. Determination of the Bicarbonate Concentration of the Blood and Plasma" *Journal of Biological Chemistry* (1922), 52: 495–499.

D.D. Van Slyke, "Determination of Urea by Gasometric Measurement of Carbon Dioxide Formed by the Action of Urease" *Journal of Biological Chemistry* (1927), 73: 695.

D.D. Van Slyke, "The Silliman Lectures 1928" *Science* (February 8, 1929), 69: 163.

D.D. Van Slyke and G.E. Cullen, "Studies of Acidosis. I. The Bicarbonate Concentration of the Blood Plasma; Its Significance and Determination as a Measure of Acidosis" *Journal of Biological Chemistry* (1917), 30: 289–345.

D.D. Van Slyke, G.C. Linder, A. Hiller, L. Leiter and J.F. McIntosh, "The Excretion of Ammonia and Titrable Acid in Nephritis" *Journal of Clinical Investigations* (1925), 2: 255–288.

D.D. Van Slyke, C.P. Rhoads, A. Hiller and A.S. Alving, "Relationship between Urea Excretion, Renal Blood Flow, Renal Oxygen Consumption, and Diuresis. The Mechanism of Urea Excretion" *American Journal of Physiology* (1934), 109: 336–374.

D.D. Van Slyke and W.C. Stadie, "The Determination of the Gases of the Blood" *Journal of Biological Chemistry* (1921), 49: 1–42.

D.D. Van Slyke, E. Stillman and G.E. Cullen, "Studies of Acidosis. V. Alveolar Carbon Dioxide and Plasma Bicarbonates in Normal Men During Digestive Rest and Activity" *Journal of Biological Chemistry* (1917), 30: 401–404.

D.D. Van Slyke, E. Stillman and G.E. Cullen, "Studies of Acidosis. XIII. A Method for Titrating the Bicarbonate Content of the Plasma" *Journal of Biological Chemistry* (1919), 38: 167–178.

D.D. Van Slyke, E. Stillman, E. Moller, W. Ehrich, J.F. McIntosh, L. Leiter, E.M. MacKay, R.R. Hannon, N.S. Moore and C. Johnston, "Observations on the Course of Different Types of Bright's Disease, and on the Resultant Changes in Renal Anatomy" *Medicine* (1930), 9: 257–386.

L.G. Wilson, Review of R.E. Kohler, *From Medical Chemistry to Biochemistry*, *Journal of the History of Medicine* (1983), 38: 462–464.

CHAPTER 3

Vitamins and the Dynamics of Molecularization: Biochemistry, Policy and Industry in Britain, 1914–1939

Harmke Kamminga

The making of the vitamin concept in the first few decades of this century relied heavily on functional ascriptions, centred on the role of specific, but chemically ill-defined nutrients in preventing specific dietary deficiency diseases. In the interwar period, efforts were made increasingly to characterize vitamins in terms of their chemical structure. In the course of these structural pursuits, vitamins were turned from hypothetical explanatory devices into molecules that could be isolated, identified, manipulated and synthesised by scientists. In the process, standard quantitative requirements for individual vitamins were defined, which provided a new rationale for nutritional and medical intervention aimed at preventing and treating diseases. Methods to produce vitamins in increasingly pure form were developed and exploited by industry, not only to manufacture standardized vitamin preparations for clinical use, but to sell vitamin preparations and vitamin-fortified foods to consumers. The promotion of vitamin supplements, in standardized and chemically defined form, served to persuade millions of people to buy these products for the benefit of their health.[1]

My aim in this chapter is to examine why the elucidation of vitamin structure came to be seen as an important scientific problem by different constituencies, with a focus on Britain. I shall draw attention to a number of scientific, medical, social, and industrial motivations which stimulated structural studies of vitamins, and outline how these different motivations were interlinked and mutually reinforced. My emphasis here being on the multiplicity of reasons underlying structural pursuits, I shall be less concerned with the detailed scientific steps whereby vitamins were turned into well-defined molecules. Several issues of policy and strategy that will be discussed encompassed a much broader domain than vitamin research. On that level, the vitamin case is presented to illustrate some general features of the dynamics of molecularization of biology and medicine in the interwar period.

I shall begin by discussing how vitamins acquired political prominence in Britain during World War I. The social and scientific importance ascribed to vitamins at this time informed government food policy, boosted vitamin research and turned vitamins into a matter of concern for the food industry. As I aim to show, the wartime formation of links between these domains provided a major impetus for the molecularization of vitamins after the war. I shall discuss the support of the Medical Research Council for vitamin research in interwar Britain, and tensions which arose between the Secretary of the Medical Research Council, Walter Morley Fletcher, and the biochemist whom Fletcher expected to take a leading role in investigating the metabolic functions of vitamins, Frederick Gowland Hopkins.

I shall argue that Hopkins came to use vitamins for his own ends, to promote biochemistry as a fundamental science of life, at the same time reinforcing a molecular perspective, not only of diet, but of all living processes. I shall then discuss how vitamins motivated the British food industry to transform its practices of quality control and product promotion and development, by recruiting research scientists. Finally, I shall point to the ways in which the different interests which motivated vitamin research in the interwar period were interconnected by, and converged on, a concern with vitamin structure.

THE POLITICS OF VITAMINS

In World War I, concern with the food supply and the threat of food shortages focussed political attention on the composition of a healthy diet, in relation to both army rations and foods available to the civilian population. In Britain especially, scientists were involved in food policy formulation at the highest levels, and they drew on new understanding of the importance of the qualitative composition of the diet, over and above its quantitative energy content measured in calories (Teich, 1995). At the same time, nutrition research to further this understanding was given high priority. Moreover, the political and scientific networks established in conjunction with these efforts, especially through the involvement of the Medical Research Committee, were of enormous influence in guiding scientific research on nutrition after the war. "Accessory food factors", or vitamins as they became known later,[2] were at the centre of these developments.

In the decade before the war, animal feeding experiments with synthetic diets of known chemical composition had indicated a vital nutritional role for hitherto unsuspected components of the diet. The demonstration that animals cannot survive on a diet of pure carbohydrates, fats, proteins, water and minerals, but that small supplements of, for instance, milk restored normal growth, was taken to indicate the presence in normal dietaries of as yet unidentified "accessory food substances" that are essential components of the diet (Hopkins, 1906; 1912). Scientific interest in these hypothetical dietary substances was heightened when they were linked tentatively to diseases that were beginning to be ascribed to dietary deficiencies (e.g., Carpenter, 1986; Codell Carter, 1977). The demonstrated role of specific foods in preventing and curing diseases such as beri-beri and scurvy stimulated efforts to isolate and characterize the protective components in these foods.

At the Lister Institute in London, for example, attempts to identify the protective ("antineuritic") factor implicated in beri-beri heralded the start, in 1910, of a renowned long-term program in nutrition research (Chick *et al.*, 1971). As part of these attempts, an animal assay was developed to determine the quantitative distribution of the protective factor in different foods. This experience was built upon during the war, when the Royal Army Medical Corps sought the advice of the Lister Institute in the face of outbreaks of beri-beri and scurvy among British troops in the Middle East and India. An animal assay for antiscorbutic activity was developed in addition to that for antineuritic activity, and extensive experiments were done to determine which foods, in a form suitable for transport to overseas troops, are most effective in preventing scurvy and beri-beri. Many of the male scientists being away at the front, most of this research was done by women scientists (many of them newly appointed for this purpose), led by Harriette Chick. Through this work, Chick and her women colleagues acquired a high scientific profile in nutritional research, which was further enhanced after the war.

Nutritional work such as that at the Lister Institute had the active support of the Medical Research Committee as well as the Food (War) Committee appointed under the Royal Society (War) Committee upon the subdivision, in 1916, of its Sectional Committee on Physiology. The Royal Society Committees, with memberships drawn from the most reputable scientific circles, were set up to organize war-related research and to advise government on policy matters. The Chairman of the Food (War) Committee, eminent physiologist E.H. Starling, was, in that capacity, also Honorary Scientific Advisor to the newly created Ministry of Food. The Committee's Secretary was Walter Morley Fletcher, former Cambridge physiologist and Secretary of the Medical Research Committee since its establishment in 1913. Among the Food (War) Committee's members was Cambridge biochemist Frederick Gowland Hopkins, by then highly esteemed for his innovative work on the physiological role of essential amino acids in dietary protein and of chemically undefined accessory food factors in the diet.

Hopkins' presence on the Committee, and the high regard in which his work was held by fellow members, ensured that accessory food factors received full attention. At the level of food policy, the Committee placed the emphasis on the importance of fresh foods, in the light of evidence suggesting that processed foods such as mechanically milled rice, white bread, dried milk and canned meats and vegetables are deficient in dietary factors essential for health (Smith,

1997; Teich, 1995). This emphasis on fresh foods, emanating from such an authoritative body, created concern among food manufacturers about adverse effects on sales of their allegedly inferior products and motivated them to set up their own research laboratories. As I shall discuss in a later section, after the war the food industry engaged in chemical research on vitamins, aimed at providing scientific justification for the promotion of its products as being, if anything, nutritionally superior to unprocessed foods.

At the level of research, the Food (War) Committee gave high priority to the determination of the accessory food factor content of different foods as measured by biological assays, but issues of etiology also received attention. In particular, the Medical Research Committee funded investigations of the etiology of rickets from different angles, with Edward Mellanby focussing on the possible dietary connection in feeding experiments with puppies (Mellanby, 1918; Medical Research Committee, 1919, chapter 5). His results provided evidence for a causal link with a fat-soluble accessory food factor in the diet, but this interpretation was opposed fiercely by a group of physiologists in Glasgow, also funded by the Medical Research Committee, on the basis of their field studies among the urban poor. In fact, the rickets controversy was to continue acrimoniously throughout the 1920s, and the disease became a focus for scientific research and experiments in public health (Smith and Nicolson, 1989; Petty, 1989).

The Medical Research Committee's active involvement in the scientific war effort had important repercussions after the war, when it was transformed into the Medical Research Council (MRC). The Committee had been set up initially for the promotion, coordination and support of tuberculosis research, according to provisions made under the 1911 National Insurance Act (Austoker and Bryder, 1989). Its intertwinement with the Royal Society (War) Committee, however, gave it a much broader role in defining medical research policy. Initially an *ad hoc* arrangement contingent on wartime conditions, this role was formalized after the war with the creation of the MRC. For the first time in Britain, there was a government funding body for the whole range of medical research, endowed with a large budget and responsible directly to the Privy Council rather than the Ministry of Health. As Secretary of the MRC, Fletcher gained enormous powers in the direction of medical research in Britain, and he built on the structures put into place during the war. Those serving on the MRC included powerful allies of Fletcher's among the scientific establishment, who shared and informed his objectives.

Among Fletcher's objectives was his ambition that the MRC should become the driving force behind fundamental medical research, with a view to building strong scientific foundations for clinical practice (Austoker, 1989). This aim was in harmony with Fletcher's own scientific background. In the Physiological Laboratory in Cambridge, he had worked on fundamental aspects of muscle physiology, in an institution which, under the leadership of Michael Foster, had fought hard for the autonomy of its research programs, independent of immediate medical concerns (Geison, 1978). Furthermore, many of Fletcher's closest advisors on the MRC were themselves engaged in basic research and intent on replacing supposedly subjective clinical judgement with supposedly objective laboratory research as the court of authority for medical matters. While Fletcher promoted fundamental research in physiology, biochemistry and pathology without demanding short-term medical applications, he nevertheless emphasised that clinical applications should be the long-term goal of this research (Austoker, 1989).

One of the areas privileged by the MRC was vitamin research. Not only was this subject deemed to be of the greatest medical and social importance, wartime efforts to coordinate research in this area had already resulted in the establishment of a special advisory committee for the promotion of research into the nature and role of accessory food factors. In 1918, the Medical Research Committee and the Lister Institute had jointly established the Accessory Food Factors Committee, with Hopkins in the Chair, Harriette Chick as Secretary, and Jack Drummond, Arthur Harden and Edward Mellanby as members. This committee remained an integral part of the MRC into the 1930s and played a central role in defining the directions of MRC-supported vitamin research. It also issued a set of highly authoritative and widely read reports on the state of knowledge of vitamins (Medical Research Committee, 1919; Medical Research Council, 1924 and 1932).

An early post-war opportunity to demonstrate the medical and social importance of vitamin work, and of the MRC's beneficial role in supporting it, presented itself with the widespread famine in Central and Eastern Europe which followed the war. Diseases such as scurvy and rickets were rife, especially among children (Figure 3.1). Under the auspices of the Accessory Food Factors Committee, Harriette Chick and Margaret Hume, later joined by more women colleagues, went to famine-struck Vienna in 1919, to investigate whether the diseases affecting the population were due to vitamin deficiencies. For three years, they studied the effects of specific dietary supplements on the prevention and cure of rickets, at a time when the etiology of this disease was still deeply contentious. The work was funded jointly by the MRC and the Lister Institute, and Chick's team was offered research facilities in the University of Vienna's children's clinic by its Professor of Pediatrics, Clemens von Pirquet.

In his clinic, Pirquet had introduced a rigorous dietary system, whereby the daily diet of individual children was related quantitatively, in terms of protein and calorie content, to each child's size. Under these conditions, it was possible to conduct controlled trials on human subjects, equivalent to systematic animal feeding experiments. By giving different groups of children different or no supplements to their controlled diet, Chick and her colleagues established convincingly that rickets can be prevented and cured by cod-liver oil. The results were sufficiently dramatic to lend credence to the vitamin deficiency theory for many of those who had previously regarded rickets as an infectious disease, including Pirquet himself (see his Preface to the report issued by the Medical Research Council, 1923; also quoted by Landsborough Thomson, 1975, p. 79). In this

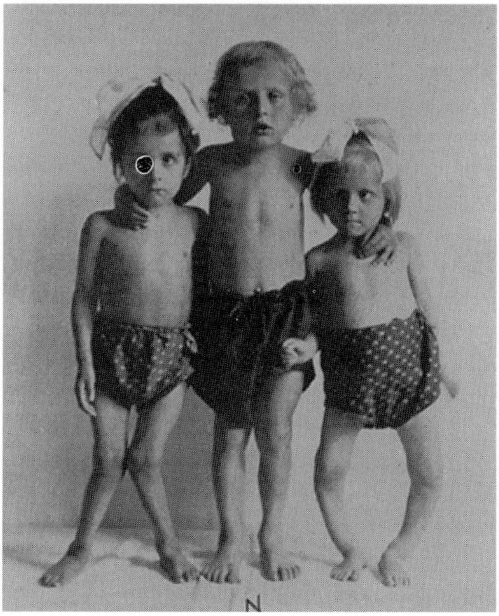

Figure 3.1. Two six-year old children with severe rickets (left and right) and a normal child of the same age (middle), Vienna, 1920.

instance, therefore, the active day-to-day involvement of research scientists in clinical practice served to persuade initially sceptical clinicians of the significance of vitamins, by means of clincial demonstration. Although the etiology of rickets nevertheless remained a contentious issue, this clinical success immediately became a favourite example used by those promoting the importance of vitamins and vitamin research.

Dietary intervention in the light of the vitamin theory thus yielded impressive therapeutic successes from an early stage onwards, notably in the case of the putative vitamin deficiency diseases — beri-beri, scurvy and rickets — associated strongly with poverty, famine and war. Such success in the treatment and prevention of socially and politically significant diseases not only raised the profile of vitamins in public health policy and strengthened the vitamin theory, but also stimulated a large scientific research effort related to vitamins.

THE SCIENCE OF VITAMINS

In interwar Britain, vitamin research was supported and promoted above all by the MRC and its Accessory Food Factors Committee. The MRC funded investigations of the etiology of deficiency diseases; the development and perfection of quantitative biological and chemical assays of vitamin activity, with a view to determining minimum requirements and setting international standards; studies of the chemical properties and structure of vitamins; and the physiological role of vitamins (Landsborough Thomson, 1975, chapter 5). I here want to sketch how these different fields of investigation had bearing on each other, with structural questions becoming the linking theme.

Questions about the chemical nature and the physiological role of vitamins had been interlinked from the very beginning, with the formulation of the notion of accessory food factors to explain the results of animal feeding experiments. Using controlled diets of systematically varied chemical composition, Hopkins showed that rats stop growing and die if they are fed an otherwise adequate diet of pure protein, carbohydrate, fat, water and minerals, while normal growth is restored by the addition of small quantities of milk (Hopkins, 1906; 1912). On these grounds, he postulated the existence in normal diets of "accessory food factors". These hypothetical substances appeared to be necessary for normal growth and survival, but they had never been detected in chemical analyses of foods, let alone identified chemically. Presumably, therefore, accessory food factors were present in minute amounts in normal dietaries. In order to consolidate the notion of accessory food factors, and to investigate their precise physiological role in metabolism, it became imperative to isolate and identify these substances. Hopkins himself tried to do so for several years, without success (Weatherall, 1994, chapter 4).

Hopkins was more successful in the identification of "essential amino acids" present in dietary protein, that is, amino acids which are required for normal metabolism, but cannot apparently be synthesised in the body from other substances and, hence, must be supplied in the diet. In fact, Hopkins had first become interested in the biochemistry of nutrition as a result of his demonstration with Edith Willcock that rats cannot survive on diets with adequate calorie content if the only protein source lacks the amino acid tryptophan (Willcock and Hopkins, 1906). He went on to identify and investigate several other essential amino acids (reviewed by Hopkins, 1916). This work, together with the notion of accessory food factors, was instrumental in shifting attention from the quantitative calorie content of diet to its composition at the level of fine chemical detail. I shall come back later to the way in which this shift was used by Hopkins in the promotion of a molecular view, not only of diet, but of the living organism in general.

That food fractions providing protection against dietary deficiency diseases contain only the tiniest amounts of protective factors is illustrated by the studies of the antineuritic factor implicated in beri-beri by Casimir Funk at the Lister Institute. After a series of laborious extractions, evaporations and other chemical operations, Funk obtained just 0.5 g of active substance from 54 kg of rice polishings (Funk, 1911). He identified his crystalline preparation as an organic base of the amine class and introduced the term "vitamine" (note 2) to describe substances required to prevent "vitamine deficiency diseases", under which he categorized beri-beri, scurvy, pellagra, epidemic dropsy and, tentatively, rickets

(Funk, 1912). However, after further elaborate chemical experiments, Funk soon found that the "vitamine" which he thought he had purified was, in fact, an impure mixture. Among the accessory food factors community, including Funk's colleagues at the Lister Institute, this set-back served as a cautionary message about the dangers of making premature claims about the chemical structure of these elusive substances.[3]

The chemist's difficulties in isolating vitamins were compounded by their instability to heat, oxidation, drying, hydrolysis and chemical reagents. Yet in the course of extensive chemical investigation in different centres, combined with animal feeding experiments, accessory food factors began to be differentiated on the basis of chemical properties. By 1920, three factors had been distinguished: those named "fat-soluble A" and "water-soluble B" by McCollum and Davis (1915) working at the University of Wisconsin, and, following similar terminology, "water-soluble C" by Drummond (1919) in London. In line with Drummond's suggestion (note 2), these became known as vitamins A, B and C, and were soon followed by vitamins D (the antirachitic factor) and E (then also known as the fertility factor). A whole string of vitamins was added in the 1920s and 1930s, in alphabetical sequence up to vitamin P, while vitamin B was subdivided into several different vitamins, first into vitamins B_1 and B_2, but eventually into twelve factors.[4]

One important incentive for studying the chemistry of vitamins was linked to the determination of the differential distribution of individual vitamins in foodstuffs, in order to work out healthy diets defined in chemical terms. Increasingly, animal experiments showed that a mildly deficient intake of vitamins did not necessarily or immediately lead to severe clinical symptoms, such as those associated with scurvy, rickets or beri-beri, but nevertheless produced abnormalities in certain tissues, organs or structures of the body which affect health in the long term (as reviewed in Medical Research Council, 1932, chapters 10 and 11). In order to define accurate standards for minimum vitamin requirements to prevent disease and optimal vitamin requirements to ensure good health, pure vitamin preparations were required.

Standardization of vitamin requirements was one of the objectives of the MRC,[5] but took on an international dimension through the activities of the League of Nations Health Organisation (LNHO). As Weindling (1995) has shown, the LNHO played an instrumental role in promoting optimal, instead of minimum, vitamin requirements. It was on the basis of international standards thus defined that John Boyd Orr (1936) concluded from his extensive investigations of dietaries that more than half the British population lived on an inadequate diet. By then, internationally agreed units of vitamin activity were measured, in bioassays, against standard samples, all held at the MRC's flagship institute, the National Institute for Medical Research in London. By 1934, there were four of these standard preparations, by then in chemically well-defined form: vitamin A in the form of ß-carotene; a standard absorption product of vitamin B_1; vitamin C as *l*-ascorbic acid; and vitamin D in the form of calciferol, obtained by irradiation of ergosterol (Burn, 1937).

Chemical studies were also undertaken for etiological reasons. In the case of rickets in particular, there was the puzzling phenomenon that not only cod-liver oil, but exposure to sunlight had a beneficial effect on patients. The effect of sunlight had been confirmed again in the investigations of Chick's team in Vienna, and while these investigations were in progress, Huldschinsky in Berlin cured rachitic children by irradiating them with a mercury lamp. At the University of Wisconsin, Harry Steenbock soon thereafter showed that many foods acquired antirachitic activity upon ultraviolet irradiation. This "Steenbock process" was patented by the University of Wisconsin (Apple, 1989; 1996, chapter 2). It was then shown, in Britain and the United States, that the irradiation of cholesterol produced small quantities of a substance with high antirachitic activity, presumed to be vitamin D.

This chemical conversion was further investigated by Otto Rosenheim's team at the National Institute for Medical Research, and by Adolf Windaus in Göttingen. These two groups, which exchanged information and materials, found that it was not pure cholesterol, but another sterol, separated from cholesterol with considerable difficulty, which produced the antirachitic activity upon irradiation. They named it ergosterol. Subsequently the substance produced by irradiation of ergosterol was isolated and named calciferol. Its remarkably high antirachitic activity was taken to indicate that this was vitamin D. The antirachitic effects of sunlight were attributed to the conversion of ergosterol ("pro-vitamin D") present in the skin into calciferol (vitamin D). Windaus received the 1928 Nobel Prize in Chemistry for his work on the structure of sterols, including ergosterol.[6] In fact, his structure of the sterol skeleton was later corrected, with Rosenheim's alternative structure receiving support from X-ray crystallographic studies by Bernal (1932). Interestingly, it was only after the Chemistry Prize to Windaus that the Nobel Committee for Physiology or Medicine awarded their first Nobel Prize for vitamin work: in 1929, Hopkins and Eijkman shared the Prize for "the discovery of vitamins", as their work of several decades earlier was then interpreted.[7]

Knowledge of the chemical structure of vitamins was also sought in order to gain insight into their mode of action in the body, and Walter Fletcher was particularly keen that the MRC should support fundamental research into the role of vitamins in metabolism. At first, vitamin function was viewed in global physiological terms: vitamins are required for normal growth, and vitamins are required to protect against particular nutritional deficiency syndromes. But functional questions typically became tied much more closely to questions about mechanisms: what was the mode of action of individual vitamins within the normal physiological process, what was the role of specific vitamins in metabolism? Fletcher held that insight into these questions of function in a more precise, local sense would be of the greatest medical importance, both for the prevention and treatment of vitamin deficiencies by highly specific chemical intervention. If MRC-supported research were to yield such crucial results, it would simultaneously provide strong legitimation for the MRC's policy of supporting fundamental research, in this case in biochemistry. In the 1920s, Fletcher continuously needed to justify this policy to the clinically orientated medical establishment, however much he and his scientific colleagues favoured it and however much he asserted that it would ultimately yield important medical benefits (Austoker, 1989).

Having the highest regard for Hopkins' earlier work on the biochemistry of nutrition, Fletcher expected that his former colleague in Cambridge and close ally on the MRC would adopt a leading role in biochemical research on the metabolic role of vitamins. To Fletcher's long-term frustration, Hopkins declined to fit in with this plan (Weatherall and Kamminga, 1996; Weatherall, 1997). In fact, Hopkins, disengaged himself from vitamin research altogether in the early 1920s. Ironically, in pursuing different, new priorities, he became a prominent propagandist for a molecular perspective on vitamins and diet.

Hopkins himself stressed repeatedly, in his writings and public lectures from the 1910s to the early 1930s, that investigations into the metabolic functions of vitamins could not even begin without knowledge of their chemical structure. This recurring message was consistent with Hopkins' own practices in biochemistry, where he always aimed to isolate and identify the molecules taking part in metabolic reactions, to elucidate the reaction mechanisms, and to investigate the control of these reactions by enzymes and their coordination within the cell (Kamminga and Weatherall, 1996).

Hopkins did not, however, encourage structural research on vitamins in his own department, with a view to investigating their metabolic role. Increasingly, research in the Cambridge Department of Biochemistry, which had been created in 1914, became focussed on fundamental biochemical research on biological oxidation and reduction and the catalytic control of specific oxidation-reduction reactions. As shown elsewhere, this shift in research was coupled strongly to Hopkins' ambition to transform biochemistry into a fundamental science of life that was to provide a unifying chemical perspective of all biological processes, across the living world. In particular, Hopkins' isolation in 1921 of glutathione, an agent with dual oxidizing and reducing properties which was found to be present in many different living organisms, stimulated the search in Hopkins' department for basic biochemical patterns common to a range of organisms, from bacteria to man (Kamminga and Weatherall, 1996).

"Fundamental" research had quite different connotations for Hopkins and Fletcher. While Fletcher promoted fundamental research with the aim of making medicine scientific, it was Hopkins' ambition to unite chemistry and all of biology. Thus, in the research pursued by Hopkins and his colleagues, medical questions, and indeed vitamins, became subordinate to a much more ambitious program of molecularizing life, through biochemistry.

The Cambridge program of biochemistry, however, remained atypical in the interwar years. Despite Hopkins' persistent stress on the underlying unity of this wide-ranging program, this unity was not appreciated widely by influential British medical scientists, who generally regarded the Cambridge research as too far removed from human and animal physiology (Weatherall and Kamminga, 1996). Hopkins countered these reservations by promoting his particular vision of life repeatedly in public. In order to give substance to that vision, however, he used therapeutic successes with vitamins and hormones as recurring illustrative examples of the social and medical need for the biochemist's special skills. In other words, Hopkins appealed to vitamins in his lectures and general writings for purposes of his own, to promote an approach to biochemistry which he defined as being concerned with "molecular dynamics in cells and tissues" (Hopkins, 1933).

At the same time, Hopkins' frequently repeated message that knowledge of chemical structure was a prerequisite for the elucidation of the basic biological functions of vitamins (among other biological molecules) gave further prominence to the problem of vitamin structure. Hopkins' pronouncements mattered, because he was regarded as the pre-eminent public authority on the biochemistry of nutrition in interwar Britain, and played an active role in the most influential scientific bodies which shaped British science policy, including the MRC and the Royal Society.

While Hopkins himself did not practise what he preached with respect to vitamins, his department was the site where Albert Szent-Györgyi first isolated vitamin C. This work began in Cambridge as an unexpected byproduct of work on biological oxidation and reduction with Szent-Györgyi's efforts to isolate and characterize a substance with strong reducing properties from the adrenal cortex (Szent-Györgyi, 1928). In view of the physiological properties of this substance, identified as "hexuronic acid", there were strong indications that it might be identical with the antiscorbutic vitamin C. Assays with hexuronic acid, undertaken at Hopkins' insistence by S.S. Zilva at the Lister Institute, failed to support this hypothesis. Szent-Györgyi persisted in his investigations, however, and eventually, after his return to Hungary, produced much larger amounts of hexuronic acid from paprika and was able to confirm its identity with vitamin C (see Weatherall, 1997, for details). The substance was subsequently renamed ascorbic acid to give expression to its activity against scurvy.

It was the Cambridge research program on mechanisms of biological oxidation and reduction that had attracted Szent-Györgyi to Hopkins' department. He later lamented the fact that his work on vitamin C had distracted him for years from his "fundamental" research in biochemistry, and described vitamins as being of greater interest to the chef than to the scientist (Szent-Györgyi, 1937, pp. 77–80). A few months after these and other disparaging remarks were published, he was awarded the 1937 Nobel Prize for Physiology or Medicine for the isolation, purification and biochemical characterization of vitamin C. In the same year, the Birmingham chemist Walter Haworth shared the Nobel Prize in Chemistry for his work on the structure and synthesis of vitamin C. His fellow laureate was Paul Karrer from the University of Zürich, who was honoured for his researches into the constitution of carotinoids, flavins and vitamins A and B.

In 1938, Richard Kuhn from Heidelberg University was awarded the Nobel Prize in Chemistry for his work on carotinoids and vitamins of the B group. (Kuhn was forbidden by the Nazi regime to travel to Stockholm to receive the award.) These awards give an indication both of the difficulties involved in structural studies of vitamins and of the importance attached to the elucidation of vitamin structure by the scientific community.

The investigation of vitamin structure, then, was pursued for different ends: to determine detailed dietary requirements, to gain further insight into the etiology of vitamin deficiency syndromes, to initiate studies of their role in intermediary metabolism, and to produce work that could gain the highest scientific accolades in organic chemistry and biochemistry. There were also strong industrial incentives, to which I turn now.

VITAMINS AND INDUSTRY

Vitamins caused major concern to food manufacturers already during World War I, when assays of foods such as those performed at the Lister Institute suggested that many processed foods were deficient in accessory food factors. The emphasis which the Royal Society Food (War) Committee placed on fresh foods in the food policies it advocated posed a serious threat to sales of processed foods and hence to profits of the food industry. To counter this threat, food manufacturers began to set up their own research laboratories after the war. Here foods were assayed and analysed, and methods explored to enrich food products with vitamins obtained from natural sources. To keep production costs of vitamin-fortified foods low, and to develop the most convenient methods of production, the technical skills of chemists and nutrition scientists were put to use in the extraction, concentration, purification and chemical investigation of vitamins. The aim was to manufacture products that could be promoted and sold as having superior nutritional value to natural foods, especially with respect to their vitamin content (Horrocks, 1993; 1995; 1997). One of the outcomes was a role for the food industry in the molecularization of vitamins.

Manufacturers of margarines were among the first to be alarmed at allegations that their products, popular because of their low cost compared with butter, were lacking in the fat-soluble substances thought to promote growth (vitamin A) and to prevent rickets (vitamin D). The major British producer was the Maypole Dairy Company, but when Lever Brothers set up the Planters Margarine Company in 1914, the two manufacturers joined forces in research into the nutritive value of their products. In 1917 (before his move away from vitamin research), Hopkins began to act as a consultant for the two margarine companies and agreed to do research on their behalf, providing that he could publish his results (Baldwin, 1961). While very little is known about Hopkins' activities as a consultant to these companies, this early link between an academic authority on vitamins and industrial food research is of interest in itself.

Hopkins also had links in this period with two food companies in Norwich, Virol Ltd and Colman's, who advertised their products citing scientific findings about their vitamin content obtained in the Cambridge Biochemical Department (Weatherall, 1994). In 1924 Leslie Harris, who had completed his Ph.D. under Hopkins, went to work for Colman's, one of the first food companies to employ its own research scientists, and set up animal feeding experiments for purposes of quality control of proprietary babyfoods (Horrocks, 1995). Harris was subsequently appointed head of the Dunn Nutritional Laboratory in Cambridge, founded with funds from the MRC and the Dunn Estate in 1927, when Fletcher had finally given up hope that Hopkins would return to vitamin research (Weatherall, 1997). This is one illustrative example of the new movement of nutrition scientists between industrial and pure research laboratories.

In the course of the 1920s, in the face of increasing competition among themselves and with the butter industry, British companies began to produce margarines fortified with vitamins A and D. Planters was the first to do so, in 1927, using vitamins extracted from liver oils (Furness, 1928; anon., 1929). In 1929, *The Industrial Chemist* reported on independent tests done by the Pharmaceutical Society of Great Britain comparing the vitamin A and D content of butter and different margarines, using biological assays. The results for butter showed significant variation while the margarines displayed constant vitamin content and were superior in antirachitic activity to all samples of butter (Furness, 1929; for details of the assays, see Coward, 1928).

The 1920s also saw the first production of vitamin preparations in Britain, both for clinical use and for public consumption as supplements to meals (Figure 3.2). The first ones were preparations of vitamin D produced by extraction from cod-liver oil, but this mode of production was superseded by the more convenient and cheaper method of producing vitamin D from irradiated ergosterol obtained from yeast. The first British product obtained this way was Radiostol, introduced to the market by British Drug Houses in 1927. It had an antirachitic activity 200,000 times higher, weight by weight, than that of cod-liver oil. The Director of British Drug Houses advocated both that this product be used by food manufacturers to fortify margarines, and that further chemical research be done to get insight into the chemical changes resulting from irradiating ergosterol (Carr, 1928).

In general, it was the pharmaceutical industry that produced vitamin preparations, although British pharmaceutical companies lagged behind German and Swiss companies in vitamin research and development. Not only was there a stronger tradition in organic chemistry in these countries, but the skills of organic chemists, both in structural analysis of natural products and in synthetic chemistry, had found a firm industrial base there much earlier than in Britain. In fact, the commercial production of synthetic vitamins in Britain began when the Swiss company Hoffmann-La Roche set up a "model plant" (Figure 3.3) in Welwyn Garden City in 1937, for the industrial synthesis of pure vitamin C and B_1 (anon., 1938).

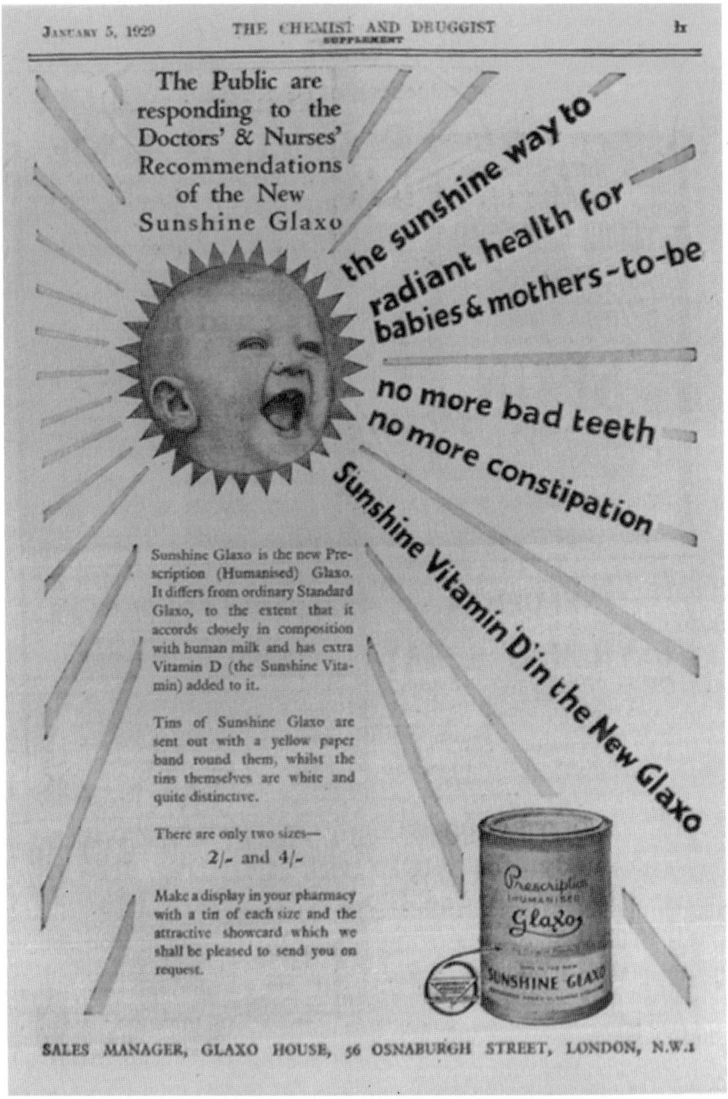

Figure 3.2. Advertisement for Glaxo milk powder fortified with vitamin D.

There was one British food company that did get involved in the production of vitamin preparations as well as vitamin-enriched foods and, through vitamins, was transformed into the pharmaceutical company now known as Glaxo (Davenport-Hines and Slinn, 1992).[8] Scientists working for the company were the motive force behind this transformation, which was coupled strongly to the molecularization of vitamins.

Figure 3.3. New "model plant" manufacturing vitamin C, Roche Products, Ltd, Welwyn Garden City.

Glaxo grew out of Joseph Nathan & Company which produced dried milk and proprietary babyfoods. In the face of poor sales of its milk powder sold under the brand name of "Defiance", the company relaunched this product in 1906 under the new name of "Glaxo". It was advertised as "the Food that Builds Bonnie Babies". The Nathan company's sales performance remained modest for some years, but a great boost to its fortunes came during World War I, as a result of a policy introduced by the new Ministry of Food, created in 1916. In 1917, the Ministry decided to provide the infant welfare centres of local authorities with milk powder, and the product it chose to distribute was Glaxo. After the war, greater competition made the company increasingly concerned about quality control of its dried milk products, and in 1919 it recruited Harry Jephcott, a young scientist with chemical and pharmaceutical training.

Jephcott was given the responsibility of ensuring for Glaxo a uniform and high standard of quality, and, because of the potentially serious commercial implications, he set out to investigate the then contentious issue of whether the drying process changed the vitamin content of milk. To help him in his investigations, Jephcott recruited Alfred Bacharach to the tiny scientific group at the Glaxo Department in 1920. Trained at the University of Cambridge, Bacharach had worked at the Wellcome Chemical Research laboratories from 1915 to 1918 and then at the Wellcome Chemical Works. Together, Jephcott and Bacharach began a series of investigations into the antiscorbutic activity, in guinea pigs, of fresh milk and of milk dried by different processes. The results for the process used in the production of Glaxo compared favourably with other drying methods, but were not quite as good as those for fresh milk. This work was deemed sufficiently significant to be published in the *Biochemical Journal* (Jephcott and Bacharach, 1921). Indeed, from the early 1920s a good number of scientific papers was published in reputable scientific journals by people working in the Glaxo Research Laboratory.

In the light of the intensive research effort of the early 1920s on the dietary connections of rickets, Jephcott began to explore the possibility of producing a Glaxo powder fortified with vitamin D. In 1923 he attended the International Dairy Congress in Washington and took the opportunity to visit major nutritional centres at the University of Wisconsin and the University of Columbia, New York. At Columbia he met Theodore Zucker, who had perfected a process for extracting vitamin D from fish-liver oil. Jephcott secured the British licence for the Zucker process and in 1924 the Nathan company began to market vitamin D concentrate in phials for drop dosage, under the name of Ostelin. This preparation, which had 2000 times the antirachitic activity of cod-liver oil, was the first standardized vitamin concentrate marketed in Britain. It was also the company's first pharmaceutical product and it sold well, helped by recommendations in the medical press, for example *The Lancet* (Davenport-Hines and Slinn, 1992, pp. 73–75).

Subsequently Jephcott obtained the licence for the Steenbock process for converting ergosterol into vitamin D by irradiation and, from 1929, Ostelin was produced in this way in the Glaxo Department using yeast as the source for ergosterol. The process had many advantages: the bulk buying of fish could be abandoned, yeast could be cultured relatively cheaply, uniformity of antirachitic activity could be secured more easily, and the product lacked the unpleasant fish taste of the earlier version of Ostelin. A further change came in 1932, when the company began to produce pure crystalline calciferol from irradiated ergosterol. This product, Calciferol GL, was four times more active than the Steenbock irradiation product. In the 1930s, the Glaxo Department also developed a range of products containing vitamin A, it sold its vitamin products on an international scale in nearly all continents, and it produced a new proprietary infant food enriched with vitamins, called Farex.

The commercial success of the company's vitamin products boosted the research department. In 1935, Glaxo Laboratories Ltd was incorporated as a private limited company, which was to "market proprietary pharmaceutical products having reasonable claims to novelty". Gradually pharmaceutical products became more important than food products as a percentage of net income. By the late 1930s, the company was doing research into the synthesis of vitamin A and iodine compounds, the isolation of bacterial antigens and pituitary hormones, and the isolation and extraction from liver of the agent that is active against pernicious anemia (later called vitamin B_{12}). In conjunction with these research efforts, the scientific reputation of the Glaxo Laboratories grew.

While Jephcott was the prime mover behind the company's move into pharmaceuticals, Bacharach was the central figure in building the company's scientific reputation. (On Bacharach, see Jephcott, 1966.) From the late 1920s, Jephcott had increasingly delegated the research on vitamins to Bacharach. It was Bacharach who perfected the biological assay for vitamin D, accurate estimations of which were essential for the standardization and quality control of the company's products. To this end, he introduced a breeding stock of homozygous rats from the Wistar Institute in Philadelphia.[9] In order to treat variations in bioassays with mathematical rigour, he introduced statistical analysis. For chemical analysis of vitamins, he also introduced newly developed physicochemical techniques in the Glaxo Laboratories, such as colorimetry in the 1920s and ultraviolet spectroscopy in the 1930s (Bacharach, 1934). Bacharach eventually became Glaxo's Chief Executive Scientific Officer, a post from which he retired in 1956.

Especially for an industrial scientist in interwar Britain, Bacharach enjoyed a high profile in the vitamin community, and his extensive personal contacts were significant in the exchange of ideas and information among academic and industrial scientists. He served on numerous committees and was an active member of the Society of Public Analysts (later the Society of Analytical Chemists), the Royal Institute of Chemistry, the Society of Chemical Industry and the Biochemical Society. Together with Harriette Chick and John Boyd Orr, he was a founder member of the Nutrition Society established in 1941 and served as its treasurer for ten years. In all these fora, he actively tried to bring together scientists working in industry, academic laboratories and governmental institutions, with a view to bringing their scientific objectives into alignment.

Bacharach also actively helped to bring science, especially the science of vitamins, into the public domain. A communist, he brought into his scientific activities the contemporary concern of the intellectual left with the "social functions of science" (Bernal, 1939). He engaged enthusiastically in workers' education and public information. In 1938, Glaxo produced a film, "Cambridge Vitamins", with the active collaboration of the Dunn Nutritional Laboratory in Cambridge.[10] Bacharach provided the spoken commentary and it is likely that he was a prime mover behind the making of the film, which stresses the social and economic aspects of nutrition as much as advances in scientific understanding of vitamins and their medical significance.

Among others, the film features Hopkins and Harris in their laboratories, experimental animals collapsing from vitamin deficiencies, and rachitic children. Furthermore, the film draws attention to the findings presented by John Boyd Orr in *Food, Health and Income* (1936) that, by internationally agreed scientific standards, over half the British population lived on an inadequate diet, and to the high cost, for those on low wages, of foods with an adequate vitamin content. In this collaborative venture between industrial and pure scientists, vitamins, by then in the form of standardized molecules, were enlisted to convey a political message about the perils of social deprivation. More strongly, vitamins were used to stress that good nutrition depends not only on the chemical composition of diet, but on the distribution of wealth.

CONCLUSIONS

In scientific investigations of vitamins in the interwar period, different interests, mediated by multiple connections between different social domains, drove each other, to converge on vitamin structure. I shall outline these interests and their interconnections, and suggest that, together, they provided a dynamic of molecularization.

Medical and social interest in vitamins was directed at the treatment and prevention of socially and politically significant diseases, and at raising public health through better nutrition. This interest connected with the scientific domain of research into vitamin function and food chemistry. It became pertinent to ask which foods contain which vitamins; to set standards for vitamin requirements; and to investigate if nutritional requirements can be met by supplementing less than adequate diets with vitamin preparations or vitamin-enriched food products. The potential economic importance of these issues also provided connections with the interests of the food industry and the pharmaceutical industry.

Industrial interest in vitamins was above all directed at making profit, by providing scientific justification for the promotion of products and by devising the cheapest means of production of vitamin preparations or of vitamin-enriched food products. Chemical knowledge and skills were required for the extraction and purification of vitamins and for the standardization of vitamin preparations. Furthermore, knowledge of the chemical structure of vitamins was needed before their artificial synthesis could be contemplated, which might turn out to be the cheapest means of production. This interest in chemical structure also connects with scientific concerns about the relation between vitamin structure and function in terms of their mode of action, to provide justification for selling vitamin products for particular purposes.

Scientific interest in vitamins was directed at the acquisition and application of knowledge about the physiology and biochemistry of nutrition. Investigations were first focussed on the global role of vitamins in normal growth and in the etiology of certain diseases, but ultimately on the specific metabolic role of individual vitamins. The conviction that the elucidation of local metabolic function could only come with knowledge about the chemical structure of vitamins was also used to promote a molecular view of diet, of particular diseases, and of the body. Vitamins thus supported and reinforced biochemical practices and strategies of investigating the body's normal and disturbed functions at the level of molecules and their reactions.

The multiple connections that I have outlined go some way towards explaining why vitamin structure came to be viewed as an important problem. This problem did not simply exist out there, waiting to be recognized, but it became a problem because people in different, but interlinked social domains decided for different, but overlapping reasons that it demanded a solution. Their interconnected interests turned vitamins from hypothetical entities confined to restricted areas of pathology, physiology and biochemistry into objects that could be manipulated by scientists investigating their structure and function, by physicians treating patients suffering from dietary deficiency syndromes, by industrialists producing profit-making commodities, and by consumers to improve their nutritional wellbeing.

ACKNOWLEDGEMENTS

I should like to thank Soraya de Chadarevian, Andrew Cunningham, Jean-Paul Gaudillière and Ilana Löwy for helpful comments on an earlier draft of this chapter.

NOTES

1 The consumption of vitamins will not be addressed in this chapter. Consumer issues are touched on by Sally Horrocks (1995) in relation to Britain, and are given prominent attention by Rima Apple (1996) in her multi-layered study of vitamins and "vitamania" in the U.S.

2 The term "vitamine" (for vital amine) was introduced by the Polish chemist Casimir Funk (1912), then working at the Lister Institute. In view of doubts about the chemical categorization, Hopkins' neutral term "accessory food factors" continued to be used widely among British authorities in nutrition, including Funk's colleagues at the Lister Institute. The word "accessory", however, was also considered misleading, in view of the indispensability for health and survival of these components of the diet (see Plimmer and Plimmer, 1922), and a whole array of terms entered the scientific literature. Eventually, Jack Drummond (1920) recommended that "vitamin" be used in order to avoid misleading chemical connotations, and that the factors "fat-soluble A", "water-soluble B" and the antiscorbutic "water-soluble C" should be renamed vitamin A, B and C, respectively. This terminology, together with the alphabetical convention, was adopted very quickly in the literature.

3 Several authors have discussed tensions between Funk and his colleagues at the Lister Institute, who appear to have regarded his claims about "vitamines" as premature and incautious (e.g., Harrow, 1955; Maltz, 1990). In this regard, it is striking how perfunctory the treatment of Funk's work is in the history of the Lister Institute written by Chick and colleagues (Chick *et al.*, 1971, pp. 144–145). Hopkins, too, was uncharacteristically censorious about Funk's contributions in many of his papers and lectures, including

his Nobel lecture of 1929 (reproduced in Needham and Baldwin, 1949, pp. 191–200).

4 The subdivision of the B vitamins was made on the grounds that chemically distinct fractions of foods with antineuritic activity had radically different physiological effects on animals. In general, different vitamins were distinguished on a combination of chemical and physio-pathological grounds, before their structures were determined. Several of the vitamins identified in the 1920s and 1930s (such as the lactating factor, "vitamin L", or the permeability factor, "vitamin P") are no longer recognised as vitamins.

5 From the beginning, Fletcher envisaged a national and international role for the MRC in the setting of biological standards, not only for drugs and vaccines, but also for vitamins, hormones, dietary requirements and even anthropometric standards in relation to the physiology of work. The broader agenda of the standardization project of Fletcher and close allies such as the pharmacologist Henry Dale and the medical statistician Major Greenwood is discussed by Desirée Cox-Maksimov (in preparation, chapters 3 and 4). See also Sturdy (this volume).

6 For details about the studies of vitamin structure by Nobel laureates in chemistry, see Farber (1963).

7 More specifically, the award was made to Eijkman for "his discovery of the antineuritic vitamin", and to Hopkins for "his discovery of the growth-stimulating vitamins". The pertinent work was done by Eijkman in the 1890s and by Hopkins in the first decade of the 20th century. In other words, the award was made long after the event. Both Hopkins and Eijkman had been nominated for the Prize on many previous occasions.

8 For my account of Glaxo's vitamin-related activities, I rely heavily on the informative company history by Davenport-Hines and Slinn (1992), although they should not be held responsible for my interpretations.

9 On the standardization of the Wistar rat, see Tocher Clause (1993).

10 An original reel of "Cambridge Vitamins" is preserved at the Dunn Nutritional Laboratory in Cambridge. The Cambridge Wellcome Unit for the History Medicine holds a copy of the video version which was produced by the University of East Anglia, Norwich.

REFERENCES

Anon., "Vitaminised Margarines: A Description of the Works of Planters Foods Ltd., Bromboro Port" *Industrial Chemist* (1929), 5: 97–104.

Anon., "Model Plant Makes Vitamins" *Food Factory Maintenance* (1938), 21: 25.

R.D. Apple, "Patenting University Research: Harry Steenbock and the Wisconsin Alumni Research Foundation" *Isis* (1989), 80: 375–394.

R.D. Apple, *Vitamania: Vitamins in American Culture* (New Brunswick, NJ: Rutgers University Press, 1996).

J. Austoker, "Walter Morley Fletcher and the Origins of a Basic Biomedical Research Policy" in J. Austoker and L. Bryder (eds), *Historical Perspectives on the MRC* (Oxford: Oxford University Press, 1989), pp. 23–33.

J. Austoker and L. Bryder (eds), *Historical Perspectives on the MRC* (Oxford: Oxford University Press, 1989).

A.L. Bacharach, "A Brief Survey of Optical Methods in Food Technology" *Chemistry and Industry* (21 December 1934), 1071–1072.

E. Baldwin, *Gowland Hopkins: A Memoir on the Discoverer of Vitamins* (London: Van den Bergh, 1961).

J.D. Bernal, "Carbon Skeleton of the Sterols" *Nature* (1932), 51: 466.

J.D. Bernal, *The Social Functions of Science* (London: Routledge, 1939).

J. Boyd Orr, *Food, Health and Income* (London: MacMillan, 1936).

J.H. Burn, *Biological Standardization* (London: Oxford University Press, 1937).

K.J. Carpenter, *The History of Scurvy and Vitamin C* (Cambridge: Cambridge University Press, 1986).

F.H. Carr, "Progress in the Fine Chemical Industry" *Industrial Chemist* (1928), 4: 6–7.

H. Chick, M. Hume and M. Macfarlane, *War Against Disease: A History of the Lister Institute* (London: Deutsch, 1971).

K. Codell Carter, "The Germ Theory, Beriberi, and the Deficiency Theory of Disease" *Medical History* (1977), 21: 119–136.

D. Cox-Maksimov, "The Making of the Clinical Trial in Britain, 1910–1945" (Ph.D. Thesis, University of Cambridge, in preparation).

K.H. Coward, "The Vitamin Content of Margarine" *Lancet* (1928), ii: 726.

R.P.T. Davenport-Hines and J. Slinn, *Glaxo: A History to 1962* (Cambridge: Cambridge University Press, 1992).

J.C. Drummond, "Note on the Role of the Anti-scorbutic Factor in Nutrition", *Biochemical Journal* (1919), 13: 77–80.

J.C. Drummond, "The Nomenclature of the So-called Accessory Food Factors (Vitamins)" *Biochemical Journal* (1920), 14: 660.

E. Farber, *Nobel Prize Winners in Chemistry, 1901–1961* (New York: Abelard-Schumann, 1963).

C. Funk, "On the Chemical Nature of the Substance which Cures Polyneuritis in Birds Induced by a Diet of Polished Rice" *Journal of Physiology* (1911), 43: 395.

C. Funk, "The Etiology of Deficiency Diseases" *Journal of State Medicine* (1912), 20: 341–368.

R. Furness, "The Oils, Fats and Soap Industries in 1927" *Industrial Chemist* (1928) 4: 16–18.

R. Furness, "The Oils, Fats and Soap Industries in 1928" *Industrial Chemist* (1929), 5: 15–17.

G.L. Geison, *Michael Foster and the Cambridge School of Physiology* (Princeton, NJ: Princeton University Press, 1978).

B. Harrow, *Casimir Funk... Pioneer in Vitamins and Hormones* (New York, NY: Dodd, Mead and Co., 1955).

F.G. Hopkins, "The Analyst and the Medical Man" *Analyst* (1906), 31: 385–397.

F.G. Hopkins, "Feeding Experiments Illustrating the Importance of Accessory Food Factors in Normal Dietaries" *Journal of Physiology* (1912), 44: 425–460.

F.G. Hopkins, "Newer Standpoints in the Study of Nutrition" *Transactions of the Chemical Society* (1916), 109: 629–649.

F.G. Hopkins, "Some Chemical Aspects of Life" *Reports of the Proceedings of the BAAS* (1933), 1–24.

S.M. Horrocks, "Consuming Science: Science, Technology and Food in Britain, 1870–1939" (PhD. Thesis, University of Manchester, 1993).

S.M. Horrocks, "The Business of Vitamins: Nutrition Science and the Food Industry in Inter-war Britain" in H. Kamminga and A. Cunningham (eds), *The Science and Culture of Nutrition, 1840–1940* (Amsterdam: Rodopi, 1995), pp. 235–258.

S.M. Horrocks, "Nutrition Science and the Food and Pharmaceutical Industries in Inter-war Britain", in D.F. Smith (ed.), *Nutrition Science in Britain: Science, Scientists and Politics in the Twentieth Century* (London: Routledge, 1997), pp. 53–74.

H. Jephcott, "Obituary: Alfred Louis Bacharach (1891–1966)" *Chemistry and Industry* (1 October 1966), no. 40: 1651–1653.

H. Jephcott and A.L. Bacharach, "The Anti-scorbutic Value of Dried Milk" *Biochemical Journal* (1921), 15: 137.

H. Kamminga and M.W. Weatherall, "The Making of a Biochemist. I: Frederick Gowland Hopkins' Construction of Dynamic Biochemistry" *Medical History* (1996), 40: 269–292.

A. Landsborough Thomson, *Half a Century of Medical Research*, Vol. 2: *The Programme of the Medical Research Council (UK)* (London: HMSO, 1975).

E.V. McCollum and M. Davis, *Journal of Biological Chemistry* (1915), 23: 181.

A. Maltz, "The Role of Language in the Discovery and Acceptance of Vitamins" (Ph.D. Thesis, University of Illinois, Urbana, 1990).

Medical Research Committee, *Report on the Present State of Knowledge Concerning Accessory Food Factors (Vitamines)* (London: HMSO, 1919).

Medical Research Council, *Studies of Rickets in Vienna* (London: HMSO, 1923).

Medical Research Council, *Report on the Present State of Knowledge Concerning Accessory Food Factors (Vitamins)* (London: HMSO, 1924).

Medical Research Council, *Vitamins: A Survey of Present Knowledge* (London: HMSO, 1932).

E. Mellanby, "The Part Played by an 'Accessory Food Factor' in the Production of Experimental Rickets" *Journal of Physiology* (1918), 52: P11.

J. Needham and E. Baldwin (eds), *Hopkins & Biochemistry, 1861–1947* (Cambridge: Heffer, 1949).

C. Petty, "Primary Research and Public Health: The Prioritisation of Nutrition Research in Inter-war Britain" in J. Austoker and L. Bryder (eds), *Historical Perspectives on the MRC* (Oxford: Oxford University Press, 1989), pp. 83–108.

V.G. Plimmer and R.H.A. Plimmer, *Vitamins and the Choice of Food* (London: Longmans, Green and Co., 1922).

D.F. Smith, "Nutrition Science and the Two World Wars" in D.F. Smith (ed.), *Nutrition Science in Britain: Science, Scientists and Politics in the Twentieth Century* (London: Routledge, 1997), pp. 142–165.

D.F. Smith and M. Nicolson, "The 'Glasgow School' of Paton, Findlay and Cathcart: Conservative Thought in Chemical Physiology, Nutrition and Public Health" *Social Studies of Science* (1989), 19: 195–238.

A. Szent-Györgyi, "Observations on the Function of Peroxidase Systems and the Chemistry of the Adrenal Cortex. Description of a New Carbohydrate Derivative" *Biochemical Journal* (1928), 22: 1387–1409.

A. Szent-Györgyi, *Studies on Biological Oxidation* (Budapest: Karl Rényi, 1937).

M. Teich, "Science and Food During the Great War: Britain and Germany" in H. Kamminga and A. Cunningham (eds), *The Science and Culture of Nutrition, 1840–1940* (Amsterdam: Rodopi, 1995), pp. 213–234.

B. Tocher Clause, "The Wistar Rat as a Right Choice: Establishing Mammalian Standards and the Ideal of a Standardized Mammal" *Journal of the History of Biology* (1993), 26: 329–349.

M.W. Weatherall, "Scientific Medicine and the Medical Sciences in Cambridge, 1851–1939" (Ph.D. Thesis, University of Cambridge, 1994).

M.W. Weatherall, "The Foundation and Early Years of the Dunn Nutritional Laboratory" in D.F. Smith (ed.), *Nutrition Science in Britain: Science, Scientists and Politics in the Twentieth Century* (London: Routledge, 1997), pp. 29–52.

M.W. Weatherall and H. Kamminga, "The Making of a Biochemist. II: The Construction of Frederick Gowland Hopkins' Reputation" *Medical History* (1996), 40: 415–436.

P. Weindling, "The Role of International Organizations in Setting Nutritional Standards in the 1920s and 1930s" in H. Kamminga and A. Cunningham (eds), *The Science and Culture of Nutrition, 1840–1940* (Amsterdam: Rodopi, 1995), pp. 319–332.

E.G. Willcock and F.G. Hopkins, "The Importance of Individual Amino-Acids in Metabolism. Observations on the Effect of Adding Tryptophane to a Dietary in which Zein is the Sole Nitrogenous Constituent" *Journal of Physiology* (1906), 35: 88–102.

CHAPTER 4

Producing Molecular Therapeutics from Human Blood: Edwin Cohn's Wartime Enterprise

Angela N.H. Creager

INTRODUCTION

One of the most celebrated advances in military medicine during World War II was the development of blood plasma fractions as therapeutic agents, largely through the leadership of Edwin Cohn, head of the Department of Physical Chemistry at Harvard Medical School. As the United States began mobilizing for war, Cohn embarked on a project to identify components of blood which could be used to deter shock in combat victims. The rapid success of his laboratory in devising a blood-derived transfusion material created a leadership opportunity for Cohn in the subsequent collaboration with the military and industry to mass-produce serum albumin from blood donated through the American Red Cross. Other plasma fractions of therapeutic value were also isolated by Cohn's research team during the war and manufactured as medical commodities. Cohn continued to work with pharmaceutical producers and clinicians after the military sponsorship ceased by exercising patents on his methods and supervising the establishment of quality control standards for industrial production.

Given the prior orientation of Cohn's laboratory towards quite abstruse protein chemistry, the involvement of his research group in developing transfusion therapeutics seemed a radical departure, the application of scientific expertise in response to a military crisis. However, Edwin Cohn had long argued for the relevance of his work to practical concerns, evidenced by his development in the 1920s of a treatment for pernicious anemia. Cohn insisted that the fundamental attributes of life derive from biologically active molecules, especially proteins.[1] He accorded the theories and tools of physical chemistry highest priority not only for pure research but for addressing medical, industrial, and military problems;

his laboratory used the same biochemical methods, such as fractionation and electrophoresis, to investigate molecules and to produce them for clinical use. This strategy, recognizable throughout Cohn's career, is what I will treat as *molecularization*.[2] Cohn's approach provides an early example of the growing reliance in the 20th century on molecular accounts of life and disease. As in the case of vitamin research (Kamminga, this volume), Cohn's activities did not respect a distinction between pure and applied research (despite his rhetoric to the contrary): They generated not only new conceptions of life but new products as well.

The growing authority of molecular answers to clinical and biological problems was not inevitable. For example, throughout the 1920s, Cohn was an active participant in the rancorous debate between physical chemists (molecularists) and colloidal chemists over the nature of proteins. The dispute concerned not only the academic status of proteins, but whether colloidal science or solution chemistry would explain and ground biology and medicine. Over the course of the 1930s colloidal chemists' claims were largely rejected by protein researchers, even as instruments originally invented by colloidal scientists, such as the ultracentrifuge and electrophoresis apparatus, were assimilated into a biochemical framework. (In those two instances, so were the inventors.) These instruments extended the range of molecularization from the biochemists' small molecules, as detected in enzyme-catalyzed reactions, to the realm of viruses, whose molecular nature was established by Wendell Stanley in 1935 (Kay, 1986). At the same time, the classification of proteins and nucleic acids as true molecules necessitated broadening the notion of a molecule to include entities held together by noncovalent bonds (such as protein subunits) as well as chemical structures linked entirely by covalent bonds. Thus the meaning of molecularization shifted over time, assimilating colloidalist notions of partially assembled particles, in ways its earliest proponents might not have predicted.

Cohn's commitment to molecularization can be seen in three roughly chronological stages. First, in the 1920s, Cohn and L.J. Henderson aimed to demonstrate through their studies of proteins that physical chemical theory was adequate to account for the behavior of even complex biological molecules if fully quantitative methods were used. Because one aspect of their critique of colloidal chemistry was that its rootedness in practical (particularly industrial) problems meant that it was inadequately grounded in theoretical physics and chemistry, they adopted a rhetoric of pure research. Cohn directed his research efforts towards resolving theoretical questions about the solubility and chemical nature of proteins, even as both he and Henderson fostered collaborative efforts with medical colleagues and physiologists.

A second stage of molecularization involved the appropriation of instruments originally developed by colloidal chemists and their deployment as machines for making molecular standards. Cohn's laboratory was among the first to possess an analytical ultracentrifuge and a Tiselius electrophoresis apparatus in the 1930s, and these instruments increasingly defined research on the behavior of proteins in the laboratory. Moreover, after Cohn became involved in blood fractionation during World War II, these instruments served as "mediating machines" between his laboratory and pharmaceutical settings, particularly in the establishment of quality control standards for molecular therapeutics (Wise, 1988). Because prior chemical requirements for molecular purity, such as crystallization, were difficult to achieve with some macromolecules, instruments such as the analytical ultracentrifuge provided important new techniques for classifying or certifying substances as pure molecules.

The Plasma Fractionation Project, centered at Cohn's laboratory during and after World War II, embodies a third phase of molecularization, the development of biological molecules as industrial products. This stage relied on an insertion of laboratory measurements into the quality control specifications for medical commodities (through the aforementioned instruments). Cohn was convinced that maintaining these standards required him to work with pharmaceutical companies after the war. These arrangements were problematic in the view of the university administrators, and Cohn was involved in complex negotiations with Harvard over how to exercise patents for his methods. In Cohn's account of the development of his laboratory's "scientific policy," published in 1952, he endeavored to justify both his applied research enterprise and his patents by contending that neither activity impeded the pursuit of fundamental research in his laboratory. This document reveals the tensions as well as the successes which resulted from molecularization as an industrial strategy.

Cohn's direct collaboration with industry was unusual among prestigious biochemists after the war, as most university researchers contributed to improving the health of Americans only insofar as they contracted to receive funding from the National Institutes of Health (NIH). During the 1950s, experimental biologists sought to increase the autonomy of their research enterprises from the demands of medicine and agriculture, particularly in the expanding postwar universities (Creager, 1996). They justified the ascendance of new disciplines on the grounds that older disciplines, such as biochemistry, bacteriology, and even genetics, were too wedded to applied concerns to discover the secrets of life. In this respect, their rhetoric inadvertently echoed that of the physical chemists such as Cohn who in the 1920s argued that colloidal science was inadequately grounded in fundamental knowledge. The heralding of a new field of molecular biology was successful not only in securing funding for disciplinary innovation but also in capturing the public imagination. In fact, the popular conception of molecular biology as a postwar success story has served to preclude historical awareness of the already prevalent uses of molecular approaches in biology and medicine prior to 1950. Within the scientific community in the 1960s, it was left to disgruntled biochemists to contend that biology had already been molecularized, and they interpreted the historical evidence to their ends (Abir-Am, 1992). Cohn's successes in developing therapeutic molecules were part of the legacy of an older biochemistry which self-interested molecular biologists would cast as representing the utility of research but otherwise outmoded, as they sought to differentiate basic research from the employment of science during the wartime mobilization.

But to the degree that postwar scientists relied on the generosity of public funding for research, they benefited from the widespread belief, supported by Cohn's entrepreneurialism, that research would yield continuing medical advances. The mobilization of science for World War II created unprecedented opportunities for the circulation of commodified molecules (not only plasma fractions, but cortisone and antibiotics) beyond university laboratories through industries, clinics, and battlefields. In response to the success of biomedical researchers in the war effort, the U.S. government not only increased scientific funding but strengthened the regulatory management for molecular medicines. Thus during the 1950s, the Public Health Service both increased the NIH extramural grant budget ten-fold and segregated its regulatory responsibilities into the Division of Biologics (Harden, 1986, p. 180; Mider, 1976, p. 851–52).[3] Cohn's wartime enterprise of producing molecular therapeutics from blood contributed to the creation of this postwar biomedical estate, even as the growing differentiation between basic and applied work made his status problematic.

EDWIN COHN'S RESEARCH ON PROTEINS IN THE DEPARTMENT OF PHYSICAL CHEMISTRY AT HARVARD MEDICAL SCHOOL

Before World War II, Cohn was noted for his repudiation of research for pragmatic ends and the valorization of "pure" research, to the consternation of some of his more clinically-minded colleagues at Harvard Medical School. However, his rhetoric of unfettered research was rooted in the long-standing debate between physical chemists such as himself and colloidal chemists, who had derived much of their credibility by directly addressing applied problems. During the 1920s Cohn endeavored to exclude colloidal scientists from the realm of biological problems, even as he subsequently capitalized on instrumentation which had been developed within a colloidalist framework. The institutional and disciplinary background of Cohn's department at Harvard illuminates his motivations for embracing molecularization as a research strategy before World War II.

The Department of Physical Chemistry was established at Harvard Medical School in 1920 to keep Lawrence J. Henderson from leaving the university. Johns Hopkins University had made a very attractive job offer to Henderson, who was renowned for "his ability for developing research along physical chemical lines in relation to physiology."[4] Henderson built his scientific reputation through his work on the acid-base equilibria in blood and his success reflected well on the utility of physical chemistry to the medical researcher and biologist (Servos, 1990, p. 68; Amsterdamska, this volume).[5] Henderson's new department was spared permanent teaching responsibilities towards the medical students, allowing its members to devote all of their efforts towards applying physical chemistry to questions in biology and medicine. The department functioned as a single laboratory which was associated with the Department of Physiology, although with "dignified independence."[6] Henderson, whose inclinations already ran towards theory over hands-on experiment, called on his former student, Edwin J. Cohn, to organize and manage the laboratory for him. Research was carried out (under the direction of Cohn) by medical students and graduate students who did voluntary projects, almost all of which were concerned with the physicochemical characteristics of physiologically active proteins.

When Cohn returned to Harvard in 1920, he had already been associated with many prominent researchers among the international community of physical chemists. Jacques Loeb, influential author of *The Mechanistic Conception of Life*, was a friend of the Cohn family and colleague of Edwin's older brother, Alfred Einstein Cohn, at the Rockefeller Institute for Medical Research. Alfred guided his younger brother to ground his interest in biology in rigorous training in the physical sciences, reflecting the strong imprint of Loeb's crusading espousal of physical chemistry among life scientists (Scatchard, 1948, p. 265; Edsall, 1961, p. 48). Edwin Cohn completed his Ph.D. in 1917 under embryologist Frank R. Lillie at the University of Chicago, but conducted much of his graduate research with Henderson at Harvard, where he investigated the solution properties of seawater. Cohn soon decided to pursue physical chemical investigations of proteins, and went to Yale to study with protein crystallographer Thomas B. Osborne until he went back to Harvard to conduct war research.[7]

After World War I, Cohn took advantage of the newly-created National Research Fellowships to study with scientists in England and Europe. Henderson guided Cohn towards tutelage under several eminent chemists in Europe, including S.P.L. Sørensen in Copenhagen, who demonstrated the molecular nature of egg albumin; Svante Arrhenius in Sweden, who formulated pH as a solution measurement; and William B. Hardy and Joseph Barcroft in England, well-known for their work on hemoglobin. When Cohn returned to the United States in 1920 to manage the Department of Physical Chemistry, he was well-prepared to undertake research in protein chemistry. The publications from the department for the next ten years were largely concerned with establishing the molecular nature of proteins, much like Loeb's research at the Rockefeller Institute during the same period (Pauly, 1987b). In an influential review article by Cohn in 1925, he synthesized Loeb's published results with those of many other prominent protein chemists, including his own, to produce a comprehensive titration curve for egg albumin. He proceeded to correlate the titration curve with what was then known of the amino acid content of the protein, the first systematic attempt of this kind (Cohn, 1925; Edsall, 1962).

Cohn's research on proteins during the 1920s, and particularly his 1925 review, must be understood within the context of then current debates about the limits of molecular approaches for understanding entities larger than small molecules (Edsall, 1962; Florkin, 1972; Fruton, 1972, "The Nature of Proteins"; Olby, 1974, Section I; Morgan, 1990). During the first decades of the 20th century colloidal chemistry held more promise for solving practical problems than almost any other field of physical science; colloidal chemists offered explanations for the behavior of complex substances such as soaps, films, and the constituents of protoplasm, especially proteins (e.g. see Fischer, 1921). The growth of industrial chemistry and biochemistry during this period provided an expanded audience of scientists interested in materials of colloidal dimension, i.e. "too small to be observed under an ordinary microscope and too large to diffuse easily across organic membranes" (Servos, 1990, p. 301). Colloidalists held that small molecules could assemble into larger aggregates, or micelles, which do not obey the laws of solution chemistry because the particles are suspended (and hence in more than one phase) rather than dissolved. The strongest advocates of this "isolationist" colloidal chemistry among American scientists were Wolfgang Ostwald, Martin Fischer, and Wilder Bancroft, who also served as editor of the *Journal of Physical Chemistry,* a major publication venue for colloidal work (Servos, 1982; 1990, pp. 300–308).

As the discipline of physical chemistry consolidated (especially in the United States) from 1900–1920, many physical chemists took issue with their fellow colloidalists' contention that solution chemistry could be replaced in many instances by empirical laws of the behavior of "gels," "sols," and colloids (Servos, 1990). While even molecularists used the term colloid (note Jacques Loeb's 1924 monograph *Proteins and the Theory of Colloidal Behavior*), they argued that colloidal materials must be fundamentally understood in the molecular terms of physical chemistry. Consequently, colloid scientists and mainstream physical chemists debated whether colloidal effects could be explained by conventional physical chemical theory, and this dispute was particularly rancorous in protein chemistry. (Under these circumstances, the naming of Henderson's unit as the Department of Physical Chemistry announced its members' allegiance.)

Physical chemists faced a difficult task in convincing other researchers that the behavior of proteins could be understood in terms of their molecular nature. Organic chemist Emil Fischer, who had postulated the peptide linkage between amino acids in proteins, contended in 1917 that the molecular weight of a genuine molecule (connected by covalent linkages such as peptide bonds) could not exceed 4000. Evidence for a molecular weight for hemoglobin of at least 16,730 had been present in the scientific literature since 1885, so the reluctance of chemists to accept substances of large molecular weight as molecules spurred the investigation of alternative notions, particularly colloids (Edsall, 1962). Also troubling to biochemists was the fact that solution theory was derived from conditions of infinite dilution, and even when dilute, protein solutions showed aberrant viscosity and osmotic pressure effects. Physical chemists who rejected the assumptions underlying the empirical laws of colloids were thus faced with the challenge of developing a more pragmatic solution chemistry which could account also for colloidal phenomena such as adsorption, charged surfaces, and unusual hydrodynamic effects. Thus the insistence by physical chemists that the realm of molecules included proteins and other biologically active materials, which I have referred to as the strategy of molecularization, created important scientific challenges.

These concerns with the limits of solution theory applied to real problems were not simply academic matters. Colloidal chemistry had gained ground in the early 20th century precisely because its practitioners claimed that industrial and medical problems were too complex to be solved through physical chemical theory, which was based on simple models. Colloidal chemists argued that their empirically-based framework provided greater applicability. Even so eminent a chemist as The Svedberg, who received the Nobel prize in chemistry in 1926 for his work on colloidal systems, appealed to practical matters in justifying his approach. Svedberg introduced his 1928 monograph, *Colloid Chemistry*, by holding out its relevance to the "main part of all living beings" as well as the "very important technical products — [of the] industries of tanning, dyeing, fermentation, cellulose products, starch, gelatin, glues, soaps, rubber, nitrocellulose, celluloid, textile products, dairying, agriculture, bread-making, asphalt, petroleum, photography, clay products, cements, in the purification of water, etc." (Svedberg, 1928, p. 17). For "the study of natural systems which appear in the organisms or as technical products of high importance," Svedberg claimed, these systems "are of such overwhelming importance that every little progress *today* is of the greatest value — we must see what can be gained *now*" (Svedberg, 1928, p. 17, emphasis in original). Svedberg's colloid chemistry promised immediate benefits to industry and medicine.

L.J. Henderson, on the other hand, vigorously defended the soundness of the theories of physical chemistry in the face of the complexities of actual industrial and clinical problems. His own research on blood provided an example of the applicability of chemical thermodynamics for understanding living systems, and he argued that to despair of the utility of an idealized physics and chemistry was to question "the theory of the regulation of breathing, one of the most splendid achievements of modern physiology, which has stood the test of aviation, of gas warfare, and of military medicine" (Henderson and Cohn, 1918, p. 860). Knowledge about proteins was a key area for demonstrating relevance to the processes of life, and each group held to a different articulation between scientific knowledge and worldly concerns. Those promoting physical chemistry (who explained that proteins were simply macromolecules, not colloids) were as eager as the colloidalists to lay claim to industrial and medical problems, but they argued that theories should not be abandoned in favor of empirical practice to accommodate complexities. The dispute between advocates of molecular versus colloidal versions of protein chemistry politicized experimental findings: Svedberg's surprise when hemoglobin appeared homogeneous in his ultracentrifuge in 1924 (rather than displaying a range of molecular weights) gave the molecular advocates an important ally. The gradual acceptance of Hermann Staudinger's term macromolecule provided a chemically palatable way to extend the realm of molecular chemistry to include proteins and other large polymers. (On Svedberg, see Pedersen, 1983; on Staudinger, see Fruton, 1972, pp. 154–155; and Olby, 1974, pp. 3–10.)

Following the precedent of his teacher Henderson, Cohn committed the Department of Physical Chemistry to relating the latest theories in physical chemistry to problems of protein structure. Cohn saw in Peter Debye's work on interionic forces, particularly the Debye-Hückel theory of 1923, a way to understand the complex interactions of proteins, amino acids, and ions, which had proven intractable by available theories of chemical equilibrium.[8] His efforts to employ theory to study protein structure involved alliances in two directions. MIT theoretical chemist George Scatchard maintained an ongoing collaboration with members of the Department of Physical Chemistry to extend chemical theory to their protein research. At the same time, Cohn took care to direct this cooperative work towards topics relevant to physiology and medicine whenever possible. By 1929, Cohn had launched a major research investigation of the physical properties of blood proteins, such as hemoglobin, fibrinogen, and immune globulins.[9]

The debate with colloidal chemists left its mark on Cohn's research endeavors beyond the emphasis on theory. The laboratory devoted years of intensive research to determine how solubility patterns of peptides and proteins are correlated with the ionic properties of constituent amino acids. These studies directly countered the colloidal chemists' adsorption hypothesis, and demonstrated the practicality of treating purified proteins as true solutions. Over the course of the 1930s, Cohn's laboratory shifted its emphasis from assessing the physical chemical behavior of amino acids and small peptides to working on the size and shape of larger proteins. The Department of Physical Chemistry added an analytical ultracentrifuge and a Tiselius electrophoresis apparatus, for investigating the size and shape as well as electrical charge of proteins, to their equipment for measuring the dielectric constant and double flow refraction of solutions. (On the significance of electrophoresis in research on protein chemistry, see Kay, 1988.) The laboratory's extensive studies culminated in the 1943 publication of *Proteins, Amino Acids and Peptides* by Cohn and John T. Edsall, which became a standard reference for protein biochemists during the next decades (Cohn and Edsall, 1943; Eisenberg, 1992).[10] As Robert Kohler has pointed out, within biochemistry there was a remarkable growth of interest in the behavior of macromolecules, and laboratories such as Cohn's Department of Physical Chemistry were viewed by 1940 as part of a frontier in the discipline (Kohler, 1982, "Epilogue").

While Harvard University funded research in the Department of Physical Chemistry for the first decade, the Rockefeller Foundation provided most of the research support for the expanding group in the 1930s, at first through its affiliation with Walter Cannon's Department of Physiology (see Cohn, 1950). In 1930 the Division of Medical Science of the Rockefeller Foundation gave Harvard Medical School a seven-year grant of $175,000 for the Laboratories of Physiology, two-thirds of which was to support Cannon's work in the Department of Physiology, and one-third of which ($10,000 per year) was to support Henderson and Cohn's research. (On Cannon and physiology at Harvard, see Laszlo, 1987, pp. 83–88.) By the time that the grant expired, the Department of Physical Chemistry had achieved more autonomy from the physiologists and Cohn had been designated Chair of the department. In 1938, the Division of Natural Science of the Rockefeller Foundation awarded a seven-year grant of $100,000 to Cohn and his department, giving the group not only a strong financial base but considerable freedom and prestige.[11]

Much has been written on the effect of the Rockefeller Foundation's science policies on the subsequent development of life science, particularly in the U.S.

(Kohler, 1976; Abir-Am, 1982; Kay, 1993). The funding of Henderson and Cohn's program of research on the physical chemistry of proteins typified Warren Weaver's efforts to advance physical, chemical, and quantitative approaches to biology. In the words of Robert Kohler, "Cohn's work was one of the showpieces of the Rockefeller Foundation's program in experimental or molecular biology" (Kohler, 1982, p. 315). Research on protein structure represented one of the most fruitful fields for such interdisciplinary innovation. Nonetheless, it should be emphasized that Weaver's vision led the Rockefeller Foundation to support a great diversity of research, furthering the causes of many different schools within chemistry and physics (Kohler, 1991, pp. 275–300). Colloidal chemists working on biological problems were funded by the Rockefeller Foundation as well as physical chemists who were committed to a molecular view of proteins; these groups of researchers held quite incompatible notions of the nature of subcellular particles. While over the course of the 1930s particular tools and methods from various approaches (ultracentrifugation, electrophoresis, radioisotope tracers) came to represent the successes of physics and chemistry in biology, often as the result of substantial Rockefeller investment, the subsequent standardization of these developments has obscured the heterogeneity of programmatic attempts to subject biology and medicine to physics and chemistry.[12]

While well-respected among protein chemists, physiologists, and biochemists, Cohn's work was not very accessible to those unversed in physical chemical theory. One report from a Rockefeller Foundation officer includes the following summary:

> The scientific level of the work in this laboratory has always been distinguished. But it is certainly true that the work over the first ten or fifteen years was pretty abstract and theoretical. Professor Cohn and his colleagues carried out a large number of difficult and precise experiments to determine the physical-chemical properties of large protein molecules. As late as 1938 the Rockefeller Foundation, making an appraisal for itself of this project, included in its statement the sentence, 'The work has been painstaking, abstruse, and likely only slowly to come to widespread recognition for its essential importance.'[13]

In addition, Cohn's approach had its detractors. While other biochemists such as S.P.L. Sørensen, C.R. Harrington, and F. Gowland Hopkins praised the high caliber of research in Cohn's laboratory, Linus Pauling questioned what value the accumulation of all the data had for fundamental knowledge about proteins.[14]

At Harvard, in part due to its heavy reliance on outside support, the Department of Physical Chemistry had a problematic relationship with the medical school. More than half of the departmental funds were derived from temporary sources (predominantly the Rockefeller Foundation), and many medical faculty members resented the support of a department which contributed neither to the teaching of medical students nor to clinical research. As Cohn wrote in 1938,

> Had I appreciated in 1920 the strong bent toward supporting primarily such research with Medical School funds as would be immediately appreciated by a faculty of medicine, I would of course not have gone to work at the Harvard Medical School.[15]

In fact, Cohn always kept his eye on medical and agricultural problems which might justify his laboratory's research on proteins and amino acids. Cohn's decision in the late 1920s to pursue the isolation of a factor effective in treating pernicious anemia, which was a departure from his protein studies, illustrates his felt need to demonstrate the medical relevance of his laboratory (Weiner, 1989). In 1926, George Minot and William Murphy at Harvard Medical School published their finding that pernicious anemia could be treated with a special diet of liver. Edwin Cohn joined the project to isolate a biochemical fraction from liver producing the same effect (the liver diet being quite unpalatable to patients). While he never identified a chemically pure active component (vitamin B_{12} was isolated in the 1940s), his fraction was produced by Eli Lilly and used clinically for several years.

During the 1930s, when Cohn's projects held no direct clinical relevance, he sought to induce industrial trade groups and companies to fund research in his department (on the advice of a Harvard Business School professor). To this end, he wrote an essay entitled "Certain Relations Between Biochemical Research, Agriculture and Industry, Biology and Medicine."[16] He appealed to the centrality of natural products, especially proteins, to problems facing industry, agriculture, and medicine. In Cohn's (opportunistic) view, "the solution of the more complex problems involving the physics and chemistry of living matter" demanded more adequate support of laboratory research on proteins. "Moreover, their characterization involves complex physical and physical chemical equipment, costly to install, and requiring a staff of highly competent, well-trained, matured investigators for their proper use. So obtained the results of electromotive force, electrophoretic, dielectric diffusion, viscosity and ultracentrifugal measurements are rapidly modifying our knowledge both of the role of these substances in health and disease and of their possible roles in agriculture and industry." Cohn's appeal did result in substantial research support from the Corn Products Industry during the 1940s.[17]

An opportunity to develop new transfusion materials in 1940 provided Cohn with a justification of his abstruse explorations of protein structure during the 1930s. Medical colleagues saw the transition in a different light. In the words of his more clinically oriented colleague, A. Baird Hastings, "If it hadn't been for World War II, Edwin Cohn might have gone to his grave continuing to measure the dielectric constants of amino acids and proteins with one more decimal of accuracy! The needs of the war literally saved him" (Hastings, 1989, p. 71).

THE LABORATORY OF WAR: TRANSFUSION TECHNOLOGIES AND BLOOD SUBSTITUTES

In the spring of 1940, as the German offensive in continental Europe worried American military leaders, the Surgeon General of the Army sought advice from the National Academy's National Research Council regarding medical technologies. The National Research Council had served as the source of scientific and engineering expertise for the American military during World War I (and earlier); following this precedent, committees of scientists were organized in

1940 to answer military medical needs in transfusion and chemotherapy (Stewart, 1947, pp. 100–102). Walter Cannon of Harvard Medical School served as head of the committee of transfusion, and he involved Edwin Cohn in helping the Navy develop a transfusion material from bovine blood.

Cohn organized his entire laboratory around the project within a few months, such that even before the Office of Scientific Research and Development began funding medical projects, the Department of Physical Chemistry was fully engaged in military concerns. His early commitment yielded unexpectedly rapid success. Cohn's protocol for fractionating plasma, devised during the U.S. preparedness period, became widely used in the industrial production of military medical supplies, and Cohn's laboratory remained the center of further research and development throughout the war. Due in large part to the authority Cohn acquired among both Navy representatives and industrial producers, biochemical methodology soon came to the fore in transfusion medicine.

The importance of Cohn's laboratory in developing plasma fractionation methods is all the more striking because it occurred at a time when whole blood transfusions were generally considered superior to the use of blood substitutes. The advantage of Cohn's method depended entirely on the medical supply policies of the military. There had been significant advances in transfusion technologies during the 1930s. After the first uses of stored blood by military surgeons in World War I, the first blood bank in the United States opened at Cook County Hospital in Chicago in 1937 (Wood, 1967). At the same time, the Republican Army involved in the Spanish civil war was using a blood transfusion service, the first of its size in a military context. Nonetheless, the challenges of providing the necessary refrigeration and transportation for whole blood on a large scale motivated the U.S. Navy to stockpile dried plasma, even though military surgeons disagreed about whether whole blood could be effectively substituted on the field. Even amassing dried human plasma was challenging due to the large numbers of blood donors required. Consequently, the Navy hoped that bovine blood, which was readily available from Midwestern slaughterhouses, would be a suitable feedstock for industrial level production of blood substitutes. The war provided the motive for Cohn to devise processes for mass-producing pure, standardized blood components.

The context of war had been instrumental in the development of transfusion technologies throughout the 20th century, and scientists had worked to develop blood substitutes during World War I as well as during World War II. During World War I military doctors began regarding shock, a leading cause of fatality on the battlefield, as a result of the depletion of blood volume. For injured soldiers, blood could be replaced by non-physiological liquids in lieu of blood, which was difficult to preserve and transport. Colloidal chemists were consulted to devise a blood substitute which would maintain normal blood volume, and gum acacia was used as a therapeutic (Janeway and Oncley, 1947, p. 444). Although researchers continued to seek alternative blood substitutes during World War II (for example, Linus Pauling worked on gelatin as a possible candidate), physiological substances were also being developed. The commercialization of vacuum-drying machines (lyophilizers) by Sharpe and Dohme in the 1930s made possible the drying of plasma, which was preferred over other blood substitutes (Kendrick, 1964). Comparing the involvement of scientists in World War I transfusion medicine with the same efforts during World War II, one notices an overall shift from the language of "colloids" to that of protein solutions, as well as a change in emphasis from blood substitutes to blood derivatives. The transition in military medicine from reliance on colloidal chemistry to physical chemistry of solutions (as assimilated into biochemistry) reveals the process of molecularization which had occurred within biomedical research in the intervening decades. By 1947, the term "colloidal" was used by scientists working on transfusion materials to refer to substances which increase the osmotic pressure of blood rather than to describe protein structure (e.g., Janeway and Oncley, 1947).

When Walter Cannon of the National Research Council and David Edsall for the American Red Cross (and former Dean of Harvard Medical School) drafted Cohn to "determine whether the plasma of animals could be made safe for human transfusion," Cohn set his laboratory on isolating the protein albumin from bovine blood (Cohn, 1948, p. 365). Medical researchers (in conjunction with colloidal chemists, actually) had established that the protein albumin is responsible for around 80 percent of the blood's osmotic pressure. Cohn's biochemical strategy was to isolate this blood component rather than to try to remove antigenic particles from bovine blood. Ten years of prior experiments in Cohn's laboratory devoted to assessing the effects of solvents, pH, ionic strength, temperature, and protein concentration on protein solubility were mobilized for the task of establishing a fractionation method for isolating serum albumin which might be reliably carried out on an industrial scale. Drawing heavily on particular methods developed in solubility studies six years earlier, the group settled on a basic framework by the late summer. Rather than relying on salt precipitation, as was conventional in laboratory purification of proteins, their method utilized ethanol at cold temperatures to separate the plasma proteins into five major fractions. The new availability of freeze-drying machines meant that this ethanol fractionation could be carried out on a large scale; a solvent such as ethanol is much easier to remove from a protein solution than a solute such as a salt. In addition to controlling the ethanol-water ratio and temperature, Cohn's method relied on regulation of three other variables to effect protein separation: pH, ionic strength, and protein concentration (Cohn *et al.,* 1940). Cohn often referred to this approach as the five-variable method, although other researchers and producers quickly termed it the Cohn fractionation method.

The re-orientation of Cohn's group away from precise physicochemical measurements of various proteins to the task of isolating a transfusion material gave the laboratory's sophisticated instruments for determining the charges and sizes of protein molecules a new mission:

> The highly specialized equipment which had been installed during the twelve years of theoretical studies — including the ultracentrifuge, the electrophoretic apparatus, and the devices for determining dielectric constant — all proved directly serviceable in this practical job of separating and testing animal products for medical use.[18]

Cohn's laboratory workers used these instruments for establishing the standards of purity for their serum albumin fraction and then for other therapeutic fractions. The successful redeployment of these technologies towards standardization and production became a new justification for the money which had been invested in the Department's research in the 1920s and 1930s. As summarized for the Rockefeller Foundation trustees, "this neat job in applied science provides a beautiful demonstration of the bread-and-butter value of theoretical research."[19] Regardless of whether plasma fractionation justified the expensive machines, the reliance on the Tiselius electrophoresis apparatus later became incorporated into the Federal licensing regulations for serum albumin as a medical product. In this respect, the biochemical equipment used to certify the molecular homogeneity of medical products carried the strategy of molecularization into industry.

While the research was initially undertaken at the request of the military through the National Research Council, early developments were funded through Cohn's Rockefeller Foundation grant. The first war-related support the laboratory received was a check for $1500 from the National Research Council in the early spring of 1941 to construct a pilot plant at Harvard for producing serum albumin on a larger scale (Cohn, 1948). A few months later the Committee on Medical Research (CMR) was constituted as a primary subdivision of the Office of Scientific Research and Development (OSRD). Funding from the OSRD for Cohn's plasma fractionation work began in August of 1941, the beginning of hundreds of thousands of dollars spent by the government on Cohn's enterprise (through a series of contracts). Early production from the Harvard pilot plant was sufficient to begin clinical trials, and also during this period Armour Laboratories in Chicago began preparing to produce blood derivatives. Bovine albumin was found to cause delayed serum sickness among some recipients when tested on prisoners (National Research Council, pp. 358–380), so Cohn shifted the emphasis to the isolation of serum albumin from human blood, which proved safe in clinical trials. When the United States entered the war on December 7, 1941, the laboratory was ready to ship human serum albumin to treat several casualties of the Pearl Harbor attack. Impressed by the results, the Surgeon General of the Navy authorized contracts for the commercial preparation of human serum albumin early in 1942 (Cohn, 1948).

The success of human serum albumin as a therapeutic posed an enormous challenge to the small national blood donation service. The American Red Cross had only recently established blood donation services in response to the military. Dr. G. Canby Robinson, the National Director of the Blood Donor Service of the America Red Cross, responded to the military's six-fold demand for blood for albumin in this way:

> The problem of collecting blood in this program is an enormous assembly line, mass production job. Many serious problems arise in the collection of this number of bleedings. The America Red Cross, as you know, is a volunteer organization. — The Plasma Program for the Armed Services was considered to be an extremely large one. Now in addition to the plasma program we (Red Cross) have been asked to collect blood from

> approximately an additional million and a half donors for the albumin program. Everyone must realize this is an enormous task. Much of the blood is collected in small localities in mobile units. — To keep the proper amount of blood going into any given laboratory is quite a task. For example: one of the laboratories wants 5,000 bleedings per week for plasma and 6,000 for albumin. It is almost an impossible task to get this number of bleedings into one laboratory (National Research Council, p. 259).

Robinson's use of industrial metaphors in describing community-based blood collection is striking, but the production language was apt. Essentially all of the donated blood went to manufacturing sites, both to prepare dried plasma and to supply the Harvard Laboratory and the collaborating pharmaceutical industries with material for the fractionation.

Cohn recognized that the human plasma proteins in the other fractions besides serum albumin carried potential medical use, and he set about isolating these by-products. Gamma globulin antibodies could be used for inoculations, isoagglutinins could provide diagnostic tools for blood typing, and blood clotting proteins could be used to treat disorders such as hemophilia and to prevent extensive bleeding from surgical procedures. The Cohn method was subsequently refined to yield five biomedical products from human plasma: serum albumin, serum gamma globulin, fibrin foam with thrombin, blood grouping globulins, and fibrin film. Dr. Joseph Stokes, Jr., who was associated with the Commission on Measles and Mumps of the Army Epidemiological Board, coordinated the clinical trials of gamma globulin in Philadelphia. Cohn also had the cooperation of physicians in Baltimore, Detroit, Durham, Omaha, Los Angeles, San Francisco, Hartford, and New York in administering the experimental blood fractions. The successful use of gamma globulins by collaborating physicians as a prophylactic against measles in the winter and spring 1942–43 provided the model for its subsequent use among U.S. troops.[20] The highest priority product continued to be normal human serum albumin, but all five blood components were prepared commercially. Between January 1942 and May 1945 the CMR, Red Cross, and Navy jointly contracted their production to seven pharmaceutical houses, Armour, Cutter, Lederle, Eli Lilly, Sharpe & Dohme, E.R. Squibb, and Upjohn (Cohn, 1948, pp. 375–76). (See Figure 4.1.)

Thus Cohn's molecular strategy for producing medicines from blood had launched a new industry, one which would not have arisen apart from the exigencies of war and the need for mobile transfusion materials. Cohn became more deeply involved in the production aspects of the blood derivatives than might have been expected for a university researcher.[21] The pilot plant which had been built in 1941 in Cohn's Harvard Medical School laboratory made it possible for his group to develop the actual industrial-scale production protocols, and it provided a prototype for subsequent plasma fractionation plants. (Figure 4.2, reproduced from Cohn's own account of the project, depicts his laboratory as a miniature factory.) Cohn's pilot plant served as the central production site until commercial production was well underway in 1943, with the

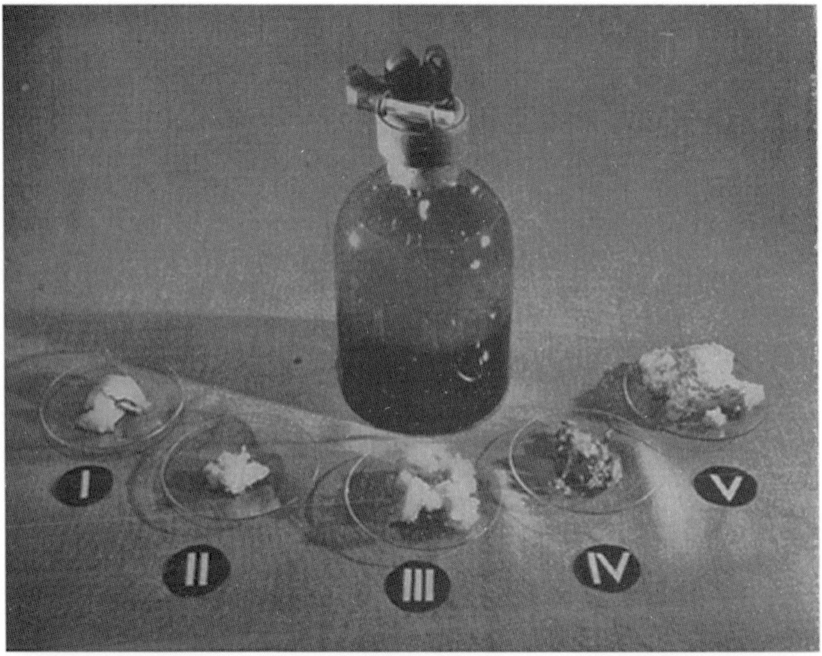

Fibrinogen and Thrombin
Fraction I and III–2

Fibrin Foam and Thrombin
Fraction I and Fraction III-2

Fibrin Film
Fraction I and Fraction III-2

Serum γ-Globulin
Fraction II

Isoagglutinins
Fraction III-I

Serum Albumin
Fraction V

Figure 4.1. Medical products from fractionation of human blood plasma.

Large-scale fractionation of human plasma proteins

Harvard pilot plant.

Figure 4.2. Juxtaposed photographs from Harvard pilot plant (left) and an industrial plant (right) showing the drying of blood proteins from the frozen state in the production of fractionation products.

Harvard researchers operating their small plant 24 hours a day to produce the serum albumin demanded by the military (Cohn, 1948, p. 375). Cohn's laboratory also took charge of training the contracted industrial producers. The pharmaceutical houses were required to send representatives from the companies for months of training under Cohn to master the procedures and begin production for the military.[22] In addition to assuring close control over production, which was very important to Cohn, this training program served to guard confidentiality of the methods, which were after all developed to give American military medicine an edge over enemies. The urgencies of war intensified Cohn's ambi-

Large-scale fractionation of human plasma proteins

Industrial plant.

Figure 4.2. continued

tious disposition; one member of his group described Cohn as possessing "the driving force and organizing ability that one might expect to find in an American corporation executive" (Janeway, 1980, p. 582; also see Lear, 1951).

At another level, the plasma fractionation technology was particularly well-suited to the management methods of a military project. Cohn organized his laboratory into groups working on each of the blood fractions, so that they divided the work and expertise just as they divided the blood.[23] This management style was easily meshed into the hierarchical structure of the Navy as well as being suited to the commodification of his fractions; researchers and producers alike were all required to report to Cohn, who was recognized by the Navy as the authoritative consultant "in all matters related to the processing method" (Cohn, 1948, p. 378). Cohn's method enabled the transformation of civilian blood into pharmaceutical commodities used to achieve transfusion of shock victims, immunization of soldiers, reconstruction of the brain covering after neurosurgery, induction of clotting in hemophiliacs, and blood typing. It was a triumph for proponents of a molecular approach to disease and pharmaceuticals. Moreover, the public credited mobilized scientists with the provision of both plasma fractions and antibiotics as mass-produced molecular medicines for the postwar era (Fulton, 1945; Neushul, 1993; also see Marks 1992).

DEMOBILIZATION AND POSTWAR SCIENCE POLICIES

Cohn's success in developing new blood-based therapeutics and his high-profile position in the multifaceted production enterprise brought him to a new level of prominence by the end of the war. Articles on Cohn and his contributions to transfusion medicine appeared in *Time* and *Life* magazines, *Collier's*, *Atlantic Monthly*, *McCall's* (a women's magazine), several newspapers, a television program, a radio program, and international publications.[24] Cohn's war work was also featured among scientific publications, with both his picture and a schematic diagram of the therapeutic aspects of his work appearing on the cover of *Chemical and Engineering News* in May 1948 (Figure 4.3). In an article which typified the acclaim surrounding Cohn's war work, a *New York Times* reporter argued that the benefits of Cohn's discoveries would overtake the high price of war:

> This war, which has caused the flowing of more rivers of blood than ever before in human history, has also led to the greatest progress in our knowledge of this vast ocean of life, its multiplicity of constituents and the mechanism of its many systems, according to Prof. Edwin J. Cohn of Harvard. The new knowledge, which is already saving countless lives of our fighting men on the battle fields, promises in the generations to come to give back to mankind more lives than the war has taken.[25]

Unlike many biologists who returned to pre-war topics of research, Cohn committed his laboratory to developing ever-better methods for fractionating blood and stockpiling transfusion materials for future national emergencies, a venture which kept him in the public spotlight. In addition, he chose to remain closely connected with the pharmaceutical houses by exercising oversight over all industrial production under patents he filed.

Despite the attendant publicity, Cohn's quest for better industrial methods of plasma fractionation was problematic for several reasons. First, late in 1943 a controversy erupted over the military's reliance on blood substitutes to treat casualties, as a prominent Army surgeon, Edward Churchill, leaked to the *New York Times*

Figure 4.3. Picture of Edwin Cohn, shortly after receiving the Richards Medal from the Northeastern Division of the American Chemical Society, on the cover of the national publication for the American Chemical Society, *Chemical and Engineering News*, May 25, 1948.

that (in his view) the disastrous losses in North Africa were due in part to over-reliance on plasma instead of whole blood. He had resisted the "oversimplified physicochemical" approach to transfusion materials from the outset of the war, urging a more widespread use of whole blood.[26] In response, by March 1944 there was a striking change in military policy towards promoting the use of whole blood over blood derivatives. As Brigadier General Douglas Kendrick reflected, "Had members of the various committees of the National Research Council been permitted to visit combat zones, as was several times requested, their recommendations would have been more closely related to the facts of military medicine" (Kendrick, 1964, p. 36). Nonetheless, behind the lines an enormous blood-processing industry had been set in motion, and the enterprising professor and at least one of the pharmaceutical houses (Cutter Laboratories) were interested in its postwar continuation.[27] The commercial development of therapeutic fractions other than serum albumin, especially gamma globulins, could justify the physicochemical or molecular orientation to biological blood products for civilian medicine.

Second, the changing ideology of research posed a different kind of problem for Cohn. With the publication of Vannevar Bush's report *Science — The Endless Frontier* in 1945 and subsequent debates in Congress about how best to foster scientific research, the apotheosis of "basic" research had commenced. Cohn's laboratory returned to their studies in protein chemistry but Cohn was unwilling to abandon his pilot plant and efforts to devise improved blood fractionation schemes. Consequently, his laboratory group was divided between those doing research on protein structure and those continuing the blood fractionation project. Some academic peers felt that Cohn never regained his stature as a researcher in biochemistry. In the words of fellow protein chemist Hans Neurath, "Cohn was approaching the peak of his great powers as a physical chemist when he had to turn his entire energies to doing practical work for the war effort."[28] Cohn's own equivocations about how his continued involvement in blood-derived therapeutics related to the framework of basic research are evident in his 1952 essay on science policy (Cohn, 1952).

A third problematic aspect of Cohn's continuing the plasma fractionation project concerned his interest in maintaining proprietary rights over his methods. Despite widely-held sentiments that academic researchers should not patent therapeutic innovations, Cohn began filing patent applications for his fractionation method during the war, in 1942.[29] This decision was strongly informed by his earlier experience with the method he and George Minot developed in 1927 for the isolation of a liver fraction effective in treating pernicious anemia (Weiner, 1989). At that time Harvard refused to allow Cohn and Minot to patent the method, even though the researchers were working with Eli Lilly to develop it as a commercial therapeutic. A researcher at Eli Lilly subsequently patented Cohn's method without informing Harvard University, preempting Cohn's legal ability to exercise control over the commercial product. Cohn subsequently argued that the "public's interest would have been better protected" had he been permitted to file patents and supervise quality control of industrial production (Cohn, 1951, p. 8).

Because the plasma fractionation method was developed before the laboratory entered into a contract with the OSRD, Cohn could legally patent his methods as a Harvard faculty member without the government being party. However, Harvard still maintained a general policy against patents being taken out on products of therapeutic value. Cohn insisted to Harvard President James Conant that the "public good" necessitated a change in policy. Given the long history of debates about the ethics of patenting by biomedical researchers in universities, Cohn had to articulate his grounds for exercising patents with great care. In order to justify this decision, Cohn presented the pharmaceutical producers, his wartime collaborators, in a new light. The fact that they would now be marketing the therapeutics for profit necessitated his involvement to protect the public good from mercenary interests.[30] Whereas the regulation of blood processing during the war by Cohn and his laboratory had been conducted in the name of national defense, the same activity was now to be pursued on behalf of the public.

Cohn's patents were initially held by the Research Corporation, a non-profit organization whose purpose was to license university-based patents and invest the proceeds in research. To oversee licensing and supervise product quality, Cohn established a Commission on Plasma Fractionation and Related Processes after the war. The commission was composed of Cohn, Dr John F. Enders, the Harvard virologist who developed a method for using tissue culture to study poliomyelitis, Dr V.A. Getting, the Commissioner of Public Health for Massachusetts, Dr Roger I. Lee, a member of the Harvard Corporation and the President of the American Medical Association, and Dr George Scatchard, MIT theoretical chemist and long-time collaborator of Cohn's. This group was to "advise the Research Corporation on the technical competence of the manufacturers who apply for licensing, to establish procedures and standards for the production and testing of products, and to authorize testing laboratories which will have the authority to check the licensees' output against the standards."[31] By 1950, Cohn began to make plans to re-acquire the patents from Research Corporation, and established the "Protein Foundation," another non-profit corporation, charged with "protecting against the improper exploitation of the public where public health and therapeutic procedures are involved."[32]

The setting of standards for blood therapeutics by Cohn, a protein chemist, had lasting consequences in the licensing of plasma fractions. At the beginning of the Plasma Fractionation Project, relevant quality standards were stipulated by the National Institutes of Health. Cohn felt these standards (particularly that for serum albumin) to be grossly inadequate; the products meeting these codes were impure when analyzed biochemically.[33] Cohn's quality standards relied on instruments which could assess the chemical purity of the proteins in each fraction. In the first Federal license application to produce human albumin, submitted by the Massachusetts State Department of Health laboratory on August 29, 1941, item 35 reads:

> Describe tests and requirements for potency or standardization. When are these tests made?
>
> Tiselius runs to show presence of albumin and globulins as well as nitrogen and ash determination. Tests made after filtration.[34]

The ultracentrifuge and electrophoresis equipment provided a way other than crystallization to establish the chemical homogeneity of macromolecules, and could mediate these standards between industrial and academic settings.

Cohn linked the commercial usage of his patents to these instruments: While the pharmaceutical producers were not required to pay a licensing fee, they were required to demonstrate the biochemical purity of their plasma fractions, which necessitated that they obtain the same equipment Cohn used. The companies also had to send a sample of every lot of each product to Cohn's laboratory for testing, as supervised by the Commission on Plasma Fractionation and Related Processes. Electrophoresis and analytical ultracentrifuge equipment became commercially available after the war (Kay, 1988; Elzen, 1986), and their proliferation in biomedical research laboratories was accompanied by their use in commercial settings for quality control. Molecularization as an industrial strategy can be seen in the reliance upon these standardized instruments to certify the safety of biologically-derived blood products. The adoption of chemical purity as a standard for pharmaceutical products was not general, however; the biochemical impurity of the Salk vaccine, for instance, was heavily criticized by some scientists in the wake of the Cutter incident, in which early batches of the vaccine from Cutter Laboratories transmitted poliomyelitis to some children.[35]

Cohn's brainchild in the early 1950s was a portable blood fractionation machine, which could be directly connected to a donor's arm in order to separate the plasma components as the blood was drawn. In this way he hoped to isolate and preserve even the most labile plasma proteins, extending the strategy of molecularization to less robust physiological agents.[36] Cohn patented his portable fractionation system and was negotiating in 1951 with a scientific instruments company (American Optical Company) to mass-produce his miniature blood fractionator.[37] He intended to oversee manufacture of a whole fleet of these machines, which were to be used at blood donation centers. As one might expect, Cohn's invention caught the public imagination. The report of the development in the *New York Times* reads:

> The world's tiniest chemical processing plant, about the size of a card table, was wheeled into a Yale University lecture room here this morning and demonstrated at the annual meeting of the National Academy of Sciences. The tiny plant, of plastic and stainless steel, performs chemical engineering steps that are virtually as complex as those of a vast petroleum refinery.[38]

The comparison of plasma with petroleum illuminates a central aspect of Cohn's venture: he had, in fact, opened up a new domain of natural products which pharmaceutical houses were eager to develop, particularly once the promise of using gamma globulins for immunization against poliomyelitis was demonstrated (Smith, 1990, pp. 193, 250; Paul, 1971, pp. 390–394). The production of therapeutics from blood, such as clotting factors for hemophiliacs, has subsequently grown into an international, highly profitable industry (Titmuss, 1971; Hough, 1978).

Cohn's entrepreneurial ventures reflected the success of a molecular approach to therapeutic problems, but scientific managers at private foundations were divided over whether the goal of producing new and improved medical products was detrimental to the advancement of basic research. Whereas Basil O'Connor, head of the National Foundation for Infantile Paralysis, championed Cohn's laboratory, one Rockefeller Foundation officer noted in 1950, "E.J. Cohn is not a first-class scientist, but is an excellent entrepreneur and promoter. His main weakness however is that he directs his laboratory so completely that the chance of their observing and finding anything 'sideways' is negligible and in this he seems therefore to resemble the worst of industrial research."[39] To the degree that criticisms of Cohn concerned his willingness to let practical ends direct his research, they represented a live issue for postwar biochemists, whose discipline had been established in the U.S. through its relevance to medicine and agriculture (Kohler, 1982). For example, in the view of virologist Wendell Stanley, in the postwar period biochemistry had "matured" beyond its juvenile servitude to medicine.[40] Likewise, self-ordained molecular biologists sought to distinguish their enterprise from more applied concerns. In the words of Salvador Luria:

> Molecular biology deals with questions of molecular structure, and therefore is biochemistry; but it is not the classical biochemistry that emerged earlier in the twentieth century out of the concerns of medical, agricultural, and industrial researchers. (Luria 1984, p. 83).

Luria offers molecular biology as a purified biochemistry, freed from the taints of industry, medicine, and agriculture. There is, of course, an ironic similarity in this rhetoric to the debate between colloidal chemists and physical chemists. Just as biochemists such as Cohn had criticized colloidal chemistry for its reliance on empirical observations over theory and its industrial orientation, phage geneticists castigated biochemists for their service to agriculture and medicine. Even as Cohn's writing during the postwar period argued for the supremacy of "untrammeled" research (see Cohn, 1952), the recognized achievement of his plasma fractionation methods in postwar medicine derived from their industrial success.

CONCLUSIONS

In surveying Edwin Cohn's career as a protein chemist and inventor of medical products, molecularization represents Cohn's valorization of chemistry as providing the decisive solutions to practical problems as well as to biological questions. Military and medical needs, as well as disputes between physical scientists in different fields, gave the strategy of molecularization its particular shape in Cohn's laboratory. On the one hand, the debate about what kind of physics and chemistry would be employed in solving medical and biological problems constrained the ways in which Cohn appealed to practical applications in the 1920s and 1930s. Because colloidal chemistry had been based from the outset in the establishment of empirical relations which might be exploited in industry and medicine, physical chemists such as Henderson and Cohn took the high road of claiming that only theoretically grounded approaches to biology and medicine were worthy. On the other hand, even as Cohn distanced himself from applied research during the interwar period, he took every opportunity to demonstrate the practical utility of his protein chemistry. By 1940, the appeal of colloidal chemistry in biology had attenuated and the context of war permitted a greater commitment to pragmatic ends.

Cohn's plasma fractions were developed in direct response to military need for a mobile, stable transfusion material. Other molecular agents important to postwar medicine and agriculture, such as penicillin and DDT, were also developed as part of the war effort. World War II thus provided an impetus for a shift from molecular representations of life to molecular products for medical and agricultural problems. Instruments which had been developed to characterize proteins and other biological materials became important for verifying the molecular purity of mass-produced therapeutics as well as in screening efforts for cancer vaccines (see Gaudillière, this volume).

The success of molecularization in military medicine had other postwar consequences. Vannevar Bush argued in his 1945 essay, *Science — the Endless Frontier,* that federal government funding of basic research in universities and institutes was essential to public welfare, national security, and continued advances in medical care. The OSRD's Committee on Medical Research took credit for having funded four of the most dramatic advances in medicine during the war: antibiotics, DDT, improved anti-malarial drugs, and plasma fractionation products (Bush, 1945, p. 49). Although the late 1940s were marked by intense debates about federal science policies (see Kevles, 1977), the successes of molecular therapeutics during World War II provided a justification for the expanded federal funding of research in the postwar era, enabling the ascendancy of the research university in postwar U.S. science (Geiger, 1993). The Atomic Energy Commission and National Institutes of Health expanded their support of university research in the postwar years (Cohn received funding from both agencies in the early 1950s), and the National Science Foundation formed a division for molecular biology by 1952. In addition, the National Foundation for Infantile Paralysis and the American Cancer Society became important research patrons. Biomedical researchers of the (macro)molecular persuasion perceived an opportunity to achieve greater autonomy from the demands of clinic and industry while seizing upon larger biological questions of the relationship between structure and function and the molecular mechanism of reproduction (Creager, 1996). The earlier success of molecular approaches to biological, medical, agricultural, and industrial problems, culminating in advances during World War II such as Cohn's plasma fractions, thus provided material and conceptual resources for the rapid expansion of research in the postwar period aimed at elucidating molecular secrets of life.

ACKNOWLEDGEMENTS

I thank Thomas Rosenbaum at the Rockefeller Archive Center and Richard Wolfe at the Rare Books Room of the Francis Countway Library of Medicine for invaluable assistance in the research for this project. I am also indebted to Professor John T. Edsall for discussing Cohn's war work with me and providing the groundwork for my analysis in his own historical writing.

NOTES

1 My understanding of Cohn agrees with Lily Kay's account (Kay 1988; Kay 1993, Interlude I) of a "protein paradigm" that guided biological research between the wars, with two qualifications. First, I do not detect an overriding interest in heredity and reproduction during this period among Cohn's network of protein chemists. Second, the adoption of instruments to visualize macromolecules was more consequential for protein research than a conceptual orientation privileging proteins, in my view. Ironically, these instruments only became standardized and widespread once the "protein paradigm" was collapsing, but they were immediately adapted to visualize other sorts of macromolecules, particularly nucleic acids. Kay's argument draws on Olby's earlier thesis that the transition from protein research to the study of DNA was a paradigm shift (Olby, 1974).

2 For simplicity, I will use the term molecule instead of macromolecule throughout the paper. Whereas physiological and organic chemists had isolated and characterized molecules from living materials since the nineteenth century, physical chemists subsequently extended the molecular realm to include polymers such as proteins and nucleic acids (Fruton, 1972), usually designating these entites of large molecular weight as macromolecules, especially after the term "colloid" fell out of fashion.

3 This reorganization was prompted by safety concerns about the Salk poliomyelitis vaccine after the Cutter incident in 1955. Cohn's therapeutics were an important part of postwar efforts to combat the polio epidemics, as one of his blood fractions offered the most promising treatment before vaccine development (Paul, 1971, pp. 390–394).

4 Letter from David Edsall to A. Lawrence Lowell, President of Harvard College, regarding the establishment of a Department of Physical Chemistry at Harvard Medical School, January 29, 1920, Deans' Files,

"Departments, HMS & HSDM: Physical Chemistry, 1920–49," 11:580, Harvard Medical School Archives at the Rare Books Room, the Francis Countway Library of Medicine (hereafter HMS Archives).

5 His most lasting contribution in biochemistry is familiar as the Henderson-Hasselbach equation, which quantifies the relationship of pH to a compound's acid-base ratio.

6 Quote from letter by David L. Edsall, op. cit., note 4. See also Laszlo (1987) and Pauly (1987a).

7 His war research with Henderson might be described as the physical chemistry of breadmaking, and it consisted of a "Scientific study of the changes that occurred when wheat was replaced by other grains less rich in protein" (Scatchard, 1948, p. 265). These studies were motivated by the shortage of wheat during the war, as the military sought recommendations on the best replacement grains.

8 Letter from Edwin J. Cohn to Walter Cannon, dated February 11, 1937, Deans' Files, "Departments, HMS & HSDM: Physical Chemistry, 1920–49," 11:580, HMS Archives. See also Edsall (1961, p. 68).

9 Cohn's research efforts and the rationale underlying each approach are well-summarized in a letter he composed to David L. Edsall, March 29, 1929, Deans' Files, "Departments, HMS & HSDM: Physical Chemistry, 1920–49," 11:580, HMS Archives.

10 Eisenberg (1992) discusses the enduring popularity of this monograph as a reference among protein chemists and biochemists in the postwar period.

11 Rockefeller Foundation Trustees Minutes, April 6, 1938, Rockefeller Foundation RG 1.1, 200D, box 141, folder 1741, from the Rockefeller Archive Center (hereafter RAC).

12 Robert Kohler (1991, pp. 358–391) argues that the development of new tools and instruments for experimental biology from 1925–35, including the ultracentrifuge, the electron microscope, radioisotopes, and X-ray diffraction, remains an underappreciated aspect of the Rockefeller Foundation's legacy in 20th-century biology. These implements proliferated during the postwar years and became standard furniture in the laboratories of molecular biologists. See also Kohler (1977), Elzen (1986), and Kay (1988).

13 Summary of history of Department of Physical Chemistry, January 26, 1944, RAC, RG 1.1, 200D, box 141, folder 1744.

14 Evaluations of Cohn's laboratory by S. P. L. Sørensen, C. R. Harrington, and F. G. Hopkins, RAC, RF RG 1.1, 200D, box 141, folder 1741; Pauling's appraisal is reported in Warren Weaver's diary, November 1, 1937, RAC, RF RG 1.1, 200D, box 141, folder 1741. On Henderson and Cohn's department in disciplinary context see Kohler (1982, pp. 314–15).

15 Letter from Cohn to J. B. Conant regarding the promotion of John T. Edsall, March 26, 1938, Cohn papers, folder "Harvard University 1934–39," HMS Archives.

16 Copy in Cohn papers, folder "Harvard University, 1934–1939," HMS Archives.

17 In a letter from Cohn to Wallace B. Dunham, May 2, 1938, Cohn wrote of his efforts to solicit support from the Corn Products Industry. Cohn papers, folder "Harvard University, 1934–1939," HMS Archives. The Corn Products Industry gave $11,250 from 1940–1945 and $60,000 from 1945–1950 to the Department of Physical Chemistry for both general research expenses and fellowships. (These amounts are listed in Cohn, 1950.)

18 "Wanted: Blood for 300,000 Transfusions," Trustees Bulletin, March 1941, RAC, RG 1.1, 200D, box 141, folder 1743.

19 *Ibid.*

20 Cohn enlisted other academic researchers as well as physicians in work on the blood fractions. Several M.D. researchers, such as John Enders at Harvard Medical School, collaborated with Cohn on various aspects of the laboratory testing, and protein chemists at Stanford, Columbia, University of Wisconsin, and Connecticut Agriculture Experiment Station aided in characterizing the isolated proteins and assessing their stability. See Cohn (1948) on the cooperation between researchers and physicians in the project.

21 Cohn's involvement in the industrial development of his blood fractions contrasts the wartime pattern set by physical scientists of designing new products for the military but handing over the challenge of production to industry (Dennis, 1994). The comparison with penicillin development is also instructive: the discoverers of penicillin and those who developed a way to mass-produce the antibiotic were quite separate groups (Neushul, 1993). For the comparative case of the industrial development of sulfa drugs, see Lesch (1993).

22 Interview with Dr Douglas Surgenor, August 5, 1993. Also see Cohn (1948, p. 377). Some of the historical figures involved in the war project have explained that the policy of having Cohn train the industrial producers served to protect the trade secrets of the process, which was classified. (Douglas Starr, personal communication, 1993.)

23 I am indebted to Steve Sturdy for this insight.

24 Copies of many of these articles are preserved in the Harvard University Archives at the Pusey Library, HUG 4290.2.

25 William L. Laurence, "War Speeds Study of Blood System," *New York Times*, September 13, 1944, RAC, RG 1.1, 200D, box 141, folder 1744.

26 Churchill, an army surgeon, told the reporter that the over-reliance on plasma in treating wounded soldiers led to higher fatality rates than necessary. See Churchill (1972, pp. 45–52) and "Plasma Alone Not Sufficient," *New York Times*, August 26, 1943.

27 "Cutter Laboratories, 1897–1972: A Dual Trust," oral history, the Bancroft Library, Berkeley, CA.

28 Quote of Hans Neurath in Warren Weaver's diary, Thursday, May 3, 1945, Rockefeller Archive Center, RG 1.1, 200D, Box 141, Folder 1745.

29 Between 1942 and 1952 Cohn and two members of his laboratory filed for 10 different U.S. patents as well as 15 foreign patents. See undated manuscript of 14 pages in folder "Patent Matters 1952," Cohn papers, HMS Archives.

30 Letter from Edwin J. Cohn to James B. Conant, President of Harvard University, November 14, 1944, excerpted from Cohn (1951, p. 14).

31 Rockefeller Foundation Trustees Bulletin, May 1, 1945, pp. 3–4, RAC, RF 200D, series 1.1, box 141, folder 1745.

32 Letter from Chester I. Barnard to Dean Rusk, President of the Rockefeller Foundation, March 5, 1953, RAC, RF 1.1, series 200D, box 141, folder 1746.

33 Interview with Douglas Surgenor, summer 1993.

34 "Application and Report on Manufacture of Normal Serum Albumin (Human), To the Surgeon General, U.S. Public Health Service, Washington, D.C.," accompanied by letter of approval from the Federal Security Agency, August 29, 1941; courtesy of Douglas Starr.

35 Wendell Stanley was among those who drew attention to the chemical impurity of the Salk vaccine, in terms of the small percentage of vaccine protein which was actually virus rather than contaminating monkey kidney protein. During the war he had developed a method to remove cellular protein from influenza virus vaccine by ultracentrifugation; like Cohn he believed that the chemical purity of medical products was an indication of their quality and safety. See "Poliomyelitis Vaccine; Hearings Before the U.S. Congress, House Committee on Interstate and Foreign Commerce, 84th Congress, Part 2, June 22 and 23, 1955," U.S. GPO, 1955, and Smith (1990), pp. 359–367.

36 Cohn received funding from the NIH for this project and in 1951 opened a new laboratory, the Blood Characterization and Preservation Laboratory for Research and Development for the National Blood Program at the Bussey Institution of Applied Biology associated with Harvard. In this way he addressed the recognized superiority of whole blood as a transfusion material over blood derivatives. Arguing that the only way to guard against the next national emergency would be to find a way to stockpile blood, his research program was to investigate how to preserve the "formed elements" of blood, red blood cells, white blood cells, and platelets. See Lear (1951).

37 Edwin J. Cohn, "Implications of New Knowledge about Blood for Transfusion Services and for the Public Health," unpublished mss., folder "Patent Matters 1952," Cohn papers, HMS Archives.

38 New York Times clipping, Cohn papers, HUG 4290.2, Pusey Archives, Harvard University.

39 WFL diary, September 15, 1950, RAC, RF 1.1, 200D, box 141, folder 1746. On Basil O'Connor's enthusiasm about Cohn, WW diary, June 20, 1950, RAC, RF 1.1, 200D, box 141, folder 1746.

40 Letter from Wendell Stanley to Vincent du Vigneaud, February 1952, carton 3, Stanley papers, University of California archives, Bancroft Library.

REFERENCES

P.G. Abir-Am, "The Discourse on Physical Power and Biological Knowledge in the 1930s: A Reappraisal of the Rockefeller Foundation's 'Policy' in Molecular Biology," *Social Studies of Science* (1982), 12: 341–382.

P.G. Abir-Am, "The Politics of Macromolecules: Molecular Biologists, Biochemists, and Rhetoric," *Osiris* 2nd series (1992), 7: 164–191.

V. Bush, *Science — The Endless Frontier — July 1945* (Washington, D.C.: National Science Foundation, 1960).

E.D. Churchill, *Surgeon to Soldiers: Diary and Records of the Surgical Consultant, Allied Force Headquarters, World War II* (Philadelphia and Toronto: J.B. Lippincott Company, 1972).

E.J. Cohn, "The Physical Chemistry of the Proteins," *Physiological Reviews* (1925), 5: 349–437.

E.J. Cohn, J.A. Luetscher, Jr, J.L. Oncley, S.H. Armstrong, Jr, and B.D. Davis, "Preparation and Properties of Serum and Plasma Proteins, III. Size and Charge of Proteins Separating upon Equilibration across Membranes with Ethanol-Water Mixtures of Controlled pH, Ionic Strength, and Temperature," *Journal of the American Chemical Society* (1940), 62: 3396–3400.

E.J. Cohn and J.T. Edsall. *Proteins, Amino Acids and Peptides as Ions and Dipolar Ions* (New York: Reinhold Publishing Corporation, 1943).

E.J. Cohn, "The History of Plasma Fractionation," Chapter 28, in E.C. Andrus (ed.) *Advances in Military Medicine*, Vol. 1 (Boston, MA: Little, Brown and Company, 1948), pp. 364–443.

E.J. Cohn, "A Brief History of the Support of the Department of Physical Chemistry, Harvard Medical School, 1920–1950," pamphlet in series *University Laboratory of Physical Chemistry Related to Medicine and Public Health* (Cambridge: Harvard Printing Office, 1950).

E.J. Cohn, "History of the Development of a Patent Policy Based on Experiences in Connection with Liver Extracts and Blood Derivatives, 1927–1951," pamphlet in series *University Laboratory of Physical Chemistry Related to Medicine and Public Health* (Cambridge, MA: Harvard Printing Office, 1951).

E.J. Cohn, "History of the Development of the Scientific Policies of the University Laboratory of Physical Chemistry Related to Medicine and Public Health: A Memorandum of the Unwisdom of Projects and Reports," pamphlet in series *University Laboratory of Physical Chemistry Related to Medicine and Public Health* (Cambridge, MA: Harvard Printing Office, 1952).

A.N.H. Creager, "Wendell Stanley's Dream of a Free-Standing Biochemistry Department at the University of California, Berkeley," *Journal of the History of Biology* (1996), 29: 331–360.

M. Dennis, "'Our First Line of Defense': Two University Laboratories in the Postwar American State," *Isis* (1994), 85: 427–455.

J.T. Edsall, "A Historical Sketch of the Department of Physical Chemistry, Harvard Medical School, 1920–1950," *American Scientific* (1950), 38: 580–593.

J.T. Edsall, "Edwin Joseph Cohn," *Biographical Memoirs, National Academy of Sciences,* Vol. 35 (New York: Columbia University Press, 1961), pp. 46–84.

J.T. Edsall, "Proteins as Macromolecules: An Essay on the Development of the Macromolecule Concept and Some of Its Vicissitudes," *Archives of Biochemistry and Biophysics, Supplement* (1962), 1: 12–20.

D. Eisenberg, "John Edsall and Protein Chemistry," *Protein Science* (1992), 1: 1–3.

B. Elzen, "Two Ultracentrifuges: A Comparative Study of the Social Construction of Artefacts," *Social Studies of Science* (1986), 16: 621–662.

M.H. Fischer, *Soaps and Proteins: Their Colloidal Chemistry in Theory and Practice* (New York: John Wiley & Sons, 1921).

M. Florkin, "The Dark Age of Biocolloidality," in *A History of Biochemistry*, M. Florkin and E.H. Stotz (eds) *Comprehensive Biochemistry*, Vol. 30 (Amsterdam: Elsevier Science Publishers, 1972), pp. 279–284.

J. Fruton, *Molecules and Life: Historical Essays on the Interplay of Chemistry and Biology* (New York: Wiley-Interscience, 1972).

J.F. Fulton, "Penicillin, Plasma Fractionation, and the Physician," *Atlantic Monthly* (Sept. 1945), 146: 107–114.

R. Geiger, *Research and Relevant Knowledge: American Research Universities Since World War II* (New York and Oxford: Oxford University Press, 1993).

V.A. Harden, *Inventing the NIH: Federal Biomedical Research Policy, 1887–1937* (Baltimore, MD: The Johns Hopkins Press, 1986).

A.B. Hastings, *Crossing Boundaries: Biological, Disciplinary, Human. A Biochemist Pioneers for Medicine* (Grand Rapids, MI: The Four Corners Press, 1989).

L.J. Henderson and E.J. Cohn, "On the Swelling of Protein Colloids. A Reply to Professor Martin H. Fischer," *Journal of the American Chemical Society* (1918), 40: 857–861.

D.E. Hough, *The Market for Human Blood* (Lexington, MA: Lexington Books, 1978).

C.A. Janeway and J.L. Oncley, "Blood Substitutes," Chapter 29, in E.C. Andrus (ed.) *Advances in Military Medicine*, Vol. 1 (Boston, MA: Little, Brown and Company, 1948), pp. 444–461.

C. Janeway, "Plasma, the Transport Fluid for Blood Cells and Humors," in M.M. Wintrobe (ed.) *Blood, Pure and Eloquent: A Story of Discovery, of People, and of Ideas* (New York: McGraw-Hill Book Company, 1980), pp. 572–599.

L.E. Kay, "W.M. Stanley's Crystallization of the Tobacco Mosaic Virus, 1930–1940," *Isis* (1986), 77: 450–472.

L.E. Kay, "Laboratory Technology and Biological Knowledge: The Tiselius Electrophoresis Apparatus, 1930–1945," *History and Philosophy of the Life Sciences* (1988), 10: 51–72.

L.E. Kay, *The Molecular Vision of Life: Caltech, The Rockefeller Foundation, and the Rise of the New Biology* (New York and Oxford: Oxford University Press, 1993).

D.B. Kendrick, *Blood Program in World War II* (Washington, D.C.: Office of the Surgeon General, Department of the Army, 1964, reprinted 1989).

D.J. Kevles, "The National Science Foundation and the Debate over Postwar Research Policy, 1942–1945: A Political Interpretation of *Science — the Endless Frontier*," *Isis* (1977), 68: 5–26.

R.E. Kohler, "The Management of Science: The Experience of Warren Weaver and the Rockefeller Foundation Programme in Molecular Biology," *Minerva* (1976), 14: 279–306.

R.E. Kohler, "Rudolf Schoenheimer, Isotopic Tracers, and Biochemistry in the 1930's," *Historical Studies in the Physical Sciences* (1977), 8: 257–298.

R.E. Kohler, *From Medical Chemistry to Biochemistry: The Making of a Biomedical Discipline* (Chicago, IL: University of Chicago Press, 1982).

R.E. Kohler, *Partners in Science: Foundations and Natural Scientists, 1900–1945* (Chicago, IL: University of Chicago Press, 1991).

A.C. Laszlo, "The Physiology of the Future: Institutional Styles at Columbia and Harvard," in G.L. Geison (ed.) *Physiology in the American Context, 1850–1940* (Bethesda, Maryland: American Physiological Society, 1987), pp. 67–96.

J. Lear, "You May Be Drafted to Give Blood," *Collier's* March 10, 1951, pp. 11–13, 58, 59.

J.E. Lesch, "Chemistry and Biomedicine in an Industrial Setting: The Invention of Sulfa Drugs," in S.H. Mauskopf (ed.) *Chemical Sciences in the Modern World* (Philadelphia: University of Pennsylvania Press, 1993), pp. 158–215.

J. Loeb, *Proteins and the Theory of Colloidal Behavior* (New York: McGraw-Hill Book Company, Inc., 1924).

S.E. Luria, *A Slot Machine, A Broken Test Tube: An Autobiography* (New York: Harper & Row, Publishers, 1984).

H. Marks, "Cortisone, 1949: A Year in the Political Life of a Drug," *Bulletin of the History of Medicine* (1992), 66: 419–439.

G.B. Mider, "The Federal Impact on Biomedical Research," in J.Z. Bowers and E.F. Purcell (eds.) *Advances in American Medicine: Essays at the Bicentennial*, Vol. 2 (New York: The Josiah Macy, Jr. Foundation in conjunction with the National Library of Medicine, 1976), pp. 806–871.

N. Morgan, "The Strategy of Biological Research Programmes: Reassessing The 'Dark Age' of Biochemistry, 1910–1930," *Annals of Science* (1990), 47: 139–150.

National Research Council, Division of Medical Sciences, *Bulletin on Blood Substitutes*, in three volumes (Washington, D.C., undated but including minutes from meetings May 31, 1940 through December 6, 1945), courtesy of D. Starr.

P. Neushul, "Science, Government, and the Mass Production of Penicillin," *Journal of the History of Medicine and Allied Sciences* (1993), 48: 371–395.

R.C. Olby, *The Path to the Double Helix: The Discovery of DNA* (New York: Dover Publications, Inc., 1994).

J.R. Paul, *A History of Poliomyelitis* (New Haven: Yale University Press, 1971).

P.J. Pauly, "General Physiology and the Discipline of Physiology, 1890–1935," in G.L. Geison (ed.) *Physiology in the American Context, 1850–1940* (Bethesda, Maryland: American Physiological Society, 1987a), pp. 195–208.

P.J. Pauly, *Controlling Life: Jacques Loeb and the Engineering Ideal in Biology* (New York: Oxford University Press, 1987b).

K.O. Pedersen, "The Svedberg and Arne Tiselius: The Early Development of Modern Protein Chemistry at Uppsala," in G. Semenza (ed.) *A History of Biochemistry. Personal Recollections*, (Amsterdam: Elsevier Science Publishers, 1983), pp. 233–281.

G. Scatchard, "The Scientific Work of Edwin Joseph Cohn," *The Nucleus* (1948), 25: 263–276.

J.W. Servos, "A Disciplinary Program That Failed: Wilder D. Bancroft and the *Journal of Physical Chemistry*, 1896–1933," *Isis* (1982), 73: 207–232.

J.W. Servos, *Physical Chemistry from Ostwald to Pauling: The Making of a Science in America* (Princeton, NJ: Princeton University Press, 1990).

J.S. Smith, *Patenting the Sun: Polio and the Salk Vaccine* (New York: Doubleday, 1990).

I. Stewart, *Organizing Scientific Research for War: The Administrative History of the Office of Scientific Research and Development* (Boston: Little, Brown and Company, 1947).

T. Svedberg, *Colloid Chemistry*, 2nd edn (New York: The Chemical Catalog Company, Inc., 1928).

R.M. Titmuss, *The Gift Relationship: From Human Blood to Social Policy* (London: George Allen & Unwin, 1971).

C. Weiner, "Patenting and Academic Research: Historical Case Studies" in V. Weil and J.W. Snapper (eds) *Owning Scientific and Technical Information: Value and Ethical Issues* (New Brunswick, NJ: Rutgers University Press, 1989), pp. 87–109.

M.N. Wise, "Mediating Machines," *Science in Context* (1988), 2: 77–113.

C.S. Wood, "A Short History of Blood Transfusion," *Transfusion* (1967), 7: 299–303.

CHAPTER 5

The Molecularization of Cancer Etiology in the Postwar United States: Instruments, Politics and Management

Jean-Paul Gaudillière

INTRODUCTION

On April 22, 1958, Dr John Heller, then director of the U.S. National Cancer Institute (NCI), was testifying on the floor of the American Congress about the state of the art in cancer research.[1] Coming to the chapter of cancer etiology, Heller focused on a very promising research domain: the study of viruses causing cancer in animals. Interrogated about the possibility of developing cancer vaccines, Heller replied:

> We think ... that in viruses we are certainly in a worthwhile area. We have gone so far as to call it a probable breakthrough. Only time will tell. We are going to support work as well as undertake it at our own place to see if we can find out the answer to the riddle [of cancer] as quickly as humanly possible. I do not know how long it will take, maybe 7 years or maybe a decade, but at least we are on an existing trail and the dogs are barking in great style, and I think we may get some very exciting game here [U.S. Senate 1959, p. 761].

Heller was not dreaming. From the mid-1960s until the late 1970s, the study of cancer viruses ranked second among the research programs at the NCI, behind chemotherapy. Discourses of public health emergency, macromolecular targets, and immunological weapons were closely associated with the viral vision of cancer causation. By the 1980s, however, after 30 years of a prosperous life, cancer viruses and cancer vaccines quietly left the stage to be replaced by molecular genetics and the promises of recombinant DNA and modern biotechnology.

Although the highly visible "Human Genome Project" is often presented as the beginning of molecular medicine, early linkages between macromolecules, medicine, and biotechnology have not passed unnoticed by historians of science and technology (Bud, 1993; Kay, 1993; Thackray, 1996). In parallel, historians of medicine in the United States have documented the fact that, after 1945, cancer became a prominent disease: a social scourge, a political issue, and a research target (Strickland, 1972; Starr, 1982; Patterson, 1987).

In this chapter, I also argue for thin dividing lines between biological research and cancer research, science policies and health policies, molecular biology and biotechnology. I will, however, focus on a rarely addressed issue, namely the part played by engineering practices, modern management, and political regulation in the rise and fall of a series of experimental models which were widely employed to articulate early promises of a medical revolution originating in molecular biology.

The word "molecularization" alludes to the discrete particles which material substances are conceived to be composed of. It is especially appropriate for pointing to the postwar (re)invention of tumor viruses, since the "molecularization" of cancer causation — if not cancer diagnosis or cancer treatment — actually consisted in realignments of biochemistry, genetics, bacteriology, and cancer studies which introduced particles to which cellular abnormality was attributed. Three stages may be distinguished. Firstly, there was the redefinition of a few tumor agents which had haunted pathology laboratories for years. In the 1940s, these entities emerged both as legitimate candidates for the role of cancer cause and as viral macromolecules. Secondly, in the 1960s, an expanding panoply of viruses isolated from chicken, rabbit, or mice tumors was transformed into genetic particles infecting cells and causing tumors in animals as well as in humans. As was the case with other infectious entities, vaccine development then dominated the strategy for control. Finally, by the early 1970s, viral DNA sequences hidden in chromosomes became the unique molecular target of cancer-inducing agents such as chemicals, hormones, or radiation. These "oncogenes" and the encoded proteins provided unifying goals for research on the diagnosis and prevention of cancer.

The molecularization of cancer etiology was a process of relocalization. Tumor viruses were displaced from bacteriology toward molecular genetics, from the outside of the body to some tiny parts of the cells. But "molecularizing" was not merely a matter of theory or culture: the basic argument of this chapter is that it was instrumental, political, and managerial in the first place. The argument will be developed in three vignettes, each one addressing a particular aspect of the postwar history of the tumor viruses.

By following the chicken sarcoma agent at the Rockefeller Institute in the 1940s, I will first show that the resuscitation of cancer viruses was closely associated with the construction of the ultracentrifuge and the electron microscope as routine means for defining and visualizing macromolecules. This will in turn highlight the influence of World War II on the development of these new practices. The second part will focus on the political molecularization of the early 1960s and the role played by the co-construction of the American "biomedical complex". The expert community then formed at the juncture of the U.S. Congress, the NCI, and the American Cancer Society was responsible for giving mammalian tumor viruses the meaning of infective hereditary particles and for launching a large-scale search for human cancer viruses and cancer vaccines. The third part will concentrate on the managerial molecularization of the 1970s when an extended network of NCI contractors made the oncogene theory possible by organizing the production, circulation, and control of biochemical and immunological tools employed to trace cancer proteins of viral origins.

MACHINES AND IMAGES: THE NEW CANCER VIRUSES

In 1942–1943, the *Journal of the American Medical Association* published contrasting lectures on cancer research given by two researchers from the Rockefeller Institute for Medical Research, James Murphy and Peyton Rous (Murphy, 1942; Rous, 1943). Rous and Murphy were former collaborators and worked on the transmission of tumors in fowls. Coming to terms with the relationship between viruses and human cancer, Murphy wrote in his paper:

> There is insufficient indication at present that viruses play any important role in the general picture; therefore no attempt will be made to discuss at length the possible relation of this group of agents to cancer. The only evidence of importance in this field is derived from the study of a group of fowl tumors which may be transmissible by filtrable agents. As no mammalian malignant tumors has yet been transmitted except by living tumor cells, the relationship to mammalian tumors cannot be considered as established. Furthermore, there is no agreement as to the nature of the transmitting agents and their relationship to the viruses (Murphy, 1942, p. 110).

In contrast, Rous argued for "dead-end viruses" as causes of cancer while other factors, such as carcinogens or hormones, were in all likelihood provocative agents. Building on "recent discoveries" about the transmission of tumors in mammals, he claimed:

> To account for the worldwide distribution of tumors ... it has been necessary to suppose that the body carries resident viruses just as it does resident bacteria, indigenous viruses, as they have been termed ... They would give no sign of their presence in most instances and would in due course be passed on to the young. But if a provocative carcinogen happened to work on the cells with which such a virus was associated, thus altering its milieu, it might undergo variation and, taking a hand in cell affairs for the first time, give rise to a tumor (Rous 1943, p. 580).

Rous and Murphy had been discussing the part played by viruses in cancer causation for two decades. Rous, a pathologist at the Rockefeller Institute, is widely known for having introduced the "filtrable agent of chicken sarcoma" before World War I. In 1911, Rous described how he could transfer chicken tumors from one animal to another by injecting filtrate of ground sarcoma. At that time most oncologists viewed the chicken system to be of little importance for the study of mammalian and human cancers. In the 1920s, this "tumor agent" was traced by Rous' collaborator, James B. Murphy, then head of a cancer laboratory at the Rockefeller Institute. Murphy was convinced that the agent was not a virus but an "endogenous", "enzyme-like" substance which could be purified from tumor cells. In the 1930s, the debate about the role of viruses in cancer at the Rockefeller Institute was revitalized by the discovery of a virus, the Shope-papilloma virus, which seemed to induce skin tumors in rabbit. Rous took over this new exemplar and resuscitated his viral hypothesis. In parallel, Murphy and his colleague Albert Claude were deeply involved in biochemical attempts to isolate an endogenous "chicken tumor agent", which they saw as a self-reproducing enzyme. The irony of the situation was, therefore, that, in 1943, Rous could not mobilize his own chicken tumor agent in support of the viral vision of cancer, but only the papilloma virus and a newly defined agent causing mammary tumors in mice. In 1947, however, Rous received the unexpected support of Claude who came to the conclusion that the "chicken tumor agent" was in "all essential properties similar to viruses".

The coming of age of the "Rous sarcoma virus" has been the topic of numerous acounts by scientists and of significant historical studies.[2] In this section, I take the history of the chicken agent as an examplar of the role played by instrumental practices such as electron microscopy and ultracentrifugation in the resuscitation of cancer viruses in the 1940s. The impact of instruments and technological developments on the history of molecular biology is a widely acknowledged problem. Historical studies have concentrated on the long-term commitment of the Rockefeller Foundation to physico-chemical instrumentation, i.e. the ultracentrifuge, the electrophoresis apparatus, the cyclotron, or the electron microscope, and hence on the reshaping of biological research (Kohler, 1976, 1992; Abir-Am, 1980). It is not surprising that the use of these new instruments was essential in defining biological macromolecules which could not be "observed" without them. Thus, Svedberg's development of the ultracentrifuge and his quasi-monopolistic use of this instrument strongly determined the way proteins were defined and investigated in the 1930s. Less obvious, however, is the extent to which the new instrumentation pressed for a new "form of experimental life". This is not only to say that the scientists who developed the "molecular" tools were more often engineers and physicists than biologists, but also to point out that these "biophysicists" were often "machine devotees" who defined strategies and problems in terms of technological frontiers which could be crossed by "scaling-up", namely by building bigger, faster working, and more precise machines. Speaking of the historical process which took the definition of viruses away from the bacteriologist's views about "minute bacteria" toward the molecular biologist's emphasis on replication and nucleic acids, André Lwoff once wrote that "viruses should be recognized as viruses because they are viruses". In contrast, one may say that instruments should be analyzed as instrumentation because they are not instruments but "technological packages" which circulate between restricted groups of producers and users. This had two consequences in the 1940s. Firstly, the molecular entities characteristics of molecular biology were operationally defined agencies whose mere existence was ascertained through an *assemblage* of people, machines, skills, and reference results. Secondly, the development of "molecular" instrumentation was strongly connected with industrial practices and shaped by the scientific mobilization of World War II. These features may be illustrated with the re-invention of the chicken sarcoma virus.

The early days of ultracentrifugation may be traced back to the 1920s (Elzen 1988; Creager, 1993). Until World War II, however, a very limited number of machines operated in the United States. One was the property of the Rockefeller Institute Department of Plant and Animal Pathology in Princeton and was operated by Wendell Stanley and his collaborators.[3] By the late 1930s, the machine was employed to purify sizeable amounts of the tobacco mosaic virus (TMV) and determine some physical properties (sedimentation rate, size, and shape) of the viral macromolecule. This work on the physical structure of viruses was intimately associated with Stanley's theory of viruses being inanimate, infective, big nucleoproteins (re)produced in the cell.

This contrasts with Claude's early use of the ultracentrifuge who employed it as a machine for fractionating tumor cells. The machine at hand in his laboratory was not an analytical machine including the optical device necessary to measure sedimentation rate and compute molecular weights. It was a (commercially available) preparative version of the instrument (Claude, 1937). This was not a tool for analyzing macromolecules, but a convenient device for preparing subcellular material. At first this material was thought to be the endogenous chicken tumor agent. As centrifugation revealed the same kind of particles in normal cells, Claude adopted the view that the particles he isolated with his machine were normal components of the cells playing little role in the actual causation of cancer (Rheinberger, 1995).

By the early 1940s and in Claude's hands, ultracentrifugation was: (1) a means to sort out and concentrate particles and aggregates of naturally variable size; (2) a means to reveal the inanimate nature of the particulate elements associated with the formation of subcellular structure, such as mitochondria, and more generally with growth, differentiation and cancer (Claude, 1941). The machine remained, however, a problematic apparatus which might well produce arbitrary classes of particles. Debates about the generation of artefacts by mechanical treatments which could disintegrate cell structures were circumvented by means of an association between the ultracentrifuge and chemical testing. Uniform results of chemical tests made differential ultracentrifugation more reliable. For instance, the similar composition of the material extracted from mouse embryo, mouse tumors, chicken embryo and chicken tumors strongly supported biological homogeneity, while positive results of tests revealing nucleic acids and phospholipids in the particles led Claude to assume "that the granular elements of the purified tissue fraction may represent mitochondria or fragments of mitochondria", since the latter were thought to be composed of "phospholipids and proteins" (Claude, 1941).

Claude started to use the electron microscope to study cytoplasmic particles long after the ultracentrifuge crossed the laboratory threshold. In 1944, Claude and Keith Porter, a cytologist working on cell cultures in Murphy's laboratory at the Rockefeller Institute, were invited to work on one of the first electron microscopes made by the RCA company. Porter was able to grow very flat tissue cultures directly on the microscope grid and to fix them with traditional dyes before drying. The pictures they obtained were, however, very dark and full of holes, so that the interior of the cells was barely visible. In 1945–46, new cutting and preparative techniques improved the quality of images, but they did not make the visual evidence unproblematic. One means to reinforce the claim that these newly visible features were actually biological structures was to put side by side pictures of the "same" entity obtained by means of light microscopy and electron microscopy (Porter, 1945). This was of limited value, however, since the most interesting particles were precisely those invisible under classical microscopes. Claude and Porter sought further certainty in a combination with ultracentrifugation. The first move was to concentrate cell extracts by ultracentrifugation in order to examine so rich a preparation that the structures observed could reasonably be attributed to mitochondria even if they were partially altered during the fixation and desiccation stages of specimen preparation (Claude, 1945). The alternative was to ground imaging on ultracentrifugal evidence. By 1946, the reality of small, electron-dense particles in the background of cell pictures was matched against the fact that spinning broken cells for one hour at 18,000 *g* isolated "microsomes", bodies that Claude had previously presented as a new class of subcellular particles (Claude, 1946). Once certified, electron micrographs stood by themselves. In 1947, Claude's final claim that the tumor agent was a virus was centered on electron microscopic pictures of chicken sarcoma cells showing numerous pairs of dark, dense, rounded bodies of a size ranging from 67 to 84 micrometers in diameter (Figure 5.1) (Claude, 1947). Claude used these images as evidence of the extraneous nature of the tumor agent — cellular bodies, i.e. "microsomes" of similar size, appeared quite different, as hollow spheres or vesicles — and of its replicative properties, the paired entities presumably resulting from the division of single particles.

The making of electron microscopic pictures into convincing evidence was therefore made possible by mobilizing macromolecular entities originating in the work with the ultracentrifuge. Yet, calibration worked both ways. As Claude put it in 1948, the electron microscope was also a "guide or check" to make sure that centrifuge fractions represented parts of the cell and not arbitrary classes of particles (Claude, 1950). The fact that the ultracentrifuge and the electron microscope did not produce traces and images that could easily be matched made cali-

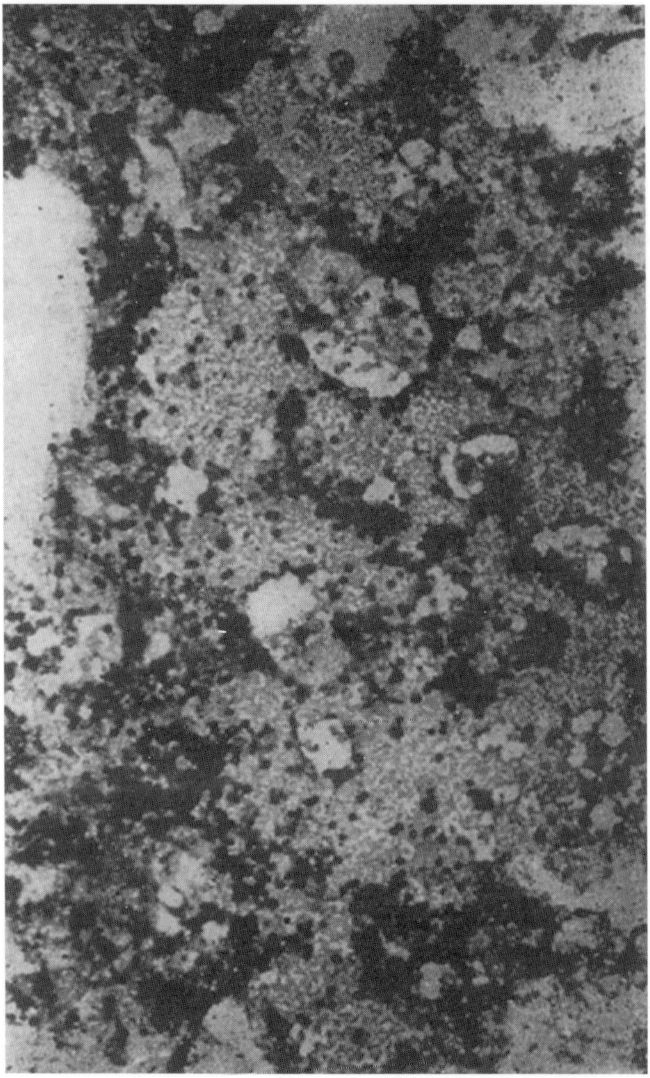

Figure 5.1. Electron microscope picture of the chicken sarcoma agent.

bration a complex achievement. The size of particles was the sole property which could be measured with both instruments. Accordingly, Claude's argument of 1947 rested on two different moves. First, in 1945, Claude calibrated electron microscopy with the ultracentrifuge: the fact that size figures for liver microsomes obtained by direct measurement of images were similar to previous ultracentrifuge computation was taken as evidence that electron micrographs revealed real cellular entities. Similar size was then a criterion for identity. Second, in 1947, Claude used the electron microscope to re-calibrate the ultracentrifuge: appearance of the virus-like particles in chicken tumor cells was taken as evidence that ultracentrifuge preparations had blurred the distinction between endogenous microsomes and extraneous viruses. Similar size was no longer a sufficient criterion for identity. Rather than "independent confirmation", this was a case of "triangulation" (Leigh Star, 1986).

So what made these two apparatus stick together and provide a basis for the characterization of macromolecules? What made the ultracentrifuge the tool of choice for determining the size of proteins and intracellular particles, while the electron microscope became the instrument for studying bacteria, viruses and infective entities? I would like to link this new "experimental game" to the war context in general, and to wartime research on the influenza virus in particular, since several key participants in the (postwar) cancer virus field were involved in the search for vaccines against influenza during World War II.

In the United States, wartime medical research was organized under two umbrellas. On the one hand, the Office for Scientific Research and Development (OSRD) headed by Vannevar Bush established a Committee on Medical Research (CMR) in 1941 for contracting research in universities and medical schools. On the other hand, in 1940, the Army and the Surgeon General set up a "Board for the Investigation and Control of Influenza and other Epidemic Diseases in the Army" which gathered together "epidemiologists, bacteriologists, chemists, and pathologists". The rationale behind the Army decision was the memory of the influenza epidemic of 1918–1919. From a military medical viewpoint, key targets in the fight against influenza were the epidemiological investigation of local outbreaks of the disease, and the production of vaccines:

> The present expansion of the Army has been accompanied by an increase in influenza and other acute respiratory diseases among troops ... In fact, the possibility cannot be ignored that the Army may again be confronted by another pandemic of influenza of the virulent type which caused such a large proportion of the total death among our troops during the last war (Board, 1941 p. 3).

Following the development of methods for producing relatively large amounts of influenza virus in fertilized chiken eggs, attention focused on "inactivated" vaccines which would be made of influenza viruses killed by exposure to heat, ultraviolet light, formadehyde, or other chemicals. Although they were poorly represented on the military board, biochemists argued that the production of vaccines would only work if based on well-characterized strains of viruses. For instance, Stanley targetted the lack of "knowledge of the fundamental biophysical properties of the influenza virus" as a major impediment in the path toward prevention and control of influenza. Purification rather than inactivation should be the priority and the ultracentrifuge should be employed. The latter plea was successful: by 1944, most groups tracing pure influenza virus were using preparative ultracentrifugation. With the support of the OSDR-CMR committee, the most extensive work along this line had been conducted in Stanley's laboratory at Princeton. Stanley's commitment to ultracentrifuge was not "pragmatism" in the sense of looking for a cost-efficient method for purifying viruses. The ultracentrifuge emerged out of his study of tobacco mosaic virus (TMV) as a way of producing macromolecules. During the war, the machine sedimented as a whole research strategy with cognitive, technical, and social aspects. In Stanley's laboratory, a pure virus was a well-behaved macromolecule which showed a regularly moving "sedimentation boundary" visible in absorption photographs of the spinning solution (Figure 5.2). This ideal of macromolecular behavior provided the background to Stanley's struggle to normalize the influenza virus and standardize influenza preparations.

In contrast, bacteriologically oriented specialists relied on (electron) microscopy as the means of choice for characterizing the influenza microorganisms. This was, for instance, the case with a team in Durham, North Carolina, led by a former collaborator of Rous, Joseph Beard. This group of experimental pathologists had little experience of ultracentrifugation and their expertise was in bacteriology and immunology. In their hands, electron microscopy was a natural complement to optical microscopy and rapidly took the upper hand in the characterization of the influenza agent. In 1944, Beard summed up their work on the identification of the influenza virus (Beard, 1944). Emphasizing the fuzziness of the sedimentation boundary obtained with the ultracentrifuge, they chose to measure the size of the virus through the electron microscope. Calibration was then based on internal controls by means of a screen with small holes of known diameter. Then, it was possible to interpret the variability of sedimentation pattern,

> from *direct observation* with the electron microscope it was *evident* that the concentrates which behaved like the virus were composed of rounded or ovoid particles of variable size … These [size] values obtained with the electron microscope were corroborated by calculations of particule size from the sedimentation velocity data obtained with the analytical ultracentrifuge (Beard 1944, p. 317).

This ordering finally led the Durham team to stigmatize Stanley's emphasis on "highly uniform sedimentation characteristics". Their practice of electron microscopy reinforced a bacteriological vision of viruses which opposed Stanley's image of the big nucleoproteic macromolecule. Beard commented on the dark rounded bodies:

> Morphologically, differentiation of structure within the particle was clearly seen in the electron micrographs … Centrally or eccentrically placed in the individual particles was a rounded area or region of increased density. Roughly the general appearance of a clearly shown virus particle was somewhat similar to that of a typical rounded cell in which there is present a nucleus surrounded by cytoplasm (Beard 1944, p. 318).

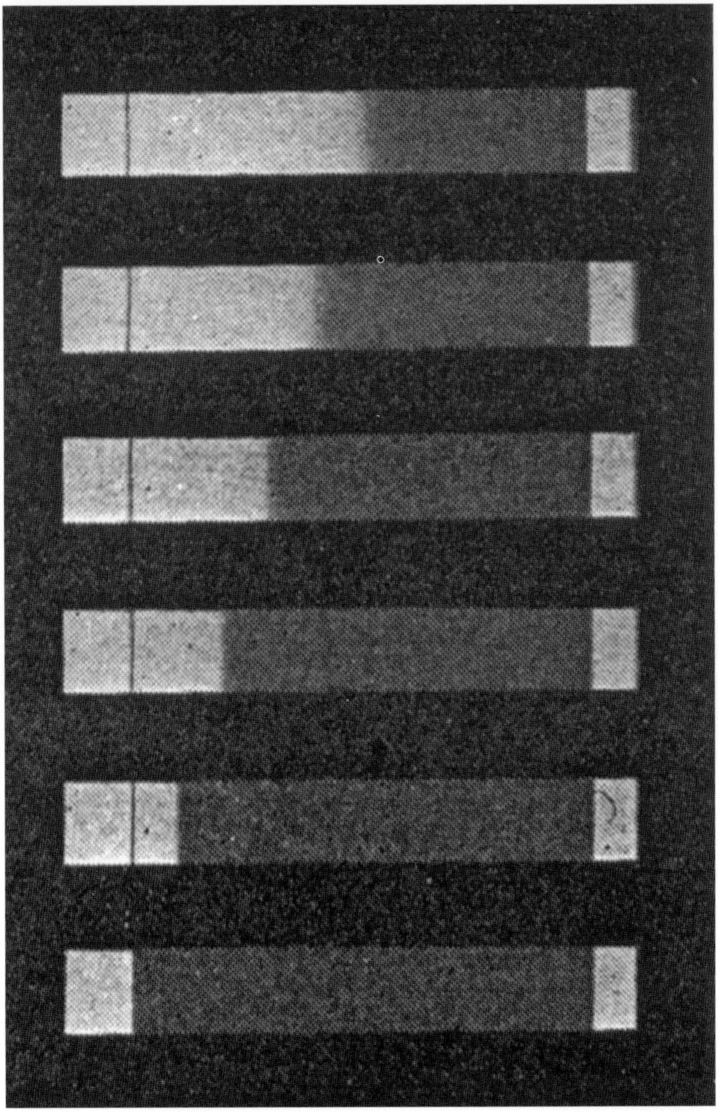

Figure 5.2. Ultracentrifugation photographs of the influenza virus.

Seeing "the virus entities themselves" revealed "variation in particle size and shape ... which would be expected from a similar examination of a pure culture of small cocco-bacilli" while "internal structure ... is not greatly different morphologically from that to be seen in electron micrographs of bacteria". All these results were "clearly incompatible with the hypothesis that the influenza virus may be molecular in nature".

One may be tempted to interpret this debate in terms of conflicting disciplinary traditions — biochemistry versus bacteriology — or alternatively for conflicting professional aims — biophysical knowledge versus preventive medicine. One should notice, however, that numerous aspects of the influenza debate were replicated in the controversy about the nature and size of bacteriophages which, in 1943–1944, opposed the physicists Max Delbrück and Salvador Luria on the one hand, to the biochemists Northrop and Stanley on the other hand (Van Helvoort, 1992). One may therefore consider that, in the 1940s, the ultracentrifuge was, in general, employed to prepare, purify, and measure (macro)molecules, while the electron microscope visualized parasitic or infective agents duplicating in host cells. The domestication of both instruments was facilitated by the scientific mobilization of the war and followed paths which maintained specific forms of usage which were stabilized by the very nature of wartime biomedical research. This does not only refer to the importance of purification within the context of rapid vaccine development, but also to enhanced collaboration between industrial instrument makers and biological research settings.[4]

As argued by Nicholas Rasmussen (1997), biological electron microscopy was, during the war, regulated by RCA and its "Committee on Biological Applications of the Electron Microscope" under the patronage of the National Research Council. The committee included the head of the RCA electronics research division, the geneticist Miroslav Demerec, two cytologists (Charles Metz and James Kempton), a radiation biologist (Caryl Haskins), a bacteriologist as chairman (Mudd), and Stanley. The role of this group of academic biologists was to supervise the work of Thomas Anderson, the RCA's electron microscope operator, evaluate research projects, allocate time slots, and check the quality of the results. Although it was challenged by Stanley, the dominant belief was that electron microscopy would only become a reliable biological instrument if it matched, confirmed, and expanded the results of classical microscopy, in other words, if it built on recognized practices in cytology and bacteriology. Hence the emphasis on producing images that would compare well with optical micrographs of inner cells, bacteria, dividing microorganisms, etc. At RCA, techniques and procedures expanded accordingly. Since electron microscopy remained a highly skilled activity practised by a small group of people who were often clients of the electronics firm and former trainees of Anderson, "RCA's right sort" of microscopy dominated virus imaging.

In parallel, as illustrated by Angela Creager's study of blood fractionation, ultracentrifuges were increasingly taken as standards for purity by virtue of their preparative function during the war (Creager this volume).[5] In the case of influenza research, this was already perceptible in 1944–1945 when Stanley had lengthy discussions with the Army Influenza Board in order to advance the use of his ultracentrifugation techniques for the production of "pure" influenza vaccines.[6] There were subtantial debates about the precise relationship between immunization and virus dosage, about adverse reactions, and about the possibility to scale-up laboratory procedures. But one thing was quickly accepted: ultracentrifugation would provide a concentrated and stable material, easy to manipulate.[7] This recognition was pushed a step further, in 1945, when the Board launched trials of the ultracentrifuge vaccine and adopted Stanley's purification procedure as reference.[8] This reinforced the formula linking purity, molecular nature, and ultracentrifuge fractions. That this preparative work paved the way for the spreading of the machine became evident in the early 1950s, when the Spinco company almost went bankrupt while selling sophisticated air-driven analytical centrifuges, to be saved only when they offered the biochemical market a preparative apparatus as easy to operate as a washing machine.

The ultracentrifuge and the electron microscope did not circulate in the same way and did not make the same viruses. Nonetheless, there were surprisingly few controversies opposing the two sets of data following the development period of the early 1940s. An explanation for this may be found in Claude's practice. Beyond the limited circle of committed biophysicists, professional interests as well as biological common sense soon pleaded against the idea of spending much time with co-calibration and precision measurement. Thus, a rather loose qualitative combination of both apparatus prevailed. This was finally reflected in an equally loose but operational concept of tumor viruses as macromolecular particles (ultracentrifugation), composed of nucleic acids and proteins (biochemical analysis), hidden in cells of the body and able to replicate (electron microscopy).

POLITICAL MOLECULARIZATION: BIOMEDICAL COMPLEX, CONGRESSIONAL HEARINGS, AND LEUKEMIA VIRUSES

Once they were resuscitated as macromolecules, what could be done with experimental tumor viruses? The response was not straightforward. Chickens are not humans and it might well have been the case that the "Rous sarcoma virus" would die of mistrust as had been the case in the 1910s. But history never repeats itself. The tumor virus of the 1940s survived by virtue of a political molecularization. This second section will focus on the policy arena and follow the development of an expert discourse on the viral causes of cancer which emerged within the American postwar "biomedical complex".

It has become a familiar claim that the events of World War II realigned government, universities, philantropic foundations and industries. It is also widely acknowledged that this change involved more than institutional reorganization. In the late 1930s and 1940s new styles of experimental work and new entrepreneurs emerged (Kevles, 1987; Gallison, 1992). The postwar "biomedical complex" has been much less investigated than the academic-military-industrial complex, but changes in the organization of the biomedical sciences were no less important. One well-known example of the new views about biology and medicine was Bush's report *Science, the Endless Frontier*, which came off the press in

July 1945 (Bush, 1945). Ironically, this report dealing with the transition from war research to science in peacetime included a new declaration of war — a war against disease. This war required an expanding body of new scientific knowledge and Bush proposed to advance the integration of biology and medicine by special schemes for coordinated attack on the development of new drugs from laboratory to large-scale production.

Historians analyzing postwar cancer research in the United States have stressed the fact that one form taken by this coordination was the formation of a tight "policy-making community" linking the American Cancer Society, the NCI, the big pharmaceutical companies, and the U.S. Congress. This nexus resulted in the formation of a system of public expertise which saw a few authoritative researchers advising policy-makers in settings including Congressional hearings, think-tanks of the Public Health Service, NCI committees, and a wide range of lay organizations. This may be illustrated with the relationship between Stanley and the American Cancer Society (ACS).

The ACS was created in 1913 under the name of American Society for the Control of Cancer. It was merely a professional body gathering elite clinicians in order to collect epidemiological data, to favor the education of general practioners, and finally to inform the general public on early detection of cancer. Until the early 1930s, the society was dominated by surgeons, and the emphasis remained on the education of professionals. Things began to change with the Depression and the New Deal. Pressing financial needs led to the re-examination of the society's strategy. The geneticist C.C. Little, then managing director of the society and director of the Jackson Memorial Laboratory, promoted lay education and collections from the general public. By 1940, there was increasing pressure on the society for a reorganization that would empower the lay activists. It was delayed until 1944 when it took the form of a take-over by a group of businessmen led by Mary Lasker, wife and later widow of the president of one of the largest advertising firm in the country, who established a foundation for medical research in 1942. By 1944, Lasker was involved in the management of the cancer society, in congressional debates on biomedical research, and in various committees pressing for reform of medical care. She convinced the society's officials to reorganize fund-raising campaigns and to use rational management and modern advertising methods.

Campaigning was then becoming a quasi-professional activity. The 1945 drive actually raised $4 million. Half of it was to be set aside for the funding of research. Thus, the cancer society was in a position to spend more money on cancer research than the National Institute of Health did. In June 1945, the society approached the National Research Council (NRC) in order to implement a broad program. The outcome anticipated the operations of the National Science Foundation (NSF). A "Committee on Growth" of nineteen members from the NRC medical division was established. Under the chairmanship of Cornelius Rhoads, then director of the New York Memorial Hospital, this body became the research committee for the cancer society in 1946. Recommendations to the society were actually made by scientific advisory panels of experts in specialized areas, i.e. experimental genetics, isotopes, chemical-biological coordination in chemotherapy, and so forth. Almost all fields of biological, chemical and physical research vaguely related to cancer causes, diagnosis or treatment were included, but molecules on the biological side and chemotherapy on the clinical side were top priorities.

The changing order within the cancer society depended on the returns of the fund-raising campaign, the natural field of expertise of lay directors. In that respect, they were successful. In 1946, the society raised $10 million, in 1947 $12 million. Collections in the early 1950s increased from $15 to 25 million per year.[9] In contrast to prewar charities, the ACS began a long-term partnership with rapidly expanding governmental research agenices. Lobbying for government funding of research, the society helped build a powerful NCI which in turn supported the society's initiatives for establishing cancer clinics within general hospitals, building new treatment facilities, launching nation-wide programs in cancer diagnosis, and organizing cancer registries.[10] Year after year, hearings on budget appropriations organized by sympathetic congressmen thus became an important setting for the display and assessment of biomedical prospects.

The new sources of funding and legitimacy were rapidly acknowledged by postwar entrepreneurs like Stanley. In 1946 Stanley received the Nobel Prize, conducted fruitful negotiations to set up a new department at the University of California at Berkeley, and was nominated member of the ACS Growth Committee.[11] In the early 1950s, Stanley's Virus Laboratory was peopled with collaborators following up work on the composition and isolation of various viruses while developing a new ultracentrifuge. Stanley's commitment to the ultracentrifuge was supplemented with electron microscopy. This resulted in important research on the polio virus.[12] Other lines of work included biochemical surveys of normal and infected cells, but there are no traces of work with tumor viruses.[13]

Stanley's original involvement in the ACS was confined to unremarkable general work as scientific advisor. Thus, in 1955–1956, Stanley played an important part in selecting and approaching would-be members of the ACS's newly established Scientific Committee. More generally, his part among numerous ACS counsellors was to back the whole field with overarching pictures of animal and bacterial model systems that could be used to investigate the role of viruses. Stanley complemented his discourse on viruses with references to the cancer problem targeted at public health officials and politicians. In 1957, as he joined the board of directors of the ACS, Stanley was for the first time invited to testify before the Senate. His plea for virus studies was based on a very general argument in favor of basic research which echoed Bush's report: "From an intellectual standpoint, what we are (after), what is the nature of life, is a most fascinating problem. Here you have a key to the study of that. And this same key, so far as I am concerned, as I have indicated, I believe contains the key to the cancer problem."[14] For Stanley, this common key was the chemical structure of viruses:

> The recent discovery by Fraenkel-Conrat [in Stanley's laboratory] that by special treatment of TMV a nucleic acid component possessing virus activity can be obtained is an achievment of the greatest significance. This finding indicates that the virus message, the ability to cause infection or to change a normal cell into a cancer cell, is actually carried by the nucleic acid molecule. This nucleic acid molecule in essence is a 4-letter word containing 1,2,3,4 or A,B,C,D, the sequence of those 4 identifying symptoms in a large molecule essentially carries the message of all of life. What we are, what we have been, what we ever hope to be (U.S. Senate 1958, p. 1411).

Since viruses could be viewed as gene-like structures, they may act as invisible ennemies "going through our bodies":

> it is not necessary to regard certain viruses as infectious agents in the ordinary sense, for some viruses appear to multiply so slowly so as to appear to be tied to the normal genetic apparatus of the cell. I am talking about the genes and chromosomes. When the cells divide and give a new set, the chromosomes and viruses go right along and almost act as part of that system. Recent experiments seem to indicate that the Rous virus causing cancer in chickens can act in this manner (U.S. Senate 1957, p. 1411).

Although he thought it "stimulating intellecually" to consider that cancer is caused by "viruslike intracellular activities", Stanley's attitude toward animal viruses remained rather cautious in the mid-1950s. Since biological assays remained the norm and no agent had been purified, the mammary tumor systems were, for instance, viewed to be too complex for fruitful work and barely mentioned in his reviews. This position changed over the periode 1958–1960 when Stanley was increasingly dragged into the public debate about leukemia and the "vertical transmission" of cancer.

This concept may be traced back to the late 1940s when a few experimental oncologists adopted a filterable agent causing mammary tumors in mice as a good model for breast cancer. Introduced in 1936, the "mammary tumor inciter" became macromolecular during the war following the path already taken by the chicken sarcoma agent. This "dead-end" virus was later widely used to illustrate the vision of cancer as originating in transmissible agents. Ludwig Gross, a cancer specialist at the Veterans Hospital, spoke of "vertical epidemics" of cancer (Gross 1946, 1949, 1958):

> 'Vertical epidemics' … designate the transmission of potentially pathogenic agents from one generation to another … The agent, though potentially pathogenic, would remain latent for a long period of time, often perhaps throughout the entire life-span of a given carrier host … Thus the host would remain in perfect health, while carrying and transmitting the seeds of the disease. In some hosts, however, particularly in those reaching middle age, the agent may become activated. Triggered by intrinsic or extrinsic stimuli, the hitherto latent and harmless agent would then change into a formidable pathogen, causing rapid multiplication of cells harboring it, and killing its carrier (Gross 1958, p. 1).

Vertical transmission implied that classical bacteriological weapons would not work easily since hidden viruses were not able to induce immune reactions in the body. Control was therefore a matter of prevention. The aim was to inhibit the entry of the virus into the body by isolation, passive immunization, or hygienic measures:

> It is possible to assume, however, that mammary carcinoma of mice does not represent a form of cancer different from breast cancer in other mammals. If this assumption is correct, the law of obligate communicability may in the future be established also for breast cancer in such animals as rats, rabbits, or dogs, or perhaps also in women. Should such a possibility materialize, the eradication of human breast cancer may become feasible by the simple method of artificial feeding of infants born to mothers having a family history of tumors (Gross 1949, p. 305).

Gross' tumor viruses were bacteriological entities laden with public health concerns and targetted at clinicians. Nonetheless, the public life of the vertical epidemics was short. By the early 1950s, Gross was no longer commenting on the new concept and concentrated on experiments. For practical matters, however, Gross selected a less studied and putatively less complex model than mouse mammary tumors. Benefiting from the availability of mice showing a high incidence of a leukemia-related disease, Gross attempted inoculation experiments to demonstrate that this disease was also caused by a cellular agent, filterable and vertically transmitted. For years, he inoculated leukemia extracts without obtaining much results and attention. In parallel, the study of the mammary tumor viruses lagged: disappointed with the variability and the complexity of the model, most experimental oncologists maintained low levels of investments in animal tumor viruses. By 1957–1958, however, the putative mouse leukemia viruses isolated by Gross and the notion of a vertical transmission of cancer experienced a new birth.

The experimental work done by Gross and the discovery of other leukemia viruses at the NCI has been discussed elsewhere. In the context of this paper, it is enough to state that, once domesticated and adopted by the NCI scientists, the mouse leukemia viruses made a great impact on officials within the agency and, soon after that, within the cancer charity. By 1958, both organizations launched a blitz campain for the support of cancer virus research. The dynamics of this campain is illustrated by the congressional hearings of the late 1950s which triggered major advances in what may be called a political molecularization of leukemia etiology. Congressional hearings for appropriating the 1959 budget were planned for April 22–23, 1958. As has become customary, drafts of the statements circulated between the NCI and the cancer society. A major point in Heller's statement was to recall the extension of Gross' work by NCI scientists and the isolation of new, more potent, strains of leukemia viruses. This was echoed by Stanley. His background statement gave a history of the field which included four major stages: Rous' discovery of 1911, Shope's work on the rabbit papilloma virus, and Gross' report of the early 1950s, now seriously confirmed and pushed at the NCI. Stanley concluded his historical review with a strong claim:[15]

> The time has come for us to assume that viruses are responsible for most kinds of cancer and to design and execute our experiments accordingly. Cancer is basically a problem in growth, and there is no reason to believe that the growth of most human cells is fundamentally different from the growth of most animal cells. Acceptance of the viral etiology of human cancer as a working hypothesis will involve a marked change in attitude on the part of many investigators, but this is necessary if the right approach and the right experimentation are to result. What we do depends in large measure upon what we think (U.S. Senate 1959, p. 1605).

Actual testimonies significantly departed from the script. They are illustrative of an expert system which brought together professional politicians, health officials, and virologists. Two issues emerged at the forefront of the discussion: the search for human viruses and the development of vaccines. In their background texts, both Stanley and Heller alluded to the observation of virus-like particles in the blood cells of leukemic mice. More important, however, was the fact that these observations implicated the use of modern instrumentation — preparative ultracentrifuges and electron microscopes. Both testimonies then stressed pressing needs for the use of the same tools in an "all-out" search for similar viruses in human tissues. Mass screening was in the air. As Stanley put it:

> Examination [by means of the electron micoscope] of the lymph node of mice dying of leukemia shows innumerable small little particles ... This inspired a few investigators to make similar examinations ... of the human lymph nodes of persons suffering from human leukemia. There again they are finding the same thing that they found in the mice. Nobody knows whether or not these little particles represent viruses, and if they do represent viruses, whether or not it has anything to do with human leukemia. But again, this is an area that is crying for exploration (U.S. Senate 1959, p. 1606).

In contrast to the search for human viruses, the vaccination perspective was not discussed in the background statements. It was first advanced during the hearings as Heller summarized early work with the mouse leukemia model:

> As in the instance of poliomyelitis and influenza, once we know the virus has been determined, isolated, and identified, it is not too fantastic to conceive that there may in time be a vaccine elaborated. We are not saying that we have such a vaccine. We are not even saying that viruses are responsible for human cancer. The evidence is strongly suggestive that this may be the case in some cancers (U.S. Senate 1959, p. 758).

A few years after the successful development and testing of the Salk vaccine against poliomyelitis, the association "virus-polio-vaccine" was a powerful incentive. Later in the day, it was further substantiated by Stanley's visual evidence. Stanley illustrated his talks about the purification and structural studies of

Figure 5.3. Stanley standing in front of electron microscopic images of the polio virus.

viruses by displaying electron micrographs (Figure 5.3). On the Congress floor, his "large and beautiful picture of poliomyelitis virus" struck an emotional chord:

> Dr. Stanley. This represents eight viruses, all at the same magnification. They are purified viruses so that you see the virus particles themselves. This one is vaccinia virus

> that is used to vaccinate or protect against smallpox. This is a crystalline insect virus. This is influenza virus, the virus which in 1918 caused the deaths of more people in 4 months in this country than we lost during all the two great world wars on the battlefield. One little virus can do that. These are the phages viruses ... that I had the good fortune to put out in crystal form. This is poliomyelitis virus which we crystallized in Berkeley.
>
> Senator Hill. You speak of the poliomyelitis virus now. If you had not first found that virus and isolated it, it would not have been possible, would it, for Dr. Salk to come forward and finally get his Salk vaccine, would it?
>
> Dr. Stanley. I think the basic knowledge of the physicial and basic biochemical, biophysical properties was useful ... This purified material is now, of course, available for, I think, a very good polio vaccine that can be uniformly made of high potency. There are some other crystals. These are some of the poliomyelitis viruses (U.S. Senate 1959, p. 1613).

The multifaceted leukemia viruses which emerged out of these policy debates nicely linked politicians, NCI officials, ACS administrators, and a few biologists. The congressional hearings of the year 1958–1964 established a strong link between the mouse leukemia viruses and classical infectious viruses through similar dialogues on purification, visualization, and vaccination. This rapidly stabilized research obviously echoed the popular vision of "crash" initiatives for the development of penicillin or polio vaccines. Thc policy issue was to commit enough money, material, and manpower into a "tumor virus coordinated attack". The scientific program was clear: (1) to study the viruses causing tumors in mice and other mammals; (2) to organize the search for human cancer viruses; (3) to produce these agents in cell culture and purify them; and (4) to produce and test vaccine preparations. The tumor viruses coming out of the congressional hearings were mere bacteriological entities whose properties resembled those of classical infectious agents. With the commitment of NCI officials and congressmen, Gross' notion of a vertical epidemic had been slightly changed, but to become more legitimate than ever before (Gross, 1964).

It would be erronous to analyze this combination of infection and tumor genesis either as a rhetorical tool employed by scientists to channel funds into basic virus research or as a constraint imposed on scientists by health officials in search of rapid pay-offs. The decision-making process which led to the allocation of important funds for studies on the role of viruses in human cancer is better analyzed in terms of converging interests and practices. The outcome was an object of research *and* policy which suited all parties. The congressional idea of "an all-out" attack on tumor viruses combined typical features of the postwar biomedical logic. First, the notion of a tumor virus "program" or "taskforce" echoed the idea popular among U.S. politicians involved in health issues that it was the government's responsibility to coordinate research and development efforts in order to find technological fixes for major diseases. As documented by historians of the NIH, this was a major push toward the creation of a whole series of disease-oriented institutes in the 1950s. Secondly, the newly isolated leukemia viruses in mice were especially suggestive to cancer specialists like Heller or Farber: they did not only suggest that a NCI-coordinated-attack would, in all probability, result in specific means for public health targets like early diagnosis and immunological prevention, but also that such a program was the ideal complement to the large chemotherapy program then run by the NCI, since the systematic search for anti-tumor chemicals concentrated on leukemia (Löwy, 1997). Finally, the mouse leukemia viruses which provided the laboratory reference for the tumor virus initiative suited the research habits of virologists like Stanley. In contrast to the mammary tumor agent which had then proved impossible to purify, they were macromolecular entities showing some ability to replicate in cell culture.

The combination of macromolecules and vaccines created by this policy-making community resulted in a significant reorganization of cancer research. The most visible aspect was the development of tumor virus studies at the NCI. First, a Laboratory of Viral Oncology was established in 1959. The new unit gathered the scientists already working on the Rous sarcoma virus, the new leukemia viruses, or the polyoma virus. The second step was the creation of a "Special Virus Leukemia Program" in 1964 with the mandate to determine whether viruses comparable to the murine leukemia viruses could be etiologically related to human leukemia and, if so, to develop immunological means of prevention. The first allocation of funding for this program amounted to $10 million. Based on the mobilization of NCI and Congress, this operation was "big biomedicine" in several ways. Firstly, as mentioned above, the study of leukemia virus was "targeted research" centralized by a government agency. Secondly, the management of the operation was based on contracts with NCI committees commissioning specific work to be done in medical schools, universities, or national laboratories. Thirdly, cancer virus research transformed a form of experimentation associated with the molecular instrumentation, large-scale screening, and scaling up into a system of public demonstration widely employed to resolve impeding conflicts about the meaning of the animal tumor viruses.

CONTRACTS AND THE MANAGEMENT OF TOOLS: FROM TUMOR VIRUSES TO ONCOGENES

One problem with the leukemia virus program, however, was its low scientific profile. Although Stanley had been an eloquent advocate for massive support of virus research in general, the research system which stabilized in the early 1960s took a different shape. Aiming at public health targets, based on bacteriological practices and large scale instrumentation, the NCI cancer virus initiative was "biologically" uninteresting. To many scientists, this seemed to be a typical bureaucratic operation, a "moonshot" in the manner of the chemotherapy program characterized as "nothing too stupid to test". Until the late 1960s, however, such comments remained marginal. In contrast, one spectacular achievement of the early days of the NCI program was the fact that prominent molecular biologists — including a few participants in the "phage group" — who viewed their activity as pure research on the fundamental properties of living organismes were attracted into the cancer domain and contributed to turning the mouse tumor viruses into "infective hereditary particles", accounting for the bacteriological notion of "vertical epidemic". This paved the way to the third phase of molecularization which took place in the late 1960s within a network of NCI contractors, namely the schift from tumor viruses to oncogenes and cancer proteins. In this section, I will show that a network of biotechnology firms and academic research centers working on cancer proteins and cancer genes emerged out of the circulation of viruses, immunological reagents, and technical procedures.

Ten days before the congressional hearings of 1958, Harry Weaver, the research director of ACS, obtained the green light for organizing a large conference on "the possible role of viruses in cancer." Stanley was the driving force behind the fifteen scientists asked to serve on the organizing committee.[16] The committee was a very heterogeneous group including historical figures in the "virus and cancer" domain (Rous, Shope, Beard), bacteriologists working on infectious viruses (Enders, Kaprowski, Puck), molecular biologists working on phage (Luria, Lwoff, Dulbecco), a specialist of cancer chemotherapy (Rhoads), but nobody working on the mouse tumor viruses. The 1959 ACS-sponsored meeting on "The Possible Role of Viruses in Cancer" seemingly juxtaposed two theoretical discourses. First, the "phage delegation" (Dulbecco and Luria) advanced the notion of "infective heredity" or of "virus infection as a class of mutations":

> In the evolution of virus-induced cancers, infective virus may be recoverable at one stage and not at another … the question arises, therefore, whether exogenous infection, or "vertical transmission" through the germ cell or the genital tract, or even de novo origin from noninfective proviruses or other cellular constituents, may be implicated for the presence of these viruses in the animals in which they are found. Especially with agents such as Gross's mouse leukemia factor, the narrow host-range specificity and the difficulty of transmission by extract suggest a remarkably inefficient adaptation to natural spread (Luria 1960, p. 678).

Central to the vision of tumor viruses as genetic particles interacting or even integrated within the nucleic acid constituents of the cell was the comparision of these viruses with phages and bacterial viruses. The comparison between tumor viruses and phages was not new — it had been employed in the 1940s by Rous or Claude — but it had deeper meanings and more direct impact on American biologists when followed by the molecular geneticists from the California Institute of Technology.[17] Thus, the cancer agents were conceived to be "elements of the genetic material of cells" which interacted with cellular genes, modified cell metabolism, and could sejourn for a long time in cells before they "reproduce and determine the formation of a specific apparatus for their own transfer to other cells". According to Luria and Dulbecco, this approach provided a vantage point from which it was no longer necessary to see the "somatic mutation hypothesis and the virus hypothesis of cancer origin as alternative and mutually exclusive."

The genetic theory of cancer causation was opposed by a well-known specialist of influenza viruses, the British bacteriologist C.H. Andrewes, who argued for the viral — exogenous, infective — nature of tumor viruses:

> I may not know very much about cancer but I have been in the virus field for 36 years, and I can confidently dispute the suggestion that tumor viruses are in a class of their own ... When [Luria] says that "virus infection is a cellular mutation" I am personally at a total loss. I don't think it even occurred to me during the 1957 epidemic of Asian influenza that we were suffering in our hundreds of thousands from cellular mutations! (Andrewes 1960, p. 690)

Andrewes' most effective point, however, was that the "genetic" hypothesis could make little profit of the numerous features revealed by bacteriologists and pathologists, for instance, variable effects in different hosts or changes of virulence observed when a virus strain was "passed" through the same species of hosts for a long time. Although practical implications were little discussed, this "virus/genes" discussion echoed alternative policy choices: structural studies of the tumor viruses and vaccine production on the one hand, characterization of both the virus genetic material and the metabolic anomaly in tumor cells with an eye on rational therapy on the other hand.

By virtue of his role as scientist-advisor and his diversified research commitments — purification and crystallization of the polio virus, biophysical instrumentaion, nucleic acids studies — Stanley was in a good position to push for a mediating role. The theoretical part of his discussion merely consisted in a juxtaposition. Stanley on the one hand defined the viral theory of cancer as a problem of alternatively "masked" and "unmasked" viruses. On the other hand, he stressed the biochemical track with respect to two angles: the control of viral replication and the enzymatic changes determined by viral nucleic acids. In conclusion, Stanley pointed to the practical developments which linked phage and tumor viruses, infection and tumorigenesis:

> Certainly the mammalian cells and the viruses are now available for extensive research in areas comparable to transduction, lysogeny and lysogenic conversion in the bacteria-bacterial virus system, and fortunately some investigators are already at work ... [Dr.Rubin] has developed a precise and efficient method for infecting and altering cells *in vitro* and is using this technique for studying the relationships between the Rous virus and host cells in a manner similar to that exploited so successfully in the study of bacterial viruses (Stanley 1960, p. 803).

Stanley was thus alluding to the entry of animal tumor viruses in his laboratory along with a student of Dulbecco, Harry Rubin, and more generally to the circulation of viruses and procedures which, in the 1960s, gradually established a network of biotechnology firms and research centers which were all NCI contractors and gathered together molecular biologists as well as virologists.

Ten years after the ACS meeting, the NCI was spending over $40 million per year on tumor virus studies. Alignment of heterogenous partners, including clinicians working in hospitals, academic virologists, NCI researchers, and company engineers, had been achieved through administrative means.[18] Contract was the standard system of management. The virus program was managed by a "Segments Chairmen Committee" which included most heads of laboratories involved in virus studies at the NCI. By 1971, 120 contracts had been established, half of them with profit-making companies. The tasks carried out by these institutions ranged from the production of research tools (cells, inbred animals, viruses), the collection of human material (tissues from biopsies, blood), to the organization of systematic screening for viral particles in human cancer. Biotechnology firms of the Washington area launched with NCI support — Microbiological Associates, Bionetics Research Laboratories, Litton Bionetics — often held many contracts, with the effect that their engineers and technicians acted as fulltime participants in the Virus Cancer Program (VCP). For example, Meloy Laboratories signed a contract to:

> propagate, concentrate, distribute Mouse Mammary Tumor Virus, perform immunological and biological assays for detection and quantitation; develop methods for propagation and detection of MMTV antigens ; conduct studies on the control of neoplasia in the susceptible murine host by vaccination with inactivated virus.[19]

Mice, viruses, immunological reagents, transformed cells, and so forth were produced within the contract framework. The production of tools was one way to normalize practices, to regulate collaboration and exchanges, to stabilize social links and finally to discipline the field. The "Segment Chairmen's Committee" made most financial and scientific decisions concerned with the collective resources. Issues included the selection of viruses, the choice and usage of animal systems, the preliminary production and testing of vaccine preparations, and the coordination of etiological studies and immunological surveys.

For instance, during one single meeting of April 1969, the committee (1) phased out long-term experiments in monkeys for cost motives and because "several years ago it was decided that an essential criteria for determining whether a virus causes cancer in man would not be the potential for inducing tumors in monkeys", but kept a smaller monkey production contract running for "providing monkeys to test the potency and safety of a vaccine when the program (will) undertake vaccine production;" (2) diminished the support to scientists at Merck who tested the therapeutic effects of some chemical analogs of tumor virus nucleic acids: "Dr Manaker questioned the relevance of the commitment of this much money to therapeutic approaches rather than to etiologic studies. It was pointed out that ... through this contract the program would have access to the purest poly I:C available"; and (3) introduced dogs and Hazelton Laboratories in the program:

> The purpose of this contract is to determine if malignancies in the canine are virus-induced as in the murine and feline systems, and if the beagle dog, or other breeds, are useful as a system for the isolation of potentially oncogenic human cancer viruses ... Following advertisement in the *Commerce Business Daily* the internal working group met to consider the proposal submitted by Bionetics and Hazelton. Following a site visit to both, the facilities at Hazelton were considered to be excellent for this work, the personnel were good and their bid was $50,000 less than Bioetics.[20]

The regulation of practices within the VCP network did not only rest on centralized control of research tools which had become emblematic of the new "molecular" virology. This point may be illustrated with the 1964 congressional hearings when the problem of negative evidence in the search for leukemia particles in human tissues surfaced.[21] Then, the new director of NCI, Endicott, responded with a statement about the collaboration between NCI and the Atomic Energy Commission:

> Blood specimens of acute leukemia patients are being examined for particles that might represent leukemia viruses ... The studies [on animal models] are done with the electron microscope on samples of highly pure virus obtained by spinning the blood at high speed in a centrifuge. At first, particles similar in shape to those found in the blood of leukemic animals were found in less that 20 percent of the specimens taken from patients. Now, with the use of a centrifuge developed collaboratively with the Oak Ridge National Laboratory of the Atomic Energy Commission, the separation of viral particles from blood is so efficient that almost all the patient specimens examined in a preliminary survey show these particles ...
>
> In anticipation of the time when human cancer virus will have been identified, the NCI is planning steps that will need to be taken to put that knowledge into use. ... In the course of protective vaccine development, virus will need to be grown in large quantities. The Institute is already supporting under contract a tissue culture operation to produce mouse leukemia virus in large quantities, with a view to developing techniques for growing human cancer virus in quantity in human tissue culture ...
>
> Plans are being made for phasing in the large culture operation in collaboration with the Atomic Energy Commission (U.S. Senate 1965, p. 1280).

The contract system favored another form of coordination between laboratories based on the development and standardization of the numerous biochemical and immunological assays employed to assess the purity, specificity, and biological nature of the cancer viruses. Such regulation rested on the activities of a "core set" of contractors maintaining specific strains of viruses and producing sera which would react specifically with the different set of antigenic proteins carried by this virus. Associated laboratories would receive virus, antigens, antisera, and assay recommendations from this "core set". Local testing would then have two results: first, specify reagents and procedures by organizing controls with known antigens locally available; second, identify new viruses or new antigens which did not show classical patterns of reaction. The latter might in turn be employed to produce new antisera and isolate new chemical fractions. Final regulation was achieved through the VCP managing committees:

> Dr. W. [Huntigton Research Center] continues to collaborate with Dr. Rowe [NIH] in the preparation of reagents and in the development of a fluorescent foci assay technique for use in the quantitation of leukemia viruses ... An extension discussion followed concerning the availability of reagents produced by this contract to all members of the program. Dr. H. [sic] pointed out the difficulty that has been experienced in many labs using commercially produced immune reagents, and, hence, the need for this contractor to prepare reagents according to specific protocols.[22]

Exchanges of materials, connected trials, and informal debates linked centers into circles of trust composed of scientists who employed mutually depending immunochemical standards composed of viruses and antisera. This immunological "networking" in turn contributed to the displacement of the molecular definition of mouse tumor viruses. Rather than being macromolecular particles visible under the electron microscope, they became aggregated proteins revealed by immunological and biochemical tests. This resulted in novel means for resolving controversies on candidate human viruses.

By the early 1970s, NCI laboratories and contractors had isolated and organized the production of viruses inducing tumors in mice, rats, cats, dogs, and monkeys. What happened to be viral particles were found in human tumor tissues, but contamination was a highly debated issue. This was a strong incentive toward the use of immunochemical tests in place of electron microscopic images. Observations of virus-like particles in human cancer tissues were viewed by chairmen of the virus program as suggestive but inconclusive evidence:

> This contract aimed at the identification and characterization of families at high risk to breast cancer. Facilities have been organized for collecting milk, serum, and tumor specimens as well as family histories from a large number of normal donor and breast cancer patients. These resources are used in the search for a human breast cancer virus ... Under a subcontract for electron microscopy, it was shown that a few unequivocal type B particles are found in the human milk. The chairmen were concerned about the size of the budget. Conclusion was that the VCP should not so heavily support a putative virus based on what appears to be inadequate, incomplete experiments.[23]

Immunochemical analysis introduced a whole range of tumor virus antigens which were considered to present host, tumor type, or group specificity. This provided opportunities for finer analysis of the biological origins of the viruses found in human tissues. Argumentation was, however, strongly constrained by established patterns of immunological collaboration. Trust — or more often doubts — about reagents, researchers, and their skills determined the fate of many a candidate virus. Take ESP-1, for instance, a virus found by electron microscopists in Houston, Texas, in a cell line established from a pleural tap performed on a young child suffering from Burkitt's lymphoma (Dmochowski, 1978). Immunological assays performed in Houston rapidly confirmed the idea that this virus shared antigens with the mouse leukemia viruses but showed immunological specificity. The contamination issue was publicly raised following the child's death from disseminated varicella. Alternative antigenic analyses were done by scientists at Meloy Laboratories, a large reagent supplier of the virus cancer program. They concluded that ESP-1 was a mouse contaminant. Controversy raged for several weeks with various sets of local reagents against similarly labelled antigens showing different patterns of reaction. Trusting the Meloy system, the chairmen of the virus program rapidly concluded that ESP-1 was an unusual but uninteresting mouse virus. The Houston workers persisted in their views, although ESP-1 quietly left the national scene. The matter was only settled two years later when both teams imported a new set of reagents developed by a German laboratory and concluded that "in spite of the various differences of biological activity, structure, and antigenicities observed in a variety of experimental systems, the major identifiable determinants of ESP-1 virus are murine." Some peculiar results could still contrast ESP-1 and the mouse leukemia viruses but they were no longer significant.

The theoretical side of this management system was an increasing emphasis on the idea of viral proteins viewed to be on the one hand tumor antigens to be purified and used for vaccination, and on the other hand transforming factors controlling key metabolic pathways in the cell. This perspective was linked to molecular biologists' views on the role of viral nucleic acids by postulating the existence of peculiar DNA or RNA sequences in the viral genome coding for these cancer proteins. The whole schema was exemplified by the "oncogene theory" advanced by two NCI researchers, R. Huebner and G. Todaro in 1969 (Huebner, 1969). The theory was an improved version of Gross' vertical epidemic. Todaro and Huebner's scheme focused on "oncogenes" controlling cellular differentiation and growth. In cells, oncogenes were defined as genetic sequences associated with endogenous RNA viruses integrated in chromosomes, transferred from generation to generation for millions of years. This case of host-parasite relationships had led to an almost complete identity between the pathogen and the cells. In the course of life, induction by different agents (chemicals, hormones, immunological disorders) could take place and activate either the "oncogene" leading to transformation of cells or the "virogene" leading to the production of virus. Huebner and Todaro used results from almost all fields of cancer research but the most important aspect was the identification of numerous hidden mouse viruses.

One final comment about the management structure of the virus cancer program and the regulation by contract is that it paved the way to the "recombinant DNA" revolution in two ways. Firstly, the close collaboration between small biotechnology companies, pharmaceutical companies, medical school laboratories, and federal laboratories which participated in the cancer virus search anticipated the "university-industrial complex" of the late 1970s.[24] Secondly, many procedures and instruments presented as classical resources for molecular biology and as essential elements of "recombinant DNA technology" emerged within the network of NCI contractors. This can be illustrated with the uses of "reverse transcriptase". This enzyme that transcribes RNA to DNA was a key element in the replication scheme attributed to tumor viruses. In 1970, assays for an enzyme catalyzing DNA synthesis from RNA in cells infected by RNA viruses were reported by D. Baltimore and H. Temin. These were bold claims, since the enzyme opposed the normal flow of genetic information.[25] Nonetheless, discussions did not focus on experimental results but on the implications of "reverse transcription" for the oncogene theory and the putative integration of viral DNA in the genome of the cell. Reverse transcriptase was rapidly normalized as a resource. Late in 1970, virus cancer program officials signed a contract with Beard, the expert in avian viruses working at Duke University. From 1971, Beard was in charge of the production of viruses which could be used as sources of reverse transcriptase.[26] Once available, reverse transcriptase circulated widely. It became a marker for RNA viruses and viral carcinogenesis. Beard's enzyme was an instrument for producing DNA molecules, a screening device for detecting viral oncogenes in tumor cells, and a research object for molecular biologists interested in the replication cycle of animal viruses.

CONCLUSION

In the early 1980s, discourses about genes, viruses, and cancer were radically reorganized. Following the dispersion of recombinant DNA techniques, a genetic theory of cancer causation replaced the viral hypothesis. This "revolution" in cancer research was then associated with the notion of belated applications of the fundamental knowledge about genes which was developed by academic molecular biologists in the 1960s and 1970s (Gaudillière, 1993). The molecularization of cancer etiology associated with the tumor viruses thus quietly disappeared from most historical accounts written by observers of contemporary cancer research. By analyzing cancer virus research in the United States after 1945, this paper suggests that the "recombinant DNA revolution" of the 1980s was only the most recent among the molecularizations of cancer etiology. Molecularization was not a single process but a series of restropectively unified "molecularization practices" which emerged in different places at different times. These can be understood as instrumental, political, and managerial innovations. Molecularization patterns have been illustrated in this chapter by following (1) the development of electron microscopes and ultracentrifuges as instruments visualizing viruses in the 1940s; (2) the congressional debates of the 1950s which established mouse leukemia viruses as policy targets; and (3) the contracts of the Virus Cancer Program of the 1960s which created the material culture within which cancer genes circulated.

Molecularization of cancer etiology was in the first place synonymous with instrumentation. From the 1940s until the 1970s, the mere existence of animal tumor viruses relied on ultracentrifugation and electron microscopy. On the one hand, tumor agents became proper macromolecular viruses because they were represented with both instruments. On the other hand, simultaneous and proper usage of the ultracentrifuge and electron microscope was defined as the ability to purify and visualize these new viruses. The role of instrumentation was not limited to this triangulation. Electron microscopes and ultracentrifuges were complex commercially produced technological devices often employed for "scaling-up" research. In this context, scaling-up was material, i.e. isolating more virus-like particles or showing numerous and detailed images, as well as metaphorical, i.e. launching "all-out" operations echoing "big science" and the organization of industrial research. This was apparent in the large screening operations of the 1960s which aimed at the identification of human tumor viruses.

Secondly, the molecularization of cancer etiology was a political process originating in the postwar "biomedical complex". As illustrated by the case of influenza virus research, well-targeted projects which mobilized a wide range of scientific, medical, and industrial partners under the lead of governmental bodies emerged during World War II. This practice was extended to cancer research by virtue of the close collaboration between the U.S. Congress, the American Cancer Society, and the NCI. A system of expertise based on public debates in the manner of congressional hearings then contributed to the establishment of projects at the boundary between federal politics, public health, and biological research, among which was the NCI cancer virus program. If one focuses on this program, one may be tempted to think that American postwar "molecular" biomedicine was "mobilized" science. The role of experts like Stanley, however, suggests that rather than conscripts working at the forefront of the NCI program — purifying viruses or screening tissues — a majority of animal virus specialists were "mobilizable": they composed the reserve and molecularized by cultivating rare skills in their own ways.

Finally, molecularization of cancer etiology was a form of scientific management modelled on industrial research practices. This is higlighted by the work of committees which coordinated research tasks within the NCI virus program. The contract system which dominated the program then combined a centralized production of resources and "inter-laboratory regulation" of practices (Keating, 1997). The contract management style favored the development of immunological tools which circulated within the networks of NCI contractors and, in turn, interwove detailed studies of tumor virus macromolecules with screening and vaccinating activities. Impeding conflicts about alternative meanings of the tumor virus approach were often resolved on the basis of trust patterns which originated in this form of standardization.

NOTES

1 U.S. Congress, United States Senate, "Labor-Health, Education and Welfare Appropriation for 1959" (Washington, DC: U.S. Government Printing Office).

2 Ilana Löwy (1990) has analyzed the changing meaning of the term "chicken tumor agent" and the reconstruction of the discovery in the late 1960s when Rous received the Nobel Prize for his 1911 findings. More recently, Rheinberger (1995) drew attention to the part played by the use of instruments producing new "representations" of viruses such as the ultracentrifuge and electron microscope.

3 On the early version of the air-driven ultracentrifuge see Elzen (1988). Commercialization remains to be analyzed.

4 In this respect, one may argue that World War II was a turning-point in the history of biological instrumentation and biotechnology. On instrumentation and industries, see J.P. Gaudillière and I. Löwy (1997).

5 In addition to the work on influenza mentioned in this paper, the purification of blood proteins was a major target (Creager, this volume).

6 W. Stanley to A.N. Richards (February 10, 1944, Wendell Stanley papers, University of California, Berkeley).

7 As T. Francis, the head of the influenza commission, put it: "The major advantage which preparation of vaccines made by centrifugation have to offer is that, presumably, a constant amount of material can be incorporated." (T. Francis to Stanley, December 21, 1945, Stanley papers).

8 War production board, standard protocol for the preparation of influenza vaccines (Stanley papers).

9 These figures compare well with the experience of the National Foundation for Infantile Paralysis which collected $6.5 million in 1943; $12 million in 1944; $25 million in 1945. To explain these collections, social historians have emphasized economical and cultural trends. Patterson (1987), for instance, listed the affluence of the postwar years, New-Deal concerns for social reform, the rising interest of a wealthy society in health issues, the successful fight against infectious disease, and finally good publicity.

10 *Cancer News* (1948), January, April, June and July issues.

11 The panel on viruses included W. Stanley, H.B. Andervont, then specialist of the mouse mammary tumor agent at the NCI, pathologist E. Goodpasture, and bacteriologists G.P. Berry and R.G. Green. (*Cancer News* (1947), January).

12 Although praised by virologists, the achievement seems to have made little impact on polio research (Smith, 1990).

13 In the 1950s grant applications to the ACS focused on improvements in tissue culture methodology.

14 U.S. Senate Committee on Appropriations for 1958, Labor-Health, Education, Welfare Appropriations (Washington: U.S. Government Printing Office).

15 Prepared Statement by Dr. W. Stanley in U.S. Senate, *Labor Health, Education and Welfare Appropriations for 1959, Hearings* (Washington: U.S. Government Printing Office, 1959). The same argument was developed in conferences.

16 D.E. Johnson to W. Stanley, April 10, 1958, W. Stanley papers.
17 Evidence for some continuity of this line of thinking is easy to find in repeated comparisons of lysogenic bacteria with tumor cells discussed, in the 1950s and 1960s, by the French molecular biologists André Lwoff and François Jacob (Galperin, 1994).
18 One should add that clinicians and patients were not involved in the virus program. They were replaced by material of human origins supplied by contracting hospitals.
19 National Cancer Institute, *Annual Report of Activities,* Fiscal Year 1968.
20 Minutes of Virus Cancer Program Segment Chairmen's Committee, April 3, 1969, NCI Archives.
21 "Satement of Dr. K. Endicott", U.S. Senate, Labor-Health, Education, Welfare Appropriations for 1965, Hearings May 22-July 1st, 1964, (Washington: U.S. Government Printing Office), pp. 1275-1294.
22 Minutes of the Virus Cancer Program Segment Chairmen's Committee, April 22, 1967, NCI Archives.
23 Minutes of the Virus Cancer Program Segment Chairmen's Committee, March 6, 1975, NCI Archives.
24 Many actors actually participated in both (Kenney, 1986).
25 "Central Dogma Reversed", Editorial, *Nature* (1970), 226: 1198.
26 National Cancer Institute, Activity Reports, Fiscal Year 1972, p. 1101.

REFERENCES

P.G. Abir-Am, "The Discourse of Physical Power and Biological Knowledge in the 1930s: A Reappraisal of the Rockefeller Foundation's 'Policy' in Molecular Biology" *Social Studies of Science* (1980), 2: 3–60.

C.H. Andrewes, "Discussion of S. Luria's Paper" *Cancer Research* (1960), 20: 689–694.

J.W. Beard *et al.*, "Ultracentrifugal, Chemical and Electron Microscopic Identification of the Influenza Virus" *Southern Medical Journal* (1944), 37: 313–320.

Board for the Investigation and Control of Influenza and Other Epidemic Diseases in the Army, *Army Medical Bulletin* (1941), 64: 1–22.

R. Bud, *The Uses of Life* (Cambridge: Cambridge University Press, 1993).

V. Bush, *Science, The Endless Frontier* (Washington: United States Government Printing Office, 1945).

A. Claude, "Fractionation of Chicken Tumor Extracts by High-Speed Centrifugation" *American Journal of Cancer* (1937), 30: 742–745.

A. Claude, "Particulate Components of Normal and Tumor Cells" *Science* (1941), 91: 77–78.

A. Claude, K.R. Porter and E.G. Pickels, "Electron Microscopic Study of Chicken Tumor Cells" *Cancer Research* (1947), 7: 421–430.

A. Claude, "Studies on Cells: Morphology, Chemical Constitution, and Distribution of Biochemical Function" *Harvey Lectures* (1950), 43: 121–164.

A. Claude and E. Fullam, "An Electron Microscope Study of Isolated Mitochondria" *Journal of Experimental Medicine* (1945), 81: 51–62.

A. Claude and E. Fullam, "The Preparation of Guinea Pig Liver for Electron Microscopy" *Journal of Experimental Medicine* (1946), 83: 499–503.

A. Creager, "Schlieren Patterns, Sedimentation Coefficients, and Subunits: How the Analytic Ultracentrifuge Structured Proteins", paper presented at the IHPSSB meeting in Brandeis, July 1993.

L. Dmochowski and J.M. Bowen, "Viruses and Human Cancer: History and Current Status of ESP-1" *Progress in Experimental Tumor Research* (1978), 22: 160–195.

B. Elzen, "Scientists and Rotors: The Development of Biochemical Ultracentrifuges" (Unpublished PhD Dissertation, University of Twente, 1988).

E. Fullam, "Reflections" *Bulletin of the Electron Microscopy Association of America* (1986), 16: 35–43.

P. Galison and B. Hebly (eds), *Big Science. The Growth of Large-scale Research* (Stanford, CA: Stanford University Press, 1992).

C. Galperin, "Virus, provirus et cancer" *Revue d'Histoire des Sciences* (1994), 47: 7–56.

J.P. Gaudillière, "Oncogenes as Metaphors for Human Cancer" in I. Löwy (ed.), *Medicine and Change* (Paris: John Libbey, 1993), pp. 212–248.

J.P. Gaudillière, "Circulating Genes: About the Production of Mice, Viruses and Cancer" in E. Mendelsohn (ed.), *The Practice of Human Genetics* (Sociology of Science Yearbook, vol. 26, Dordrecht: Kluwer, forthcoming).

J.P. Gaudillière and I. Löwy (eds), *The Invisible Industrialist: Manufactures and the Construction of Scientific Knowledge* (London: Macmillan, 1997).

L. Gross, "The Possibility of Exterminating Mammary Carcinoma in Mice by a Simple Preventive Measure" *New York State Journal of Medicine* (1946), 46: 172–176.

L. Gross, "The Vertical Epidemic of Mammary Carcinoma in Mice" *Surgery Gynecology and Obstetrics* (1949), 88: 295–308.

L. Gross, "The Aetiology of Cancer and Allied Diseases" *British Medical Journal* (1958), 5087: 1–5.

L. Gross, *Oncogenic Viruses* (New York: Pergamon Press, 1964).

T. Van Helvoort, "The Controversy between J.H. Northop and M. Delbrück on the Formation of Bacteriophage: Bacterial Synthesis or Autonomous Multiplication?" *Annals of Science* (1992), 49: 545–575.

R. Huebner and G. Todaro, "Oncogenes of RNA Tumor Viruses as Determinants of Cancer" *Proceedings of the National Academy of Sciences, U.S.* (1969), 64: 1089.

L. Kay, *The Molecular Vision of Life* (New York: Oxford University Press, 1993).

R.E. Kohler, *Partners in Science* (Chicago: University of Chicago Press, 1992).

P. Keating and A. Cambrosio, "Interlaboratory Life: Regulating Flow Cytometry", in J.P. Gaudillière and I. Löwy (eds), *The Invisible Industrialist* (London: Macmillan, 1997).

M. Kenney, *Biotechnology: The University Industrial Complex* (Yale: Yale University Press, 1986).

D.J. Kevles, *The Physicists. The History of a Scientific Community in Modern America,* (Cambridge: Harvard University Press, 1987).

R.E. Kohler, "The Management of Science: The Experience of Warren Weaver and the Rockefeller Foundation Programme in Molecular Biology" *Minerva* (1976), 14: 279–306.

R.E. Kohler, *Partners in Science* (Chicago: University of Chicago Press, 1992).

S. Leigh Star, "Triangulating Clinical and Basic Research: British Localizationists" *History of Science* (1986), 24: 29–48.

I. Löwy, "Variances in Meaning in Discovery Accounts: The Case of Contemporary Biology" *Historical Studies in the Physical and Biological Sciences* (1990), 21: 87–121.

I. Löwy, *Between Bench and Bedside* (Cambridge, MA: Harvard University Press, 1997).

S. Luria, "Genetic Concept of Virus Infection" *Cancer Research* (1960), 20: 677–688.

J. Murphy, "An Analysis of the Trends in Cancer Research" *Journal of the American Medical Association* (1942), 120: 107–111.

J.T. Patterson, *The Dread Disease: Cancer and Modern American Culture* (Cambridge, MA: Harvard University Press, 1987).

K.R. Porter, A. Claude and E. Fullam, "A Study of Tissue Culture Cells by Electron Microscopy" *Journal of Experimental Medicine* (1945), 81: 233–246.

N. Rasmussen, *Picture Control* (Stanford, CA: Stanford University Press, 1997).

H.-J. Rheinberger, "From Microsomes to Ribosomes: 'Strategies' of 'Representation'" *Journal of the History of Biology* (1995), 28: 49–89.

P. Rous, "The Nearer Causes of Cancer", *Journal of the American Medical Association* (1943), 122: 573–581.

D.F. Shaughnessy, *The Story of the American Cancer Society* (Unpublished PhD Dissertation, Columbia University, 1957).

J. Smith, *Patenting the Sun* (New York: Anchor Books, 1990).

P. Starr, *The Social Transformation of American Medicine* (Basic Books, 1982).

W. Stanley, "Some Chemical, Medical, and Philosophical Aspects of Viruses" *Science* (1941), 93: 145–151.

W. Stanley, "Virus-Induced Neoplasia" *Cancer Research* (1960), 20: 798–804.

S.P. Strickland, *Politics, Science and the Dread Disease* (Cambridge, MA: Harvard University Press, 1972).

A. Thackray (ed.), *Private Science: Biotechnology and the Rise of the Molecular Sciences* (Philadelphia, PA: University of Pennsylvania Press, in press).

U.S. Senate, *Labor Health, Education and Welfare Appropriations for 1959, Hearings* (Washington: U.S. Government Printing Office, 1959).

CHAPTER 6

Following Molecules: Hemoglobin Between the Clinic and the Laboratory

Soraya de Chadarevian

In his well-known paper of 1949 "Sickle Cell Anemia, a Molecular Disease" Linus Pauling and his collaborators at the California Institute of Technology described a new type of hemoglobin (Pauling *et al.*, 1949). This new hemoglobin was characterized by its electrochemical behavior in the Tiselius apparatus. Under certain conditions the hemoglobin of sickle cell anemic patients migrated to one pole and the hemoglobin of non-sickling individuals to the other. Pauling's team suggested that the different electrochemical behavior of sickle cell hemoglobin as compared to "normal" hemoglobin was due to a small structural change in the globin of the molecule. This alteration would allow the molecules under certain circumstances to stack onto one another, rendering the red blood cells rigid, to the point of bursting them. With this explanation Pauling interpreted the known clinical picture of sickle cell anemia as a "molecular disease".

The Californian team not only identified a new type of hemoglobin, the first after the description of fetal hemoglobin as a globin variant. They also offered a genetic interpretation of their findings. In their studies they had included the hemoglobin of individuals whose red blood cells, under low oxygen pressure, took on the characteristic crescent forms of sickled blood, but who did not suffer from anemia. This condition was known as sicklemia or sickle cell trait. Pauling's group showed that in the Tiselius apparatus the hemoglobin of these individuals behaved like a mixture of normal and sickle cell hemoglobin. This finding suggested to them that sickle cell anemia, which was long known to be an inherited disease, depended on the homozygous condition of a recessive gene, while sicklemia was due to a heterozygous condition of the same gene. These conclusions confirmed results recently presented by human geneticist James Neel from Ann Arbor, but were apparently formulated independently (Neel, 1949).

For its color and ready availability as well as for its striking physiological characteristics hemoglobin had long been a preferred research object for biochemical and biophysical investigations. After the war the efforts of human geneticists had converged on blood and its components as a research topic which was amenable to exact experimentation. In the investigations on sickle cell anemia physico-chemical and genetic approaches converged in new and unexpected ways.

Following the work of the research group at Caltech a whole series of new or "abnormal" hemoglobins was described in the laboratory.[1] These were identified through analysis of patients or through population surveys which were also exploited for anthropological or population genetic considerations. Anemic conditions were redefined as "hemoglobinopathies" and the electrophoretic characterisation of hemoglobin was introduced as routine diagnostic procedure in the clinic. Abnormal hemoglobins also became an important laboratory tool for biochemical, structural and genetic research.

By following the laboratory and clinical career of abnormal hemoglobins I want to investigate how scientists and clinicians came to share a new research culture, centred around molecules. In particular, I will focus on the figure of Hermann Lehmann, first Professor of Clinical Biochemistry at Cambridge, who functioned as a mediator between the clinic and the laboratory throughout his career, and on the role of his reference collection of abnormal hemoglobins in the circulation of tools and practices between the two institutional sites.

Pickstone, distinguishing several types of knowledge which succeed one another historically, but are retained in later stages, has stressed the importance of collections or a natural history kind of knowledge for "analytical" work in science and medicine (Pickstone, 1993, 1993a and 1994). He has also pointed to the difficulty of reconstructing these practices (Pickstone 1993a, p. 36). As if in confirmation of Pickstone's perspective, a leading biochemist described Lehmann and his colleagues as "in some way the biochemical counterparts of the classical naturalists, with rare hemoglobins as their quarry". Like naturalists, they travelled to remote parts of the world to find new hemoglobin variants. However, "unlike the classical naturalist Dr Lehmann then finds out exactly in which chemical aspect his new kind differs from the other known species ..." (Young, 1966, p. vi).

The special interest of Lehmann's collection for our theme of molecularization is that it incorporated clinical knowledge as well as a series of laboratory technologies aimed at the description and classification of molecules. In this paper I will investigate the interrelationship of these two kinds of representations and their respective uses in the clinic and the laboratory. The field as a third point of reference present in the collection will also enter my considerations, even if more marginally.

My study will entail a critique of the simplistic model according to which basic biological research acquired in the laboratory is applied to the clinic. It will also throw new light on the development of molecular biology.[2] I shall claim that the development of the new science relied from the beginning on material resources and functional knowledge from and social and institutional links with the clinic. The clinical knowledge of abnormal hemoglobins proved essential for interpreting protein functions in molecular terms and the move to molecular genetics relied crucially on the structural study of abnormal hemoglobins. Abnormal hemoglobins gained such a central role for the study of molecular structure-function relationships that some concerned with the history of that research have claimed that hemoglobin, rather than DNA, had "made" molecular biology (Wintrobe, 1980, pp. 26–27, also Wintrobe, 1985, p. 353). Francis Crick has remarked that man is generally not "the ideal experimental animal", but "proved quite useful in molecular biology because of his hemoglobin, mainly because there are so many physicians looking at so many patients" (Crick, 1969, p. 178).

The new molecular practices aimed at describing and characterizing the various hemoglobins and the related pathological conditions did not translate into therapeutic breakthroughs. Rather, the laboratory success of sickle cell hemoglobin contrasted with the public failure to recognize the importance to combat the disease which affected mainly blacks (Michaelson, 1972). Despite this failure, since the 1950s, hematological textbooks heralded Pauling's and later Vernon Ingram's work on sickle cell anemia as path-breaking advances in the field. Their work confirmed reductionistic hopes of finding simple explanations for complex clinical pictures. It was used to point to the power of molecular approaches in the study of diseases and to argue for the importance of linking the clinic and the research laboratory.

RE-DESIGNING POSTWAR BIOLOGY

According to a much publicized story, Pauling embarked on the research project on sickle cell anemia following a casual conversation with William B. Castle, professor of medicine at Harvard Medical School in 1945 (Conley, 1980, p. 339).[3] Castle spoke of "a disease [Pauling] had never heard of before" (Rich in Olby, 1970, p. 82) and of the observation that the blood of the patients affected by this disease was birefringent. This, so the story goes, suggested to Pauling at once that the pathology was due to a modification of the hemoglobin molecule and he set out to show that this was the case.

Like all discovery stories, this one requires some analysis. Some important details can be recovered from an attentive reading of Pauling's own account (Pauling, 1970 and 1980). Castle and Pauling met because they were both members of the Medical Advisory Committee, one of four committees set up by Vannevar Bush, Director of the Office of Scientific Research and Development, to assist him in drawing up recommendations about governmental policies in medical and scientific research after the war.[4] The encounter occurred when Pauling, together with Beadle, was formulating a gigantic new research project for the life sciences at Caltech to be submitted to the Rockefeller Foundation (Kay, 1993, pp. 225–233). It was an attempt to re-design peacetime biological research trying to secure the support, the organization and new alliances of war-time research. The program was centered on the physico-chemical study of proteins, the structure of which was seen to be the key to understanding all life functions. Pauling had worked on the structure of hemoglobin before. He was interested in the structural basis of the sygmoidal oxygen dissociation curve of hemoglobin and had studied the bonds and orientation of the iron atoms. Also the contention that he had never heard of sickle cell anemia before has to be treated with caution. The pathology, first described by the American physician James Herrick in 1910, had presented a problem regarding the recruitment of black American soldiers during World War II, and some research on sickle cell anemia was pursued during the war.[5] Pauling served on the committee where these research contracts were discussed. Taken together these points indicate that the "chance discussion" fell on well prepared ground. It should also be remembered that a connection between the sickling phenomenon and hemoglobin was already postulated in the 1920s (Conley, 1980, p. 332).

To pursue his project Pauling enlisted three postdoctoral students, two of whom were medics and one a physical chemist. Trying to track down a difference in the hemoglobin molecule, Pauling's team resorted to the tools of choice for molecular analysis, the ultracentrifuge and the Tiselius electrophoresis apparatus. Pauling's laboratory was among the few which by the early 1940s was equipped with both a Tiselius apparatus and an ultracentrifuge. The use of these instruments required skill and experience. But by the late 1940s commercial prototypes of both instruments were available and an increasing number of laboratories engaged in molecular protein research.[6] Hemoglobin had already been studied by both methods. In Svedberg's early investigations with his instrument hemoglobin had represented a key case to suggest that proteins were not colloids, but molecules of a definite size, and to establish the ultracentrifuge as a tool for molecular analysis (Elzen, 1986, pp. 640–646).[7] In the early 1940s, a team at the Department of Pathology of the University of Pittsburgh, connected to the local Children's Hospital, had studied fetal and adult hemoglobin with both the ultracentrifuge and a modified Tiselius apparatus and had reported an electrophoretic difference between the two molecules (Andersch *et al.*, 1944). Pauling and his colleagues were probably aware of this work, even if they did not refer to it in their article.

The team at Caltech first subjected the hemoglobin samples of sickle cell anemic patients and of normal adults to ultracentrifugation. Unable to find any difference in the behavior of the hemoglobins with this approach, the investigators turned to electrophoresis. Under certain conditions a dramatic difference between the two molecules showed up. The team also reported that the hemoglobins of normal adult white and negro individuals were found to be indistinguishable (Pauling *et al.*, 1949, p. 544). This was a necessary control experiment, but it is also a reminder that blood was used to characterize races and that sickle cell anemia was a condition known to affect only blacks (Figure 6.1).

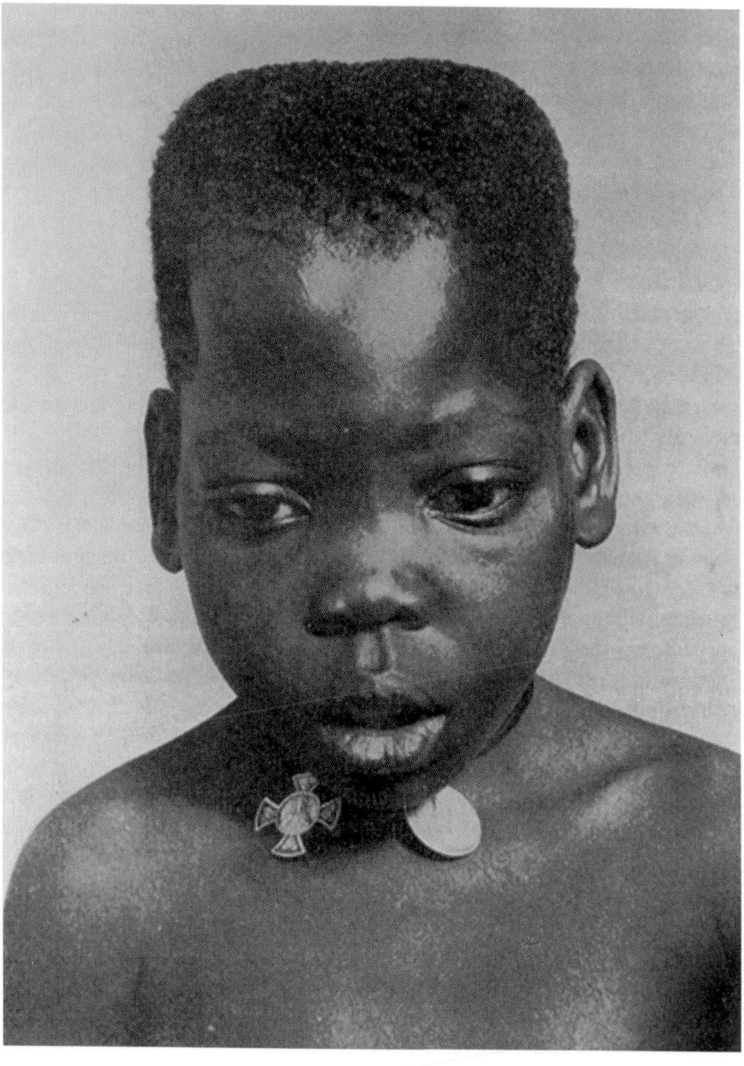

Figure 6.1. Child with sickle cell anemia. The bossing of the skull is one of the clinical symptoms of the disease.

In Pauling's own and the popular scientific press's enthusiastic promotion of the work, sickle cell anemia was to validate the claim that "life is basically an affair of molecules" and that basic molecular research would eventually benefit medicine (Gray, 1951, p. 59).[8] As Lily Kay has argued, this happened at a time when the Rockefeller Foundation was changing its vision of social control through molecular control to one of medical prevention (Kay, 1993).[9]

But the work on sickle cell anemia at Caltech also invites another reading. Well before molecular knowledge of proteins benefitted medicine, laboratory research mobilized medical resources. Pauling received the problem, the functional knowledge and the research material from the clinic. He relied on the collaboration of medics in his team and justified his project in medical terms. The work did not spur research on therapy, but a hunt for further abnormal hemoglobins, at Caltech and elsewhere.

For sickle cell hemoglobin, Caltech marked only the beginning of a long laboratory career. The functional and structural difference to normal hemoglobin turned sickle cell hemoglobin and other hemoglobin variants into important tools for molecular structure-function analysis. Here my story moves to Cambridge, England.

THE LABORATORY CAREER OF SICKLE CELL HEMOGLOBIN

Max Perutz had been investigating the crystallographic structure of hemoglobin in the Physics Department in Cambridge since before the war. He himself has often told the beginning of his research which became a life-long commitment (Perutz, 1987). A graduate in chemistry, he had moved from Vienna to Cambridge to study X-ray crystallography under Bernal in 1936. This was just a few years after Bernal and Dorothy Crowfoot (later Hodgkin), had taken the first X-ray picture of a crystalline protein, thus opening up a new field of research. Hemoglobin was suggested to Perutz as research object by Felix Haurowitz, then at Prague, for the interesting shift in crystal structure between the oxygenated and reduced form of the molecule which was observable under the microscope. The first crystals of horse hemoglobin were given to him by Gilbert Adair from the Physiological Laboratory in Cambridge.

In Cambridge there was an active network of researchers working on different aspects of hemoglobin. Much of this work was attributed to the "enthusiasm, energy and inspiration" of Joseph Barcroft who worked in the Physiological Laboratory (Roughton and Kendrew, 1949, ix).[10] Barcroft had done pioneering work on hemoglobin and the oxygen dissociation curve in the 1910s and had published the standard textbook on the subject (Barcroft, 1914).[11] During World War I he was appointed Chief Physiologist at Porton Down, the British defence centre for chemical warfare where he worked as the only civilian scientist on medical aspects of gas poisoning. Building on his earlier work he found that certain poison gases as well as shock led to oxygen deficit in the circulating arterial blood. His activities at Porton were thus tightly linked to his physiological research before the war. After the war Barcroft chaired the Medical Research Council Haemoglobin Committee, which was founded on his instigation in 1919 to encourage and sponsor research on hemoglobin.

Research on hemoglobin received a particular impetus during World War II, when much other research came to a halt. As Angela Creager has shown in her contribution to this volume, in World War II, blood and its products assumed scientific, medical, political and commercial importance. In Britain as in the United States a blood transfusion service was built up and research on the different blood components, on blood groups and other compatibility factors was pursued intensively (Mellanby, 1943). By the end of the war, no less than six Cambridge laboratories were actively engaged in hemoglobin research. An annual meeting was held under Barcroft's chairmanship to draw together the different workers in the field. The third such meeting planned for 1947 was turned into an international conference to commemorate Barcroft's death which had occurred in that year (Roughton and Kendrew, 1949).

Perutz was able to pursue his research on hemoglobin during the war, albeit with some interruptions, on a fellowship of the Rockefeller Foundation which Lawrence Bragg, the Cavendish Professor of Physics in Cambridge, was able to secure for him. It was one of the few lines of investigation which was continued in the Cavendish during the war mobilization.[12] By the end of the war, John Kendrew, a physical chemist who had served on high-level posts in operational research during the war and, on a mission in South East Asia, had been converted by Desmond Bernal to protein crystallography, joined Perutz as a PhD student. Again on the advice of Barcroft, he started working on the structural differences of adult and fetal sheep hemoglobin. Later he moved on to study the structure of myoglobin, the oxygen carrier in muscle, which presented a related, but less complex structure than hemoglobin. As with his first project, the expectation was that his work could provide clues for deciphering the highly complex X-ray diffraction picture of hemoglobin.

In 1947 Perutz and Kendrew's crystallographic work was put on a more secure footing when they were set up as a unit funded by the British Medical Research Council. Their work fell under the general heading of "biophysics", a field which attracted important funding after the war.[13] If we ask for a particular reason why the Medical Research Council funded the Cambridge unit, it should be remembered that a few years before, Dorothy Crowfoot Hodgkin and Barbara Low's crystallographic analysis had provided the conclusive proof for the structure of penicillin. Even if this knowledge was not immediately used for the synthetic production of penicillin, a program which was abandoned at the end of the war, the work marked an important achievement of X-ray crystallography and raised its recognition as a tool for structure determination of complex molecules. It will certainly have played an important role in the Council's decision to fund crystallographic work on other biologically and medically important molecules.

Once funding was assured, more researchers joined the Cambridge unit and the research projects diversified. Most famously, in 1953, Watson and Crick, working together in the unit, presented their model of DNA structure. But throughout the 1950s the structural study of proteins, in particular of hemoglobin and myoglobin, remained a central research topic of the unit.

In 1950 Perutz went on a first extended trip through the United States. Friedrich Eirich from the Polymer Institute of the Polytechnic Institute in Brooklyn, and Chandler Stetson of the Rockefeller Institute drew Perutz's attention to the "interesting problem" of sickle cell hemoglobin and provided some specimens which Perutz arranged to be sent to Cambridge (Perutz and Mitchison, 1950). For Perutz this represented a particularly important outcome of his journey.[14] Together with Murdoch Mitchison, a Cambridge zoologist, and Alfonso Liquori, a visitor in the lab, he tried to crystallize sickle cell hemoglobin in order to study the structural difference with normal hemoglobin. Again the hope was that the comparison of the two diffraction patterns would produce some clues for the resolution of the structure of the molecule. The three researchers succeeded in crystallizing the arterial form of sickle cell hemoglobin but found that it was indistinguishable from the normal hemoglobin. They failed to crystallize the venous (oxygen-free) form but discovered that its solubility was drastically lower than that of normal venous hemoglobin. This observation explained why sickle cell hemoglobin precipitated in the red blood cells on loss of oxygen and redissolved on oxygen uptake. The hope that the study of the hemoglobin variant would provide clues for the structure determination of the hemoglobin molecule, however, remained unfulfilled (Perutz and Mitchison, 1950; Perutz, Liquori and Eirich, 1951).[15]

In the mid-1950s, Anthony Allison, an Oxford geneticist, visited the Cambridge unit. His work was instrumental in providing evidence that carriers of sickle cell anemia were more resistant to malaria, a claim first advanced by Haldane for thalassemia (Haldane, 1949) and much less enthusiastically received than Pauling's work (Allison 1954).[16] Allison came to Cambridge because, at that time, the unit housed the best X-ray equipment for protein research and comprised the people most skilled in protein crystallography. Allison wanted to take X-ray pictures of sickled hemoglobin and prove its aggregated form. His experiments were unsuccessful, but he left some sickle cell hemoglobin behind. Considering the difficulties involved in securing blood samples of sickle cell anemic patients, this was a precious gift.[17] Both Max Perutz and Frances Crick, if for different reasons, suggested to Ingram, an organic chemist who had recently joined the research group, that he apply the protein fingerprinting techniques he had developed to the problem of the difference between sickle cell and normal hemoglobin.[18] While Perutz was mainly interested in the structural differences of the two molecules, Crick was eager to show that genes determined the amino acid sequence of proteins. If an inherited defect like sickle cell anemia resulted in a change of the amino acid sequence or, for that matter, of the composition of the respective protein, it would prove the point. Crick, together with Ingram, was investigating lysozymes of different sources to find evidence for the hypothesis, yet had not met with success (Crick, 1990, pp. 103–105). Earlier chromatographic studies performed in Pauling's lab did not show a definite difference in the amino acid composition of normal and sickle cell hemoglobin. Yet Ingram's refined technique of analysing protein digests by combining paper chromatographic and electrophoretic separation methods justified a new attempt.

Ingram's experiments resulted in the well-known conclusion that the difference of the two hemoglobins was due to the substitution of a single amino acid

(Ingram, 1956 and 1957).[19] This was the first direct experimental evidence that an inherited defect was in fact laid down in the amino acid sequence of a protein. Pauling's experiments had established a first link between the biophysical properties of hemoglobin and an inherited disorder. But only Ingram's experiments turned hemoglobin variants into a powerful tool for genetic analysis on a molecular level and more generally for molecular structure-function relationships. Crick reckoned that Ingram's experiments also did most in stirring the interest of protein chemists in genetics (Crick 1990, p. 106).

The combination of structural, biochemical and genetic approaches became central for the definition of molecular biology and for the local research culture which was developed at the Laboratory of Molecular Biology in Cambridge, the negotiations for which started just around that time. Abnormal hemoglobins became a crucial research tool for the work pursued in the unit. Samples of the hemoglobins used by the researchers in Cambridge came from Hermann Lehmann's "capacious freezer" at St. Bartholomew's Hospital in London (Ingram, 1989, p. [148]).

COLLECTING AND CATALOGUING

As a fellow-hematologist remarked, Hermann Lehmann was "more than anybody responsible for cataloguing the enormous wealth of variations in the hemoglobin mutants and setting our base of understanding of the population genetics, anthropology, and clinical diversity of the structural variants".[20]

Lehmann, a refugee from Germany, had a medical degree from the University of Basel (1934) and a PhD in Biochemistry from Cambridge (1938). His interest in blood diseases and his first experience with surveys for epidemiological studies dated from his service time in the Royal Army Medical Corps in India during the war. While still in Cambridge, he had investigated the relation between chlorophyll and iron in green leaves. His experience with the handling of organic iron complexes caused him to be released from general medical duties and to be drafted to the General Headquarter (India Command) Anemia Investigation Team to help in the investigation of the hypochromic anemia that commonly affected the troops. Released from military duties, he spent three years in Uganda as Colonial Medical Research Fellow for Malnutrition and Anemia. Lehmann's work involved screening hundreds of African school children and adults to investigate the causes of various forms of anemia and their relation to other diseases. In the course of these investigations, Lehmann and his collaborator observed that sickle cell trait was equally distributed in anemic and non-anemic Africans (Lehmann and Milne, 1949). In a second paper published in the same year, Lehmann together with Raper, government pathologist in Kampala, was the first to show that different tribes and races in Africa had different incidences of sickle cell trait.[21] The authors pointed out that the chronic anemia developed by a few of those individuals with sickle cells in their blood, rather than the inherited abnormality itself, had claimed the attention of most medical investigators.[22] Although also their interest was primarily medical, they reckoned that some of their findings had a bearing on anthropology. For this reason, they had chosen to present their results in *Nature*. They interpreted the possession of the sickle cell trait as an additional physical character which allowed them to distinguish certain tribes which had been classed together on a linguistic basis (Lehmann and Raper, 1949). Lehmann's first papers on sickle cell trait and its distribution in African populations appeared in the same year as Neel's paper on the inheritance of sickle cell anemia and as Pauling's findings on sickle cell hemoglobin.

On his return to Britain, Lehmann was first appointed consultant pathologist in the District General Hospital, Pembury, Kent and then lecturer in chemical pathology at St. Bartholomew's Hospital Medical College in London. In his London years (1951–1953), Lehmann developed his interest in sickle cell anemia, sickle cell trait and sickle cell hemoglobin and expanded it to embrace other abnormal hemoglobins. A crucial part of this endeavour was the development of appropriate analytical techniques.

In his first surveys in Uganda, Lehmann had determined the incidence of sickle cells by observing a blood drop under a coverslip sealed with vaseline for 48 hours. This method was described by Victor Emmel in 1917.[23] Very soon, however, Lehmann adopted a new method recently described by two American researchers. This test consisted in incubating a drop of blood with a drop of an oxygen-consuming broth culture of coli bacteria. Sickling was demonstrable in five to ten minutes. This quick bacteriological test greatly facilitated survey work (Lehmann and Raper, 1949). Searching for a still better alternative, Lehmann, on his return from Africa, visited Rose Schneider in Galveston, Texas, to discuss with her the possibility of identifying sickle cell hemoglobin containing red blood cells immunologically (Dacie, 1988, p. 415). Immunological techniques were commonly used for blood typing. The tests were quick and easy to perform and highly specific. However, it turned out that the different hemoglobins were too similar to be distinguished by immunological methods.[24]

Electrophoresis had proved a powerful tool for the search of new hemoglobin variants. The Tiselius apparatus, however, even if commercially available since after the war, was still a cumbrous and costly machine (Kay, 1988). The development of low-cost paper chromatographic and electrophoretic methods offered new opportunities (Figure 6.2).

Lehmann was introduced to the technique of paper electrophoresis by Elspeth Smith, a graduate biochemist and research assistant in his department at St. Bartholomew's. Smith was applying the technique to the separation of serum proteins, but it could be applied with equal success to the separation of hemoglobins. In the note they published on the subject, Lehmann and Smith did not fail to point out that the separation technique, which could run on battery power, could be carried out in any laboratory in the tropics (Lehmann and Smith, 1954) (Figure 6.3). In 1954, Lehmann also visited Pauling's laboratory at Caltech on a Rockefeller Foundation Travelling Fellowship to learn directly

Figure 6.2. Simple equipment for filter paper electrophoresis.

Figure 6.3. Laboratory equipment used in survey of abnormal hemoglobins in Western Uganda. The use of high tension batteries permitted survey work to be carried out without mains electricity.

from the experience of the American researchers with separation techniques (Dacie, 1988, p. 432). At Caltech, especially Harvey Itano was still active in the field of hemoglobin studies, while concurrently serving as surgeon at the National Cancer Institute.

By that time several other abnormal hemoglobins had been identified, with Itano leading the field. Soon Lehmann made his own first contribution announcing the finding of a new hemoglobin variant in the course of a survey of blood samples from West Africans (Edington and Lehmann, 1954). The new hemoglobin, named hemoglobin G according to the convention of designating hemoglobins by letters of the alphabet in order of their discovery, was described by its electrophoretic mobility (Figure 6.4). This was only the first of a long series of hemoglobin variants identified by Lehmann on surveys he conducted in all parts of the world (Figure 6.5).

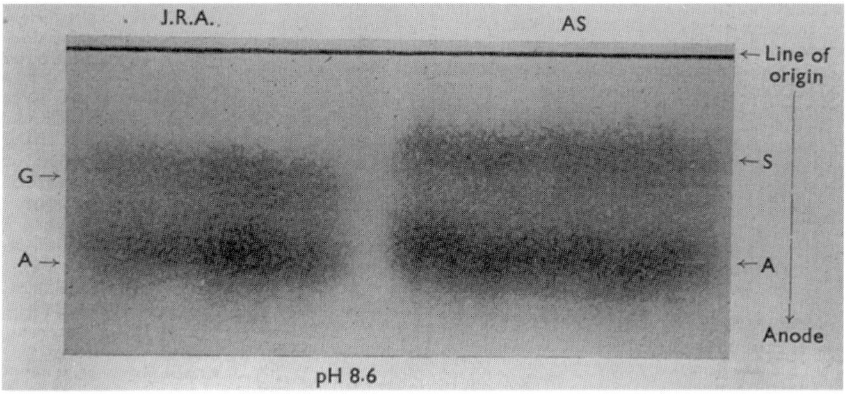

Figure 6.4. Filter-paper electrophoresis of blood samples of the West African J.R.A. and of a sickle cell trait carrier (A and S hemoglobin).

If his first survey work was conducted with the aim of identifying the causes of widespread anemic disorders in the Indian troops and in African populations, his later surveys inscribed themselves in the tradition of physical anthropology.[25] Since the work of the Hirszfelds in the 1910s, physical anthropologists investigated the blood groups and serum proteins of human populations.[26] Especially in the 1950s Lehmann collaborated closely with the Blood Transfusion Research Unit and the Serological Population Genetics Laboratory in London, and a blood-group serologist from one of these groups accompanied him on some of his trips.[27] For his studies on the distribution of hemoglobin variants in different world populations Lehmann is considered the founder of molecular anthropology. However, he never lost clinical interests from sight. Significantly, in the publication in which he, together with his collaborator Richard Huntsman, summarized his investigations on hemoglobin, the anthropological studies were presented side by side with a discussion of the symptoms and treatment of diseases

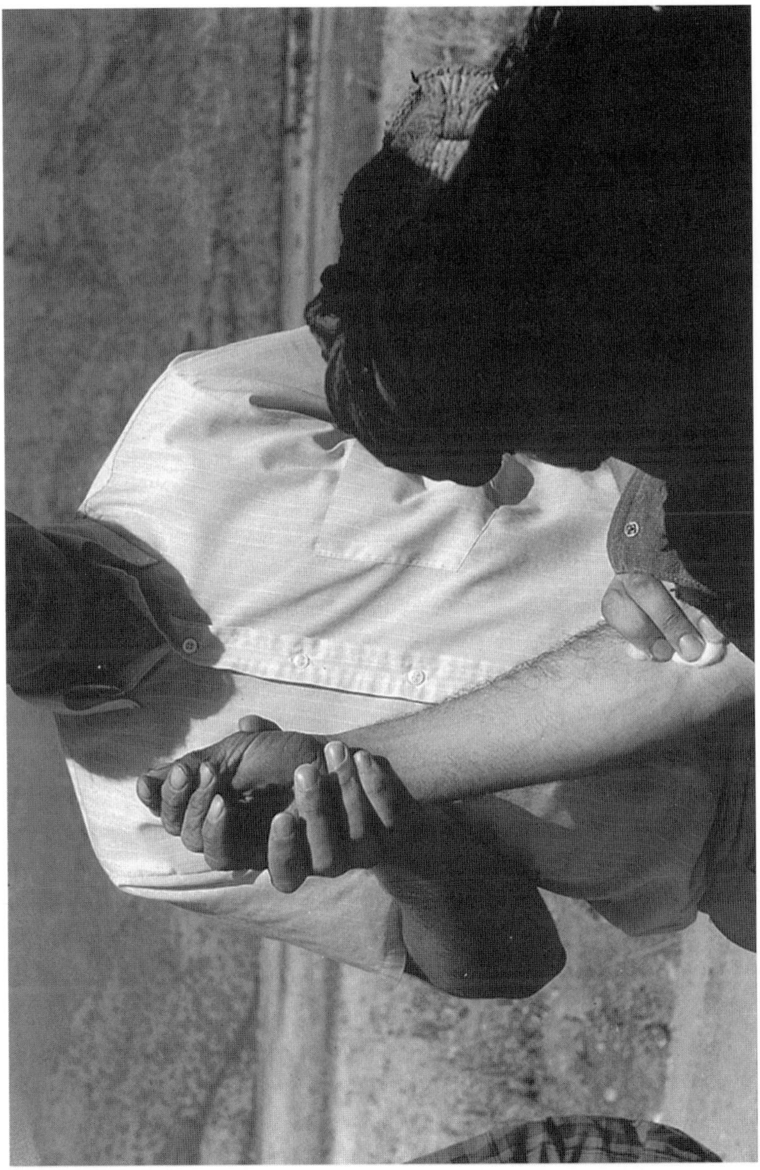

Figure 6.5. Blood sampling in a survey in Kurdistan, 1969.

connected to abnormal hemoglobins and with the laboratory methods used for their identification (Lehmann and Huntsman, 1966). Lehmann and Huntsman were also instrumental in setting up an abnormal hemoglobin unit in Zambia. This unit provided a diagnostic service for sickle cell anemia for hospitals and clinics in Zambia and beyond, and maintained the only comprehensive sickle cell anemia clinic in the country. In addition, the staff of the unit accomplished survey work for sickle cell hemoglobin and other abnormal hemoglobins (Barclay and Noak, 1970).[28] This institution was in many ways exemplary for the kind of interventions Lehmann was promoting more generally in countries where hemoglobinopathies represented an important public health problem, even if it was not always perceived as such. As chairman of various WHO scientific expert groups on the subject he tried to alert public opinion and governments to the problem and to convince them to make money available for laboratory equipment, training courses for medical personnel and the necessary infrastructure.[29]

Lehmann had started collecting and characterizing blood samples since his survey work in Uganda. Only from 1954 onwards, however, did he characterize the samples by the electrophoretic behavior of their hemoglobins. The characteristic electrophoretic mobility was indeed the most important, if not the only difference between the hemoglobin variants, and the only means to identify and separate them in the laboratory (Figure 6.6). The talk of "slow" and "fast" hemoglobins highlights the extent to which electrophoresis defined the different hemoglobins. Some hemoglobins, for instance hemoglobin S and D, showed the same electrophoretic behavior, but could be distinguished through the sickling test or through their different solubility.

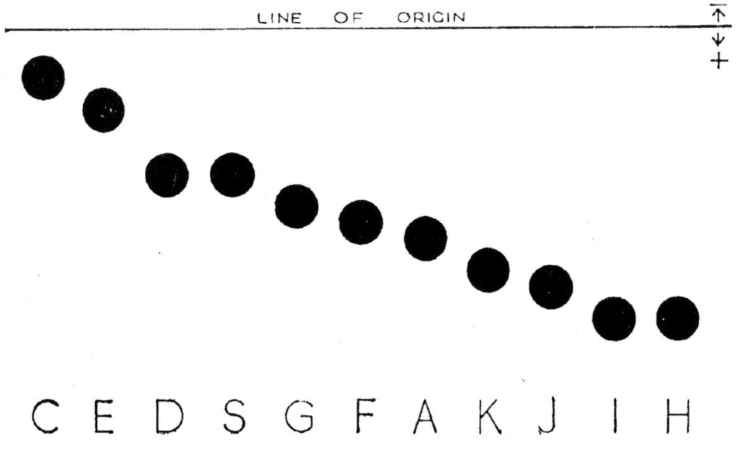

Figure 6.6. Scheme of electrophoretic mobility of variants of human hemoglobin.

Hemoglobin variants came to define various forms of anemic disorders, referred to as hemoglobinopathies, or, more specifically, as Hemoglobin C, D, E, etc. disease. These syndromes, together with detailed descriptions of the laboratory techniques for their identification, made their entrance in clinical textbooks in the mid-1950s.[30] A few years later electrophoresis was widely used in hematological laboratories.[31]

"Abnormal" hemoglobins were inherited, as ascertained by family studies, even if rare cases of spontaneous mutations seemed to occur. Yet not all "abnormal" hemoglobins caused anemia, particularly not in heterozygotes for an abnormal gene.[32] To include all conditions Lehmann suggested the term "hemoglobinoses", some of which would then be "hemoglobinopathies" (Lehmann, 1957, p. 5).

Lehmann's collecting activity had started on an informal basis. In 1963 it was institutionalized more formally, when he became head of a newly founded Medical Research Council Unit for Abnormal Haemoglobins established at St. Bartholomew's Hospital Medical College in London. This recognition of his work was soon followed by his move back to Cambridge, where he was first appointed University biochemist to Addenbrooke's Hospital and later became first professor of clinical biochemistry at Cambridge. The Medical Research Council unit, now transferred to Cambridge, combined research on abnormal hemoglobins with clinical service work. The latter included clinical diagnosis and reference services but also extended to advice on treatment and genetic counselling.[33] In 1965 the World Health Organisation designated the unit as an International Reference Centre for Abnormal Hemoglobins. By that time Lehmann's collection counted more than one hundred abnormal hemoglobins.[34]

For the Medical Research Council Lehmann always had to explain again that the identification of such a wealth of hemoglobin variants was not "merely a matter of collecting and cataloguing".[35] In his second progress report in the early 1970s, Lehmann conceded that the long parade of new hemoglobin variants may induce the reader to think that "this is a pursuit resembling more that of a factory rather than of a scientific laboratory."[36] To counter this impression, Lehmann pointed to the use of the collection of abnormal hemoglobin variants as a research tool. This, more than its use as a reference tool for medical diagnosis or population surveys, justified the enterprise for the Medical Research Council. Legitimation, however, worked also in the opposite sense. The researchers using hemoglobin variants to study structure-function relationships justified their work in terms of its medical implications. In addition, the identification and description of the abnormal hemoglobins relied on molecular technologies. To the extent to which these descriptions were used to classify diseases, molecular technologies were introduced, if not in the cure of patients, into the work of clinical pathologists. Thus, before the advent of therapies based on the new molecular technologies, Lehmann's collection of abnormal hemoglobins became an important tool for the circulation of knowledge and technologies between the clinic and the laboratory and for the construction of a common laboratory culture. Insofar as the collection mediated between different professional groups which, even if sharing tools and practices, made different uses of them, we can also refer to it as "boundary object" (Star and Griesemer, 1988).

MOLECULAR PATHOLOGY

After the successful determination of the structural difference between sickle cell and normal hemoglobin, Ingram resolved to stick to the tools of his success. Applying his fingerprinting techniques to other hemoglobin variants, Ingram hoped to explore the structural and genetic relationships of other abnormal hemoglobins. As mentioned already, Lehmann's extensive collection of blood samples provided the material for his experiments. The exchange of material, however, marked only the beginning of a productive collaboration between Ingram and Lehmann. Ingram initiated Lehmann, who was much impressed by Ingram's "brilliant" work, in the techniques of fingerprinting (Lehmann, 1959, p. 959). Together with Seymour Benzer, visiting from the United States, they published a paper on the fingerprints of hemoglobin D, which the new technique showed to exist in at least three varieties, each one carrying a chemical change in a different part of the hemoglobin molecule (Benzer, Ingram and Lehmann, 1958).[37]

These findings exacerbated the already existing problems regarding the naming conventions for abnormal hemoglobins. They did, however, also point to a possible solution to the nomenclature problem. The convention of designating hemoglobins by letters of the alphabet in order of their discovery was viable as long as variants of the molecule were discovered in small numbers. The discovery of ever more hemoglobins by different workers in the field had created several problems. The experience that new techniques could demonstrate that hemoglobins thought to be identical were in fact different, presented a further complication. In addition, Ingram's work on the point mutation of sickle cell anemia hemoglobin suggested that there was an almost indefinite number of permutations and possible variants of hemoglobins — far outnumbering the letters of the alphabet. Lehmann now suggested that eventually it would be possible to designate the hemoglobins by their amino acid sequence (Ager *et al.*, 1958, p. 6). From the mid-1960s onwards it became, indeed, common practice to use fringerprinting to identify new hemoglobin variants and to designate them by the amino acid exchange which characterized them. Thus, in 1965, Lehmann and his collaborators announced the finding of "a new hemoglobin, D Ibadan (beta-87 threonine — lysine)" (Watson-William *et al.*, 1965).[38] By that time the complete amino acid sequence of hemoglobin had been worked out by researchers in München, New York, Pasadena and Cambridge, Massachusetts.

Lehmann's use of fingerprinting techniques was not a one-sided exchange, or merely the application of a new laboratory technique to the cataloguing of hemoglobin samples and to clinical diagnostics.[39] Rather, the extensive knowledge of actual amino acid substitutions in human hemoglobins, now integrated into the collection, suggested new uses of the collection as a molecular tool.

An important demonstration came in 1965. On the basis of fifty-nine known substitutions in hemoglobin variants, it could be proved that the genetic code which had been established for bacteria and viruses was also valid for humans. All amino acid substitutions proved to be the outcome of a single mutation in the coding gene. At first there appeared to be one exception to this rule. However, on re-examination, and as predicted by Francis Crick, it was found that an error had occurred in the determination of the amino acid substitution (Beale and Lehmann, 1965). This was the first time the genetic code was used as an aid in protein analysis. The study of hemoglobin variants gave further important clues on mutation sites and on the structure and regulation of genes.[40] It was the only mammalian system which lent itself to the study of genetics on a molecular level.

Lehmann's extensive knowledge of amino acid substitutions in hemoglobins combined with his clinical knowledge of the various pathologies was put to new use once Perutz had built his first atomic model of (horse) hemoglobin in the oxygenated form (Figure 6.7).[41] For Perutz "everything fell into place" when the known amino acid substitutions could be built into the model and the structural changes could be interpreted in functional terms, on the basis of the known clinical characteristics of the different hemoglobin variants.[42] Twenty years after Pauling had introduced the notion of a "molecular disease", Lehmann and Perutz published a paper entitled "Molecular Pathology of Human Hemoglobins" (Lehmann and Perutz, 1968). Lehmann and Perutz showed that the hemoglobin molecule was insensitive to replacements of most amino acid residues on its surface, but very sensitive to any small alteration of internal non-polar contacts. Replacements near the heme groups and at the contacts between the different subunits which composed the hemoglobin molecule affected respiratory functions. The two authors displayed their findings in long tables relating clinical symptoms and abnormal hemoglobin properties to structural effects of amino acid replacement as well as in detailed stereochemical drawings.[43] While the title of the paper announced that clinical symptoms could be explained in molecular terms, the reverse was also true: being able to relate clinical symptoms to atomic structure, Perutz gained decisive clues on the normal mechanics of the molecule. The difference in function known from the clinic furnished the knowledge which was necessary for correlating structure to function, but which structural work alone could not provide.

Once a model of deoxyhemoglobin was also available, Perutz and Lehmann were able to offer a more detailed stereochemical interpretation of the altered oxygen affinity of some of the hemoglobin variants. Again, manipulation of the model with known amino acid substitutions, combined with the knowledge of clinical symptoms, offered important clues for the understanding of the cooperative mechanism of hemoglobin function (Morimoto *et al.*, 1971).[44] From then on the values of oxygen affinity were routinely incorporated in the description of new hemoglobin variants.[45] Perutz's hemoglobin model was used extensively by all members of Lehmann's unit to relate changes in the amino acid sequence of

Figure 6.7. Atomic model of horse oxyhemoglobin used by Perutz and Lehmann for their work in 1968.

hemoglobin to the three-dimensional structure of the molecule. Lehmann himself seemed to be "completely at home in the molecule, as though in the architecture of a large and complicated building".[46]

It should be noted that Perutz's work on the molecular mechanism of hemoglobin function was contested. A point which aroused particular concern was the fact that most of the functional data derived from analysis of mutant hemoglobins (Moffat, 1980–1981). Another point which is rarely mentioned is that for the particular case of sickle cell hemoglobin, the exact location in the three-dimensional model did not suggest straightaway a molecular explanation for the dramatic physiological effect of the mutation (Perutz, 1971). The exact molecular mechanism of the sickling of red blood cells is still debated today. It certainly presents itself as a much more complicated process than the crude mechanical explanation originally proposed by Pauling and his team.[47]

Perutz nevertheless rejoiced that what he had done had become medically important.[48] Yet Perutz and Lehmann's joint paper closed on a more sober note. "The data presented here", the authors concluded, "do not hold out any hope that the lesions in mutant protein molecules could be repaired directly" (Lehmann and Perutz, 1968, p. 909). According to the authors genetic repair was imaginable, but still utopian. Less remote seemed the possibility to replace the organ which produced the abnormal hemoglobin. For the time being, however, the only hope lay in preventing conception of homo- or also heterozygotes carrying strongly pathogenic mutations, in other words genetic counseling.[49] Knowledge of the molecular mechanism of hemoglobin function was not necessary to this end.

BUILDING A COMMON LABORATORY CULTURE

I have argued that Lehmann's collection of hemoglobin variants was an important mediator for the exchange of tools and practices between the clinic and the laboratory. It may be objected that what I have presented is an asymmetrical account which pays much more attention to the uses of the collection by laboratory scientists. This may be an artefact of the sources I have consulted or which are generally more readily available. However, it is also true that only a few of the hemoglobin variants produced clinical effects which presented a significant problem of public health.[50] These, we have seen, have for a long time been scandalously neglected. When they were finally acknowledged, their management was often doubtful and aroused many misgivings (Michaelson, 1972 and Duster, 1990). Twenty years after Pauling had heralded sickle cell anemia as the first molecular disease and as a model case for the application of biological knowledge to medicine, he publicly suggested that a symbol tatooed on the forehead of the person should indicate the possession of the sickle cell gene. This measure would help young people carrying the same sign not to fall in love with one another and not to procreate children. Thus no children with the disease would be born (Pauling, 1968, p. 269).[51]

Therapies based on the new molecular technologies used to classify hemoglobin variants are still not in sight. Seen in this light the emphasis of my account on the use of the hemoglobin collection by research scientists is not an artefact, but an important indicator of how resources and interests were divided and used and of the limitations of molecularization as a strategy. Rather than providing fundamental knowledge for medical applications, laboratory researchers used clinical interests and resources to promote their own work. This may not come as a surprise, even if in the history of molecular biology this point is not often made. In the decades following World War II, molecular biologists like nuclear physicists and other branches of science profited from the appeal of fundamental research and this rhetoric has found its prolongation in the dominant historiography of the field.

But new molecular techologies were incorporated in the collection. This we have seen for paper electrophoresis, for fingerprinting and for Perutz's hemoglobin model to which variants were related. The very growth of the collection depended on the incorporation of new technologies which made finer differentiations possible. And the collection was not only used as a research tool, but also as a reference tool for clinical diagnosis. In fact, many of the variants derived from blood samples which were sent for advice from other clinical laboratories to the unit housing the collection. Despite his collaborations with "fundamental" researchers, Lehmann himself never lost clinical concerns from sight. He was always interested in working out the clinical effects of the abnormal hemoglobins. Part of the "fascination" of the study of abnormal hemoglobins for Lehmann consisted exactly in the variable clinical manifestations to which they gave rise. (Lehmann and Huntsman, 1966, p. 263).[52] He continued seeing patients and had an important share in devising and assessing treatments for sickle cell anemia, even if these were not targeted at the molecular level. He was also personally involved in campaigns to alert public opinion and governments to the public health problem of some hemoglobinopathies. With the exception of his collaborative papers with Ingram, Perutz and other research scientists, he mainly published in medical journals. In fact most of his publications represented a mixture of laboratory data, clinical descriptions and family analysis. In this way Lehmann's work (like that of others who worked in the field of hematology) and his collection represented a continuous articulation of the clinic and the laboratory. Passing from the sickling test to incorporating a number of new molecular technologies, the collection became itself a tool for molecular investigations. Using it for the classification and diagnosis of diseases, pathologists and medics participated in the same molecular research culture. The study of the distribution of abnormal hemoglobins, finally, also molecularized the study of human populations.

In the mid-1970s, Lehmann's unit was disbanded. By that time, electrophoresis and fingerprinting of hemoglobins had become standard technologies and did not justify any longer the existence of a special research unit.[53] The study of abnormal hemoglobins and the term molecular pathology it inspired, were celebrated as having shown the whole field of medicine the way into the 21st-century (Bernard, 1992, pp. 9 and 50–51). But long before the advent of new therapies based on molecular interventions or genetic medicine, research on abnormal hemoglobins had served as a model system for an understanding of the mechanism of inheritance and for molecular structure-function relationships which defined the burgeoning new field of molecular biology.

ACKNOWLEDGMENTS

Earlier versions of this paper were presented at the conference on "Molecularizing Biology and Medicine" in Cambridge (July 1994), at the session on "Post-World War II Biomedicine" at the meeting of the International Society for the History, Philosophy and Social Studies of Biology in Leuven (July 1995) and in a seminar of the INSERM U158 in Paris (June 1996). I would like to thank the participants for many useful comments and especially for helping me to develop the clinical part of my argument. Surely much more remains to be done. Special thanks go to Jean-Paul Gaudillière, Harmke Kamminga, Ilana Löwy, Paolo Palladino and John Pickstone, who have all commented on written drafts of the paper. I gratefully acknowledge support for this work by the Wellcome Trust.

NOTES

1 The description of the first abnormal hemoglobin after sickle cell hemoglobin resulted from a collaboration between Pauling's colleague Itano and Neel. As a medical researcher Neel had access to patients, but had no ready access to a Tiselius apparatus which had proved a powerful tool for hemoglobin studies. Soon after reading the publication of the Caltech group, Neel suggested to Itano that they join forces. The new hemoglobin found in one of Neel's patients was named hemoglobin C for being the second type of abnormal hemoglobin to be known, the letter "A" being reserved for the normal hemoglobin (Neel, 1994, pp. 45–46; Itano and Neel, 1950). Neel also contributed to the study of the distribution of sickle cell hemoglobin in African populations (Neel, 1994, pp. 46–54). Before embarking on his studies on sickle cell anemia, Neel was the leading geneticist in the Atomic Bomb Casualty Commission genetic project in Japan.

2 In contrast to other contributions to this volume, this chapter deals with developments more directly relating to the establishment of molecular biology as a new disciplinary formation. As I argue elsewhere, in Britain molecular biology was defined as an independent research field only in the late 1950s (de Chadarevian, 1994). The first institution (worldwide) to carry the name was the Medical Research Council Unit of Molecular Biology in Cambridge which adopted it in 1957. Until 1962 the unit was housed in the Physics Department in Cambridge and the work pursued in the unit was generally referred to as "biophysics", a loosely defined field

which attracted important funding and carried a strong cultural appeal in the postwar era. Note that many investigations which later were regarded as key developments of molecular biology, most prominently Watson and Crick's presentation of the double helical structure of DNA, took place before molecular biology was defined as an independent research field, but were retrospectively re-appropriated by molecular biologists.

3 According to Castle's recollections, his encounter with Pauling took place on an overnight train between Denver and Chicago (Conley, 1980, p. 339). Pauling recalled that his conversation with the Harvard professor of medicine occurred in the Century Club in New York (Pauling, 1970, p. 1011 and 1980–1981, p. 4). Both reconstructions stress the fortuitous character of the encounter.

4 The work of these committees led to the publication of Bush's *Science, the Endless Frontier* (1945).

5 On sickle cell anemia as a "true 'discovery' of the 1920s", especially regarding the hereditary and racial characterisation of the disease, see Wailoo (1991). Research on sickle cell anemia since 1910, including Pauling's contribution, is critically reviewed in Feldman and Tauber (in press). According to the authors, advances in protein research and especially Beadle and Tatum's one gene-one enzyme hypothesis set the stage for the molecular understanding of the disease and its genetics by Pauling and Neel.

6 On the commercialization of the ultracentrifuge and the Tiselius apparatus after World War II see Elzen (1986) and Kay (1988, pp. 68–69).

7 On the dispute between molecularists and colloidalists in the interwar period see Creager's contribution to this volume.

8 Gray, America's most prominent science writer, was on the Rockefeller Foundation staff.

9 On the success of Pauling's notion of molecular disease with biomedical researchers and in the press see Paul and Edelson (this volume).

10 On Barcroft see Roughton (1949).

11 In his investigations Barcroft had shifted from working with blood cells to using hemoglobin solutions because they were easier to handle. Only when Svedberg subjected hemoglobin to tests in his ultracentrifuge, however, did hemoglobin become a substance with a definite molecular weight.

12 On Perutz's wartime experience as "enemy alien" and his later involvement in the mission "Habakkuk", see Perutz (1989, pp. 101–136). The pursuit and conditions of his hemoglobin research during and just after the war can be gleaned from Bragg's correspondence with the Rockefeller Foundation and from Foundation officers' diary entries on their visits to the Cavendish; see folder 561–564, box 43 and folder 565–570, box 44, series 401, Record Group 1.1., Rockefeller Foundation Archives, Rockefeller Archive Center, North Tarrytown, New York.

13 The first and biggest biophysics group receiving funding from the Medical Research Council was John Randall's group at King's College, London.

The complex negotiations leading to the Medical Research Council's commitment to Randall's project reveal both the difficulties and the opportunities for postwar research proposals in a situation in which new research budgets were allocated and competencies and policies still had to be defined. On biophysics in postwar Britain, see de Chadarevian (1994).

14 Perutz to Pomerat, 18 May, 1950; folder 566, box 44, series 401, RG 1.1, Rockefeller Foundation Archives, Rockefeller Archive Center, North Tarrytown, New York. Perutz arranged for further samples of sickle cell blood to be sent to him from Ibadan in Nigeria. On the costs involved in organizing the transport in ice-filled thermos flasks, see letter of 20 October 1950 from Perutz to the Accounts Branch of the Medical Research Council; file FDI/426, Medical Research Council Archives, London.

15 According to Judson's reconstruction, Perutz's "diversion" into sickle cell anemia proved nonetheless crucial. It brought Perutz a letter from Austen Riggs with information which indirectly suggested to him a new and eventually successful way of pursuing his investigations of hemoglobin structure (Judson, 1979, p. 551).

16 Haldane's suggestion as well as Allison's work on the distribution of sickle cell anemia were based on the notion of "balanced polymorphism", a key term of the Oxford School of genetics founded by E.B. Ford, with whom Allison was closely connected. Evolutionary explanations of this kind complicated the neat connection of gene and disease postulated by Pauling with his notion of sickle cell anemia as a "molecular disease" (Kay, 1993, p. 259). The postulated link between malaria and the distribution of the gene for sickle cell hemoglobin made it also necessary to rethink the anthropological considerations based on the geographical distribution of the gene (Edington and Lehmann, 1956). On the Oxford School, its expansion to Liverpool and its contribution to blood group research and medical genetics in Britain see Zallen (in press). To back up his findings gained in population surveys and in directed tests of African children, Allison artificially induced malaria in two comparable groups of fifteen African volunteers, experiments which he later professed to regret (Allison, 1954 and Kevles, 1985, p. 363).

17 When Ingram's research on the chemical difference of sickle cell hemoglobin started to show promising results (see below), Crick wrote to Sydney Brenner, who had moved back to South Africa, asking for fresh supplies of sickle cell anemia blood (Judson, 1979, p. 328).

18 Ingram had refined techniques first developed by Fred Sanger in the Biochemistry Department in Cambridge. For the uses of Sanger's protein sequencing and fingerprinting techniques by protein crystallographers and molecular geneticists in Cambridge and the interactions between these groups see de Chadarevian (1996). For Ingram's own account of his research on sickle cell anemia which turned into a lifelong commitment to the study of abnormal hemoglobins, see Ingram (1989).

19 Later Ingram reckoned that he had been lucky that the substitution of glutamic acid with valine in sickle cell hemoglobin was one which showed up in the fingerprint. Substitutions which did not result in a charge difference were much more difficult to trace in a fingerprint; interview with V. Ingram, Cambridge, Mass. 20 July, 1993.

20 David Weatherall, quoted by Dacie (1988, p. 431).

21 This work was done independently of considerations regarding genetic polymorphism, but later provided indirect support for the hypothesis that the incidence of the sickle cell gene in various populations was linked to the degree of endemic malaria. The surveys could thus be used for both epidemiological and anthropological or population genetic considerations (see below).

22 When Lehmann first came to Africa all patients who were anemic and showed a positive sickling test were classified as having sickle cell anemia. Lehmann's work helped to distinguish between sickle cell trait and sickle cell anemia. The diagnosis of sickle cell trait in apparently healthy, i.e. non-anemic, individuals could only be made through systematic blood sampling. The diagnosis was nonetheless medically important. I am not aware of disputes between more clinically orientated medics and hematologists regarding the diagnosis of carriers of the sickle cell gene. Hematologists themselves were traditionally very laboratory orientated and used to base their classifications on signs of non-pathological conditions.

23 On the role of Emmel's test in the making of sickle cell anemia and in establishing laboratory techniques as crucial diagnostic tool in clinical practice see Wailoo (1991).

24 In the early 1980s, radioimmunological methods to identify hemoglobin variants were successfully devised. On Schneider's collaboration with Lehmann and her contribution to hemoglobin research, see Schneider (1981).

25 Besides blood slides and samples, the documentary material brought home by Lehmann from his trips included portrait photographs of people from the tribes he was surveying. Unfortunately this material, which presents rich evidence of the material culture of the surveys, is only very partially preserved and very often not inscribed or dated. Some of this material is currently held at the Wellcome Unit for the History of Medicine in Cambridge.

26 On the change from the cephalic to the biochemical index in physical anthropology and the eugenic uses of these new classification system, with special reference to France, see Schneider (1990, pp. 208–229). According to Schneider, in the long run, the study of blood groups was crucial for moving physical anthropology away from comparative anatomy towards the study of gene frequencies.

27 A.E. Mourant, the director of the Serological Population Genetics Laboratory, was also director of the Blood Group Reference Laboratory housed at the Lister Institute in London. In 1952, the Laboratory was des-

ignated as International Blood Group Reference Laboratory by the World Health Organisation. See below for a parallel development of Lehmann's research unit.

28 The operation of the unit and the way it was viewed by the local population calls for further investigation. For an account of indigenous perceptions of conditions relating to sickle cell anemia see Edelstein (1986).

29 See e.g. *Hemoglobinopathies and Allied Disorders.* World Health Organisation Technical Report Series No. 338 (Geneva: World Health Organization, 1966) and *Treatment of Hemoglobinopathies and Allied Disorders.* World Health Organization Technical Report Series No. 509 (Geneva: World Health Organisation, 1972).

30 Wintrobe introduced a new chapter on abnormal hemoglobin syndromes in the fourth edition of his authoritative textbook (Wintrobe, 1956). Others followed suit; see e.g. Bernard and Bessis (1958); Gruchy (1958) and Britton (1963). Bernard and Bessis stressed emphatically that the study of abnormal hemoglobins had opened up and clarified an important chapter in hematology. Their book addressed specifically medics who were not specialized in hematology.

31 More specialized hematological laboratories, however, were rare above all in many African countries where hemoglobinopathies and especially sickle cell anemia represented a major public health problem. Proper diagnosis was important to distinguish inherited forms of anemia from those due to malnutrition, malaria or intestinal parasites. Still in the mid-1970s Lehmann lamented: "We are with the hemoglobinopathies still at the stage of diagnosis where the medical and auxiliary professions are prepared to use microscopes but are still overawed by chemical techniques — indispensable in the study of hemoglobinopathies — such as electrophoresis. It took some time to convince governments that a simple electrophoretic setup costs as little or less than a small microscope" (Schmidt *et al.*, 1974, p. 133). However, also in Western countries there was a long time gap between the introduction of electrophoresis in clinical laboratories and the recognition of hemoglobinopathies, especially sickle cell anemia and thalassemia, as public health problems in certain countries.

32 With the number of known hemoglobin variants growing, Lehmann became convinced that many of these had to be considered "unusual" rather than "abnormal" (Watson-Williams *et al.*, 1965, p. 7). However, it was also reported that apparently symptom-free carriers of abnormal hemoglobins showed increased susceptibility to certain drugs or other diseases. Hemoglobin variants which also in homozygotes did not lead to (apparent) clinical symptoms, as in the case of hemoglobin G, could only be identified through population surveys. However, additions to Lehmann's collection came also through samples of patients with unusual anemic disorders sent to the laboratory for analysis.

33 See "Progress Report 1963–1967 on the Abnormal Haemoglobin Research Unit, University Department of Biochemistry, Cambridge" (file E131/44) and "Progress Report 1967–1972" (file E131/61), Medical Research Council Archives, London.

34 By the early 1980s, Lehmann's collection counted more than 350 variants, more than a third of which were identified by the Cambridge unit.

35 "MRC Abnormal Haemoglobin Unit. Report of Visiting Subcommittee". MRC 73/161, Annexe 1 (8 November 1972); file E131/61, Medical Research Council Archives, London.

36 "Progress Report 1967–1972", Annexe 2, p. 5.

37 Benzer's fine-grained mutation maps of bacteriophage T4 had proved the linearity of genetic information. In Cambridge, Benzer, Crick and Brenner were trying to use Benzer's genetic system to show that genes and proteins were co-linear. These experiments made use of similar fingerprinting techniques developed by Fred Sanger, yet using radioactive markers (see de Chadarevian, 1996). The collaboration between Lehmann and Ingram continued even after Ingram left Cambridge, England for Cambridge, Mass. in 1958. See Ingram and Lehmann (1960) for a further common communication by the two researchers. Plans for future collaborations were also mentioned in Lehmann's application to the Medical Research Council in 1962; see "Application by Dr. H. Lehmann for the Establishment of a Research Unit on Abnormal Haemoglobins and Related Characters at St. Bartholomew's Hospital Medical College, London", MRC.62/1180. TMRB.62/60; file E131/3, Medical Research Council Archives, London.

38 Note the superimposition of old and new nomenclatures.

39 For routine analysis in hospitals, paper electrophoresis remained the method of choice. Problematic samples, however, could be sent for further analysis to specialized laboratories like Lehmann's. On this point see e.g. Lehmann and Huntsman (1966, p. 264).

40 The various forms of thalassemia provided particular fruitful research material in this context.

41 Human and horse hemoglobin were indistinguishable at the resolution of Perutz's hemoglobin model.

42 Interview with M. Perutz, Cambridge, 25 June 1992.

43 Significantly, clinical symptoms included observations obtained at the bedside (e.g. cyanosis, splenomegaly) as well as classifications based on laboratory data (e.g. inclusion body anemia, hemolytic anemia). Even before paper electrophoresis of abnormal hemoglobins became a standard procedure in hematology, hematologists relied heavily on microscopic observation and laboratory tests for the description and classification of diseases. How this was achieved calls for more detailed historical investigation, but see e.g. Wailoo (1991).

44 Perutz's work on the stereochemical changes in the hemoglobin molecule was used by Jacques Monod to elaborate the concept of "allostery" in the mid-1960s. This is another instance of the role of hemoglobin as model molecule (Creager and Gaudillière, 1996).

45 "Progress Report 1967–1972 on the Abnormal Haemoglobin Unit", Annex 2, MRC 73/161, p. 6; file E 131/61, Medical Research Council Archives, London.

46 Mourant on Lehmann according to Dacie (1988, p. 435). See also "Progress Report 1963–1967", p. 10.

47 For current research in the field see e.g. Bookchin and Lew (1996).

48 Interview of M.F. Perutz by H.F. Judson, November 1987. Video Recording. The Biochemical Society, London.

49 On the introduction of screening programs for sickle cell anemia and thalassemia see Michaelson (1972) and Kevles (1985, pp. 255–258).

50 "Abnormal Hemoglobins" *Medical Research Council Annual Report April 1965–March 1966* (London: Her Majesty's Printing Office, 1966), pp. 37–44.

51 Pauling presented this clearly eugenic measure as a moral requirement aimed at eliminating a source of suffering from the world. Pauling suggested the identification of heterozygotes should be made compulsory, while limiting the number of progeny of heterozygotes should be achieved through a process of education and the provision of information. In this respect Pauling did not argue in a strictly eugenic manner, since from a population genetic point of view the procreation of heterozygotes is more deleterious than the birth of a sickle cell anemic child who would most probably die before reaching the age of procreation. I am grateful to Diane Paul for making the text of Pauling's notorious statement available to me.

52 Against the objection that where no treatment is involved, one cannot properly speak of clinical practice, it should be remembered that the diagnosis and prognosis of a disease has always formed an important aspect of the work of the clinician. In a rather resigned mood, an authoritative researcher in the field of the thalassemias (which are based on insufficient production of adult hemoglobin) reckoned: "Perhaps some of the basic knowledge is not entirely wasted. By knowing the pathophysiology well, the clinician is able to organize his supportive management of the disease much more logically even though there is still no cure for any of the thalassemic disorders" (Weatherall, 1980, p. 406).

53 Lehmann continued research on a personal grant for another five years. He then made sure that his laboratory equipment as well as the collection of abnormal hemoglobins were transferred to St. Bartholomew's Hospital in London, where Adrian D. Stephens was running a unit which was still active in the field. When the London group was made redundant by the NHS hospital reform in 1994, no funds for the refrigerating costs necessary

to keep Lehmann's collection in good conditions could be secured. To Stephens' regret, the collection was therefore discarded. I am indebted to Dr Stephens for this information.

REFERENCES

J.A.M. Ager, H. Lehmann and F. Vella, "Hemoglobin 'Norfolk': A New Hemoglobin Found in an English Family with Observations on the Naming of New Hemoglobin Variants" *British Medical Journal* (1958), 2: 539–541.

A.C. Allison, "Protection Afforded by Sickle Cell Trait Against Subtertian Malarial Infection" *British Medical Journal* (1954), 1: 290–294.

M.A. Andersch, D.A. Wilson and M.L. Menten, "Sedimentation Constants and Electrophoretical Mobilities of Adult and Fetal Carbonyl-Hemoglobin" *Journal of Biological Chemistry* (1944), 153: 301–305.

G.P.T. Barclay and M. Noak, "An Abnormal Hemoglobin Unit in Africa" *Medical Journal of Zambia* (1970), 4 (Supplement): 238–241.

J. Barcroft, *The Respiratory Function of the Blood* (Cambridge: Cambridge University Press, 1914).

D. Beale and H. Lehmann, "Abnormal Hemoglobins and the Genetic Code" *Nature* (1965), 207: 259–262.

S. Benzer, V.M. Ingram and H. Lehmann, "Three Varieties of Human Hemoglobin D" *Nature* (1958), 182: 852–853.

J. Bernard and M. Bessis, *Hematologie clinique* (Paris: Masson & Cie, 1958).

J. Bernard, *La légende du sang* (Paris: Flammarion, 1992).

R.M. Bookchin and V.L. Lew, "Pathophysiology of Sickle Cell Anemia" *Hematology/Oncology Clinics of North America* (1966), 10: 1241–1253.

C.J.C. Britton, *Whitby and Britton. Disorders of the Blood. Diagnosis, Pathology, Treatment, Technique* (London: Churchill, 1963).

V. Bush, *Science, the Endless Frontier* (Washington: U.S. Government Printing Office, 1945).

C.L. Conley, "Sickle Cell Anemia — the First Molecular Disease" in Wintrobe (1980), pp. 319–371.

A.N.H. Creager and J.-P. Gaudillière, "Meanings in Search of Experiments or Vice-Versa: The Invention of Allosteric Regulation in Paris and Berkeley (1959–1967)" *Historical Studies in the Physical and Biological Sciences* (1996), 27: 1–89.

F.H. Crick, "Molecular Biology and Medical Research" *Journal of the Mount Sinai Hospital* (1969), 36: 178–188.

F.H. Crick, *What Mad Pursuit. A Personal View of Scientific Discovery* (London, Penguin: 1990).

S.J. Dacie, "Hermann Lehmann. 8 July 1910–13 July 1985" *Biographical Memoirs of the Fellows of the Royal Society* (1988), 34: 405–449.

S. de Chadarevian, "Building Molecular Biology in Post-War Britain", paper presented at the Mellon Workshop "Institutional and Disciplinary Contexts of the Life Sciences (Mid to Late Twentieth Century)", Massachusetts Institute of Technology, Cambridge, Mass., April 1994.

S. de Chadarevian, "Sequences, Conformation, Information. Biochemists and Molecular Biologists in the 1950s" *Journal of the History of Biology* (1996), 29: 361–386.

T. Duster, *Backdoor to Eugenics* (New York and London: Routledge, 1990).

S.J. Edelstein, *The Sickled Cell. From Myth to Molecules* (Cambridge, MA: Harvard University Press, 1986).

G.M. Edington and H. Lehmann, "Hemoglobin G. A New Hemoglobin Found on a West African" *The Lancet* (1954), July 24, 173–174.

B. Elzen, "Two Ultracentrifuges: A Comparative Study of the Social Construction of Artefacts" *Social Studies of Science* (1986), 16: 621–662.

S.D. Feldman and A.I. Tauber, "Sickle Cell Anemia: The Historical Genesis of the First "Molecular Disease" *Bulletin of the History of Medicine* (in press).

G.W. Gray, "Sickle Cell Anemia" *Scientific American* (August 1951), 185: 56–59.

G.C. Gruchy, *Clinical Hematology in Medical Practice* (Blackwell: Oxford, 1958).

J.B.S. Haldane, "The Rate of Mutation of Human Genes" in G. Bonnier and R. Larsson (eds), *Proceedings of the Eighth International Congress of Genetics* (Lund: Berlingska Boktryckeriet, 1949), pp. 267–273 (issued as supplementary volume of *Hereditas*).

J.A. Hunt and H. Lehmann, "Abnormal Human Hemoglobins. Hemoglobin 'Bart's': Foetal Hemoglobin without æ-Chains" *Nature* (1959), 184: 872–873.

V.M. Ingram, "A Specific Chemical Difference Between the Globins of Normal Human and Sickle Cell Anemia Hemoglobin" *Nature* (1956), 178: 792–794.

V.M. Ingram, "Gene Mutations in Human Hemoglobin: The Chemical Difference Between Normal Human and Sickle Cell Hemoglobin" *Nature* (1957), 180: 326–328.

V.M. Ingram, "A Case of Sickle Cell Anemia: A Commentary" *Biochimica et Biophysica Acta* (1989), 1000: 147–150.

V.M. Ingram and H. Lehmann, "The Globin Molecule in Different Human Hemoglobins" in *Proceedings of the Seventh Congress of the European Society of Hematology, London 1959* (Basel: Karger, 1960), volume II, pp. 1021–1027.

H.A. Itano and J.V. Neel, "A New Inherited Abnormality of Hemoglobin" *Proceedings of the National Academy of Science* (1950), 36: 613–617.

H.F. Judson, *The Eighth Day of Creation. Makers of the Revolution in Biology* (New York: Simon and Schuster, 1979).

L. Kay, "Laboratory Technology and Biological Knowledge: The Tiselius Electrophoresis Apparatus, 1930 to 1945" *History and Philosophy of the Life Sciences* (1988), 10: 51–72.

L. Kay, *The Molecular Vision of Life. Caltech, The Rockefeller Foundation, and the Rise of the New Biology* (New York: Oxford University Press, 1993).

D. Kevles, *In the Name of Eugenics. Genetics and the Use of Human Heredity* (New York: Alfred A. Knopf, 1985).

H. Lehmann, "Hemoglobin and Its Abnormalities" *The Practitioner* (Symposium on Diseases of the Blood) (1957), 178: 198–214.

H. Lehmann, "Classification and Identification of Human Hemoglobines" *Proceedings of the Royal Society of Medicine* (1959), 52: 959–963.

H. Lehmann and R.G. Huntsman, *Man's Hemoglobin. Including the Hemoglobinopathies and Their Investigation* (Amsterdam: North Holland Publishing Company, 1966).

H. Lehmann and A.H. Milne, "The Sickle Cell Trait in Relation to Hemoglobin Level and Anemia" *East African Medical Journal* (1949), 26: 247–254.

H. Lehmann and M.F. Perutz, "Molecular Pathology of Human Hemoglobin" *Nature* (1968), 219: 902–909.

H. Lehmann and A.B. Raper, "Distribution of the Sickle Cell Trait in Uganda, and Its Ethnological Significance" *Nature* (1949), 164: 494–495.

H. Lehmann and E.B. Smith, "Separation of Different Hemoglobins by Paper Electrophoresis" *Transactions of the Royal Society of Tropical Medicine and Hygiene* (1954), 48: 12.

E. Mellanby, "Medical Research in Wartime" *British Medical Journal* (1943), 2: 351–356.

M.G. Michaelson, "Sickle Cell Anemia: An 'Interesting Pathology'" *Ramparts* (1972), 52–58.

K. Moffat, "Structure Function Relationships in Hemoglobins: Scientific Aspects" *Texas Report on Biology and Medicine* (1980–1981), 40: 191–195.

H. Morimoto, H. Lehmann and M.F. Perutz, "Molecular Pathology of Human Hemoglobin: Stereochemical Interpretation of Abnormal Affinities" *Nature* (1971), 232: 408–413.

J.V. Neel, "The Inheritance of Sickle Cell Anemia" *Science* (1949), 110: 64–66.

J.V. Neel, *Physician to the Gene Pool. Genetic Lessons and Other Stories* (New York: John Wiley & Sons, 1994).

R. Olby, "The Origins of Molecular Biology at Cambridge and Caltech" in *Proceedings of the Conference on the History of Biochemistry and Molecular Biology* (Brookline: American Academy of Arts and Sciences, 1970), pp. 60–95.

L. Pauling, H.A. Itano, S.J. Singer and I.C. Wells, "Sickle Cell Anemia, a Molecular Disease" *Science* (1949), 110: 543–548.

L. Pauling, "Abnormalities of Hemoglobin Molecules in Anemia" *Harvey Lectures* (1955), 49: 216–241.

L. Pauling, "Reflections on the New Biology. Foreword" *UCLA Law Review* (1968), 15: 267–272.

L. Pauling, "Fifty Years of Progress in Structural Chemistry and Molecular Biology" *Daedalus* (1970), 99: 988–1014.

L. Pauling, "The Normal Hemoglobins and the Hemoglobinopathies. Background" *Texas Reports on Biology and Medicine* (1980–1981), 40: 1–7.

M.F. Perutz, "Hemoglobin: Genetic Abnormalities" *New Scientist and Science Journal* (June 1971), 50: 762–765.

M. Perutz, "The Birth of Molecular Biology" *New Scientist* (1987), No. 1561: 38–41.

M. Perutz, *Is Science Necessary? Essays on Science and Scientists* (London: Barrie & Jenkins, 1989).

M.F. Perutz, A.M. Liquori and F. Eirich, "X-Ray and Solubility Studies of Hemoglobin of Sickle Cell Anemia" *Nature* (1951), 167: 929–931.

M. Perutz and M. Mitchison, "State of Hemoglobin in Sickle Cell Anemia" *Nature* (1950), 166: 677–679.

J.V. Pickstone, "Ways of Knowing. Towards a Historical Sociology of Science, Technology and Medicine" *British Journal for the History of Science* (1993), 26: 433–458.

J.V. Pickstone, "The Biographical and the Analytical. Towards a Historical Model of Science and Practice in Modern Medicine" in I. Löwy (ed.), *Medicine and Change: Historical and Sociological Studies of Medical Innovation* (Paris: Les Editions INSERM, 1993a), pp. 23–47.

J. Pickstone, "The Place of the Analytical/Comparative in Nineteenth-Century Science, Technology and Medicine" *History of Science* (1994), 32: 111–138.

F.J. Roughton, "Joseph Barcroft 1872–1947" *Obituary Notices of the Fellows of the Royal Society* (1949), 6: 315–345.

F.J. Roughton and J.C. Kendrew (eds), *Hemoglobin. A Symposium Based on a Conference Held at Cambridge in June 1948 in Memory of Sir Joseph Barcroft* (London: Butterworths Scientific Publications, 1949).

R.M. Schmidt, T.H.J. Huisman and H. Lehmann (eds), "Standardization of Laboratory Reagents and Methods for the Detection of Hemoglobinopathies" in *Proceedings: International Conference on Standardization of Diagnostic Materials, June 5–8, 1973* (Atlanta: U.S. Department of Health, Education and Welfare, 1974), pp. 131–159.

R.G. Schneider, "Hemoglobins We Have Known" *Texas Report on Biology and Medicine* (1980–1981), 40: 479–487.

W.H. Schneider, *Quality and Quantity. The Quest for Biological Regeneration in Twentieth-Century France* (Cambridge: Cambridge University Press, 1990).

S.L. Starr and J.R. Griesemer, "Institutional Ecology. 'Translations' and Boundary Objects: Amateurs and Professionals in Berkeley's Museum of Vertrebrate Zoology 1907–1939" *Social Studies of Science* (1988), 19: 387–420.

K. Wailoo, "'A Disease *sui generis*': The Origins of Sickle Cell Anemia and the Emergence of Modern Clinical Research, 1904–1924" *Bulletin of the History of Medicine* (1991), 65: 185–208.

E.J. Watson-Williams, D. Beale, D. Irvine and H. Lehmann, "A New Hemoglobin, D Ibadan (ß-87 Threonine — Lysine), Producing No Sickle Cell Hemoglobin D Disease with Hemoglobin S" *Nature* (1965), 205: 1273–1276.

D.J. Weatherall, "Toward an Understanding of the Molecular Biology of Some Common Inherited Anemias: The Story of Thalassemia" in Wintrobe (1980), pp. 373–414.

M.M. Wintrobe, *Clinical Hematology* (London: Henry Kimpton, 1956).

M.M. Wintrobe, *Blood, Pure and Eloquent. A Story of Discovery, of People and of Ideas* (New York: McGraw-Hill, 1980).

M.M. Wintrobe, *Hematology. The Blossoming of a Science. A Story of Inspiration and Effort* (Philadelphia: Lea and Febiger, 1985).

F.G. Young, "Preface" in Lehmann and Huntsman (1966), pp. v–vii.

D.T. Zallen, "From Butterflies to Blood: Human Genetics in the U.K." in M. Fortun and E. Mendelsohn (eds), The Practices of Human Genetics. *Sociology of the Sciences Yearbook*, volume 19 (Dordrecht: Kluwer, in press).

CHAPTER 7

The Struggle Over Metabolic Screening

Diane B. Paul and Paul J. Edelson

> The ailment is rare, but its importance is not to be measured in terms of numbers alone ... The study of PKU promises to help uncover the causes of other hereditary disorders, particularly the basis for other types of mental deficiency. The disorder thus exemplifies many so-called molecular diseases, and the rewards that may accrue from the ceaseless probing of these points on the far frontiers of modern bio-medical research. (Harold M. Schmeck Jr, *The New York Times*, 1961, p. 8).

INTRODUCTION

Newborn screening for phenylketonuria (PKU) is today among the most routine of medical procedures. In many countries, including the U.S., Canada, the nations of Western and most of Eastern Europe, Israel, Egypt, Japan, Australia and New Zealand, virtually all infants are tested for this autosomal recessive disease. Unless the parents receive a positive result, they may not be aware that a test has been done, even in those American states where they have the legal right of refusal.

The incidence of PKU in the U.S. is estimated at about one in every 10,000 births. When mass screening began in the early 1960s, it was thought to be even lower; about one in 25,000. A 1939 survey by the American physician George Jervis found that 0.79% of patients in fourteen institutions for the mentally retarded were diagnosed with PKU (Jervis, 1939). A 1962 Children's Bureau census identified 399 children with PKU who had been admitted to programs for the mentally retarded during the preceding five years, on average 80 cases per year.

How and why did such a relatively rare condition become the focus of so much attention? The answer might seem to be obvious. Most untreated individuals with PKU are profoundly retarded and suffer from other physical and behavioral abnormalities; some are unable to walk or talk or control their bowels or bladder. In the 1950s, a number of studies indicated that restricting infants' dietary intake of phenylalanine could prevent or mitigate retardation and other symptoms associated with PKU in the same way that galactosemia, an even rarer metabolic disorder, might be treated by substituting a soy bean-based substitute for milk. (Medical genetics texts of the 1950s often accorded galactosemia more attention than PKU).

These studies led some hospitals to test infants for PKU. The diagnosis was made on wet diapers using a strip of filter paper that immediately turned green when a drop of ferric chloride solution was added. But as this urine screening test was unreliable until the age of six to eight weeks, after the infant had been discharged from the hospital and possibly after he or she had already suffered some degree of irreversible brain damage, it did not work well enough for purposes of early identification of affected infants.

In 1960, a new test was developed, which was more sensitive than the urine test and could be administered while the infant was still in the hospital. In its initial form, the heel of the newborn was pricked and the resulting blood dried on discs of filter paper. The paper was placed on the surface of an agar culture plate containing an analogue of phenylalanine, ß-2-thienylalanine, and spores of *Bacillus subtilis*. While normally these bacteria are capable of synthesizing their own phenylalanine, the presence of ß-2-thienylalanine blocks a crucial metabolic step in this process and the bacteria cannot reproduce. When a filter paper disc impregnated with blood from an affected individual is placed on the medium, however, the high levels of phenylalanine which it contains diffuse into the medium, and are available to the bacteria to overcome their phenylalanine starvation, producing a zone of growth surrounding the disc.

Two years after this test became available, Massachusetts became the first state to mandate PKU screening. By 1965, 32 states had enacted screening laws, all but five making the test mandatory; by 1975, virtually all newborns in the U.S. were being tested. In some states the laws passed with neither hearings nor floor debates and by acclamation or voice vote without dissent.

Before the new test had been validated or the effects of dietary treatment assessed, the PKU program was being cited as a model for the screening and prevention of other metabolic disorders. Thus at a March of Dimes symposium in 1967, the detection of PKU was characterized as "a model system for the application of genetic screening to public health" (Honeyman and Gabrielson, 1968, p. 100). Although a five-year collaborative study of treated phenylketonuric children for purposes of evaluating dietary therapy only began in 1967, Katherine Oettinger, then Chief of the U.S. Children's Bureau, characterized PKU screening in a December 14, 1965 press release as "a breakthrough in the prevention of mental retardation," noting that it "has paved the way for similar screening procedures for other inborn metabolic errors." Before the decade had ended, a variety of other rare metabolic conditions were being detected with the same filter paper blood specimen employed for PKU screening (Guthrie, 1968; 1992).

Newborn screening for PKU is often characterized as a triumph of applied human genetics and as the program that linked genetics to public health (see Paul, in press). Even in the 1960s, it was claimed that PKU, "because it can be both detected and prevented, has opened the public health door to genetics and genetic screening" (Honeyman and Gabrielson, 1968, p. 100), although neither the test nor treatment for PKU was (or is) genetic. Charles Scriver reflects the common perspective that every test of an inherited disorder involves genetics when he writes that, by 1970, "newborn screening for PKU had become one of the most widely applied genetic technologies in human society" (Scriver, 1994). But PKU screening was not a genetic technology as that term is generally understood today. Guthrie's test was employed to identify affected newborns, not parent carriers from whom the disease might in future be inherited.

In the 1930s, Lionel Penrose suggested that PKU heterozygotes were more prone to senile dementia, leading some scientists to hope that the condition could be detected early and carriers prevented or discouraged from mating with each other (Haldane, 1941, p. 139; see also Munro, 1947). Carrier detection finally became a reality in 1956, when the geneticist David Yi-Yung Hsia and colleagues at Harvard Medical School found that clinically asymptomatic heterozygotes could be identified through the use of a phenylalanine tolerance test. (With half the normal enzyme activity, carriers are unable to efficiently metabolize the additional load of phenylalanine, which can be detected in the blood). The use of that technology would indeed have constituted genetic testing in the stricter sense. But while some geneticists expressed hope that carrier detection plus counseling might represent "an important method of control" of the disease, that prospect was never realized (Honeyman and Gabrielson, 1968, p. 103; see also Hsia, 1968, pp. 83–84; Sutton, 1961, p. 93). The loading test was not very reliable and attracted little interest among high-risk individuals or their physicians.

Newborn screening, on the other hand, was quickly routinized. This may appear a logical result of the invention of a reliable and cost-effective test combined with the discovery that exclusion of dietary phenylalanine prevents retardation in PKU. That is indeed the standard account. But at the time mass screening was adopted, neither the reliability of the test nor the efficacy of the PKU diet had been proven. Researchers in the field of human metabolism argued that the legislated screening was premature and that some infants might even be harmed by the diet. They protested that not enough was known about the disease, the test, or the proposed treatment to justify large-scale legislated screening programs. Moreover, there was no effective therapy — then or now — for some of the other "inborn errors of metabolism" (or as they were increasingly called, "molecular diseases") that were soon piggy-backed onto the system for PKU testing. To its critics, these programs represented a vast and uncontrolled experiment being perpetrated on a naive and unsuspecting public.

Thus the road to mandatory screening, while short, was far from smooth. From a distance, its introduction in the 1960s may appear the inevitable result of advances in diagnosis and therapy. But a closer examination shows that there was no inexorable bridge from the science to the screening programs. Screening was the product of a well-crafted political campaign pursued in the context of a more general shift from social toward biological approaches to mental retardation. Given its low incidence, PKU did not directly touch many lives. But the political constituency for PKU screening was much larger than the number of affected individuals and their families might suggest. That constituency quickly overwhelmed the opposition from researchers (and also from physicians concerned about malpractice suits). In this chapter, we explain how scientific and political trends intersected to produce enthusiasm for a molecular approach to the problem of mental retardation.

THE SCIENTIFIC BACKGROUND TO SCREENING

Phenylketonuria was first described in 1934 by the Norwegian physician and biochemist Ivar Asjbørn Følling, who recognized the condition as one of Archibald Garrod's "inborn errors of metabolism." When Følling added ferric chloride to the urine of two severely retarded children (then a standard procedure to detect ketones in the urine of diabetics), a deep green color developed. The substance responsible was found to be an abnormal metabolite, phenylpyruvic acid. To learn if other mentally-retarded patients also excreted this acid, Følling collected samples from 430 patients and demonstrated the green color in eight. He hypothesized that phenylpyruvic acid was produced when patients were unable to metabolize phenylalanine normally (Følling, 1994, pp. 5–6). In 1939, George Jervis confirmed that an excess of phenylalanine was associated with the condition Følling called "oligophrenia phenylpyruvica" and identified the cause as an inability to convert phenylalanine to tyrosine (Jervis, 1939; Jervis *et al.*, 1940). He also demonstrated that this inability results from a defect in the liver enzyme phenylalanine hydroxylase (Jervis, 1953).

Among Følling's original ten patients were three pairs of siblings, suggesting a genetic basis for the condition. In 1935, a pedigree analysis suggested to Penrose that the disease (which he and biochemist Juda Quastel renamed phenylketonuria) was probably inherited in a recessive manner. Penrose's hypothesis was soon confirmed by Jervis, who analyzed 213 institutionalized patients and concluded that the condition was due to a single autosomal recessive gene (Jervis 1939). Jervis and his colleague Richard Block also proposed that affected babies be fed a diet from which the phenylalanine had been removed through a filtering process (Laurence, 1939; Jervis and Block, 1939). At the time, their proposal did not arouse interest among specialists in mental retardation, who apparently thought it far-fetched (Brecher and Brecher 1959, p. 109). In Britain, Penrose had also experimented with low-phenylalanine diets in the 1930s. But he was discouraged from pursuing the experiment by Frederick Gowland Hopkins (then Britain's leading biochemist) on the grounds that it would cost 1000 pounds sterling per week to feed one patient on the synthetic phenylalanine-free diet (Kevles 1985, pp. 177–178).

However, in 1951 the idea of dietary therapy was revived by the English biochemists Louis I. Woolf and David G. Vulliamy at the Hospital for Sick Children in London. They suggested that high levels of phenylalanine or one of its intermediary metabolites circulating in the blood was responsible for the symptoms of PKU and suggested that, "if the amount of phenylalanine and the breakdown products could be reduced, normal cerebral function might result" (Woolf and Vulliamy, 1951, p. 151). Using charcoal to extract the unwanted phenylalanine from their foods, Woolf tested the theory on three small children; all showed some improvement. In the early 1950s, groups associated with Horst Bickel (1954) at the Children's Hospital in Birmingham, England, with Marvin Armstrong and Frank Tyler (1955) at the University of Utah College of Medicine, and with Frederick Horner and Charles Streamer (1956) at the University of Colorado Medical School also reported cognitive and behavioral improvement of children treated with a low-phenylalanine diet. Although some other groups were less successful, and the first retrospective statistical study confirming the benefit of dietary therapy only appeared in 1960 (Knox, 1960), the positive reports generated great excitement.

That enthusiasm is easy to understand. The field of mental retardation was then considered "the ultimate in therapeutic hopelessness" (Cooke, 1964, p. 392). Lamenting physician disinterest in the field, one neurologist conceded that it was in part "a reflection of the undeniable fact that, in the majority of instances of mental retardation, medical treatment has little to offer" (Masland, Sarason, and Gladwin, 1958, p. 26). Now it seemed that an understanding of biochemistry might actually lead to a cure — and not just for PKU. Even in the 1950s, when treatment (mostly with young children) had only begun and preliminary results were mixed, dietary therapy in PKU was already being cited as a successful model for other metabolic disorders.

POLITICAL BACKGROUND TO SCREENING

During the late 1950s, both private and governmental organizations in the U.S. began to rethink their traditional emphasis on the social approach to retardation. Of particular importance in this reorientation was the role of the parents' group, the National Association for Retarded Children (NARC). Organized in 1950, the NARC had been very effective in lobbying Congress and the Children's Bureau of the Department of Health, Education, and Welfare (HEW) to expand their support of programs for the mentally retarded (Edelson, 1994, pp. 7–8).

The NARC's initial interest was largely in expansion of educational, social, and rehabilitative services. That emphasis coincided with the approach favored at HEW, where it was thought that many parents of the retarded "would be happy to take care of their children in their own homes, if necessary supporting services were available to them in their own communities." But Representative John Fogarty, chair of the subcommittee on health of the House Appropriations Committee, thought the money would be best used in a search for biological, and especially hereditary, causes of mental illness (Edelson, 1994, pp. 4–5).

At the same time, the NARC's own position was shifting. In the mid-1950s, it commissioned a survey of the field in connection with a program aimed at discovering causes of and ways of preventing mental retardation. Two reports, one on prevention and the other on psychological and cultural factors in mental subnormality, were published in 1958. The author of the report on prevention was Richard Masland, a young neurologist and neuroscientist, and a member of the NARC's Scientific Advisory Board. He suggested: "the time is ripe for a large-scale attack on the genetically determined defects" through the application of new techniques to the study of basic metabolism and to the study of patients suffering from mental deficiency (Masland *et al.*, 1958, p. 50).

These developments converged with the election of John F. Kennedy as U.S. President. The President's sister Rosemary had long been institutionalized for mental retardation and the subject was one of great personal interest to the entire Kennedy family. In 1961, President Kennedy announced a major new initiative in this area. He stressed that five million Americans were mentally retarded — a number far larger than the number of persons suffering from diabetes, polio, and other diseases which were the focus of intense public concern and the targets of major government expenditures. Moreover, facilities for the care of mentally retarded persons were vastly overburdened. State institutions averaged 367 patients above their rated capacities while for the 160,000 patients in public institutions, there were only 500 full-time physicians. Thus there existed a huge imbalance between the scope of the problem of mental retardation and the efforts to address it. On current educational and rehabilitative approaches, "the central problem remains unsolved, for the causes and treatment of mental retardation are largely untouched" (Kennedy, 1962, p. 496). The President promised to seek to double the amount spent by the National Institutes of Health in this field. He also announced the appointment of a Presidential Advisory Commission on Mental Retardation headed by Leonard Mayo, Executive Director of the Association for the Aid of Crippled Children. The panel was charged with appraising the adequacy of existing programs and recommending future action.

The Commission's appointees included Elizabeth Boggs, the NARC's Research Chairman and proponent of the scientific approach to retardation, and its proposals strongly reflected the NARC's perspective. Thus the panel's staff urged the Advertising Council to circulate information about the PKU field trials in states without legislated screening, a proposal that even Children's Bureau officials doubted was appropriate (Theodore D. Tjossem to Arthur J. Lesser, July 9, 1964). When the Commission's recommendations were published, they included a statement that newborn screening and the management of metabolic disorders was an "important" step in preventing mental retardation, and recommended that such programs be expanded although at this time the only U.S.

experience with newborn screening programs involved the unreliable ferric chloride urine test, which missed many cases. Thus the attractiveness of such approaches lay not in their specific accomplishments, but rather in the hopefulness generated by the idea that science could be mobilized to rescue America from the large and apparently intractable problem of mental retardation (Edelson, 1994). As yet, hope was nearly all that science could offer.

At just this time, Robert Guthrie, a microbiologist and non-practicing physician, developed a cheap, sensitive, and simple bacterial inhibition test. Guthrie had a retarded son and a niece who had been diagnosed with PKU at 15 months, too late to benefit from therapy. This personal interest in the disease led him to devise an assay for monitoring the blood phenylalanine level of children on low-phenylalanine diets. It soon became clear that the same test could be used to measure blood phenylalanine levels in newborns, with the possibility of beginning dietary therapy before irreversible brain damage had occurred. It was also found that the blood or serum sample did not need to be fresh, making it possible to process samples in batches at central laboratories to which test samples could be regularly mailed (Guthrie, 1973, pp. 229–30; 1992, pp. 6–7).

Soon after inventing the test, Guthrie proposed both to the Children's Bureau and to Elizabeth Boggs of the NARC that it be field tested for suitability in a national screening program for all newborns. The Children's Bureau agreed. In late 1961, it commenced a field trial involving over 400,000 infants in 29 states. A number of researchers expressed concern that the Bureau was already committed to a screening program based on the Guthrie test — a charge indignantly denied by Bureau staff. A regional medical director protested: "So far as I know, no member of the Bureau staff questions the need for further testing of the method. At this point it is not intended as a recommended screening program" (Lucille J. Marsh to Franklin D. Yoder, June 4, 1962). But by the time the trial ended in 1963, the Bureau had aggressively taken up the screening cause. While it was forbidden by statute from promoting state legislation, its slogan became "Test Every Newborn for PKU."

RESPONSE TO THE GUTHRIE TEST

Announcement of the Guthrie test was greeted in a number of quarters with elation. Even before the field trial had ended, and in spite of the rarity of the disease, it was hailed as a major discovery. The American Medical Association included the Guthrie test in its 1962 year-end report on several major medical breakthroughs, along with the cracking of the genetic code (*New York Times*, 1963; Wehrwein, 1963). PKU was now discussed everywhere — in newspapers and magazines ranging from *Good Housekeeping*, *Better Homes and Gardens*, and *Life*, to *Today's Health*, *Scientific American*, and the *Saturday Review*. The significance accorded the disease is reflected in the first awards for outstanding achievement presented by the Joseph P. Kennedy, Jr. Foundation in 1962: the

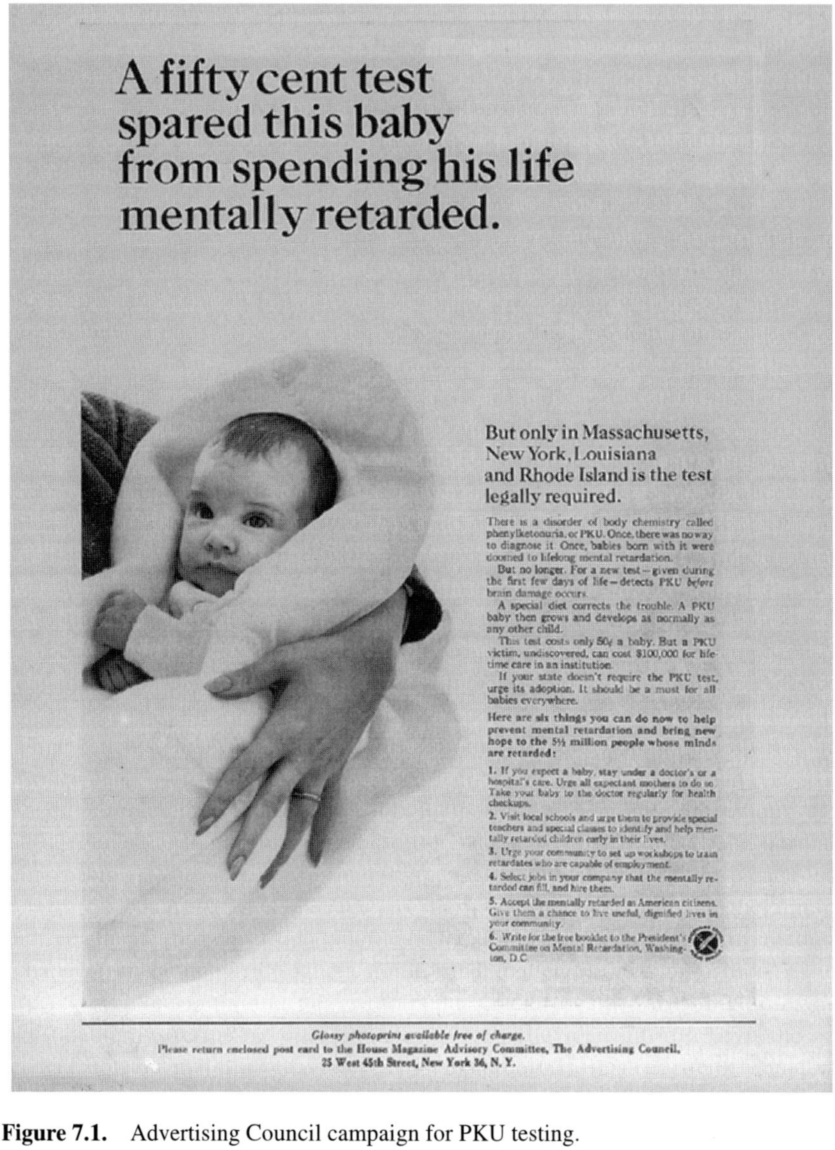

Figure 7.1. Advertising Council campaign for PKU testing.

recipients were Joe Hin Tijo, Jerome Lejeune, Murray Barr — and Følling. The Advertising Council campaign featured ads with the heading: "A fifty cent test spared this baby from spending his life mentally retarded." The following text appeared to the right of a picture of an obviously normal infant (Figure 7.1):

> But only in Massachusetts, New York, Louisiana and Rhode Island is the test legally required.
>
> There is a disorder of body chemistry called phenylketonuria, or PKU. Once, there was no way to diagnose it. Once, babies born with it were doomed to lifelong mental retardation.
>
> But no longer. For a new test — given during the first few days of life — detects PKU before brain damage occurs. A special diet corrects the trouble. A PKU baby then grows and develops as normally as any other child.
>
> The test costs only 50c a baby. But a PKU victim, undiscovered, can cost $100,000 for lifetime care in an institution.
>
> If your state doesn't require the PKU test, urge its adoption. It should be a must for all babies everywhere.

Screening advocates regularly emphasized the cost effectiveness of the program when lobbying the federal or state legislatures. Arthur Lesser, Director of the Children's Bureau Division of Health Services, testified to a subcommittee of the House Committee on Appropriations in 1961 that PKU testing had saved the federal government $2 million. Some cost-benefit claims simply compared the cost of a single test to the cost of lifetime institutionalized care. More sophisticated analyses compared the expense of laboratory testing and evaluation and treatment for metabolic disorders with the assumed expenses to the state of institutionalized care (typically for twenty years) for a proportion of the affected infants and the costs of medical and hospital care for the non-institutionalized. On this basis, Massachusetts officials estimated a saving of $364,662 per year (Massachusetts Department of Public Health, 1974). The expense of laboratory testing was sometimes equated with the unit cost of the test rather than the cost of identifying one affected individual — perhaps because the latter would include the cost of retesting the large number of false positives intrinsic to screening for very rare conditions. (In the early years of screening, the false positive rate was 95 percent; that is, there were nineteen false positives for every true one). And the costs of treatment were equated with the expense of providing the special dietary formula rather than the panoply of medical and social services usually required for effective therapy (Figure 7.2).

In some cases, reporters and politicians vastly overstated the significance of PKU as a cause of mental retardation. Thus Senator Joseph Montoya asserted: "The Department of Health, Education, and Welfare has reported that there are approximately 5.5 million mentally retarded in the United States today. Many of those are a result of phenylketonuria and mental retardation and could have been

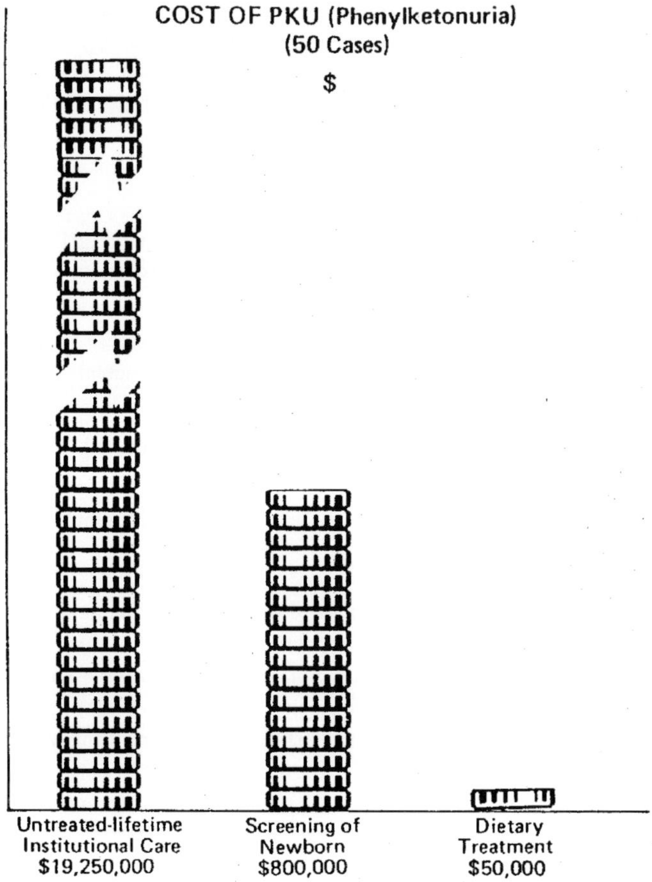

Figure 7.2. Cost of life-time institutional care of PKU patients, compared with cost of screening newborns and of dietary treatment.

prevented if detected in infancy. Most of the state training schools for the mentally retarded are overcrowded and have long waiting lists for admission" (Congressional Record, 27919, Oct. 21, 1965; see also Brecher and Brecher, 1959, p. 32).

However, most commentators understood that PKU was a rare disease, whose prevention would not begin to solve the problem of overcrowding in institutions for the mentally retarded. They viewed the Guthrie test as a major medical breakthrough because they considered PKU a model for the prevention of other diseases. If it could be prevented, why not the other inborn errors of metabolism? Indeed, why not mental disorders of other types? Thus after noting that fewer than 500 individuals with phenylketonuria are born each year, one author wrote:

> But the relatively low incidence of PKU is no measure of the attention it is currently receiving, especially from mental health experts. Their interest is easily explained: Never before has mental retardation been proved to have a known organic cause. The discovery of such a cause in the case of PKU has exciting implications. It suggests that, in time, certain other mental ailments including schizophrenia and manic-depressive psychosis — may be found to have similar roots. And, naturally, the discovery of organic origins for a disease tends to hasten its cure and prevention (Greer, 1966, p. 25).

Commentators often implied that cures were just around the corner or even already at hand. Referring to galactosemia, cystic fibrosis of the pancreas, glycogen storage disease, and idiopathic hyperlipemia, one author wrote: "These diseases can now be readily diagnosed and controlled by changes in diet" (Pompian, 1961, p. 61). According to a January 14, 1963 report in *Newsweek*: "With their new knowledge about the chemical processes underlying heredity, scientists are rapidly approaching the control of such disorders as PKU, not just by diet, but altering the faulty gene itself" (p. 70).

These discussions reflect the success of biochemist Linus Pauling's project to view diseases — including mental diseases — in molecular terms. By the 1960s, PKU was often referred to as a "molecular disease," a term coined by Pauling in 1949 to characterize sickle-cell anemia, a condition whose molecular basis he had described (Pauling, 1949; see also Kay, 1993; and de Chadarevian, this volume). The term soon gained a wide currency and broad meaning; by the 1960s it was often employed synonymously with "inborn error of metabolism" (Fraser, 1961, p. 59; Penrose, 1967, p. 18).

While not all commentary was positive, the few skeptical voices were heard only after screening was already near-universal (e.g. *New York Times*, May 4, 1968, p. 1). Many researchers would have preferred a more cautious approach to screening but they were also very cautious in their criticism. They did not complain to the press or legislators. In general, they kept their doubts to themselves. The Committee on Nutrition of the American Academy of Pediatrics criticized many aspects of the metabolic screening programs in a July 30, 1965 report to the Children's Bureau. Its members argued that:

> (1) The objectives and ways and means for implementing such wide scale screening programs remain to be evaluated as do the methods for following through of patients with heritable metabolic diseases detected by such programs.
>
> (2) The beneficial effects of good dietary management of any of the heritable metabolic diseases ... have yet to be proven unequivocally. Under the circumstances, the committee felt that caution was indicated in launching widespread screening programs until practical therapeutic programs, the effects of which are predictable in the majority of children participating, have been designed.

(3) It must be demonstrated with certainty that any normal children who receive these restrictive diets through inadvertence or because of misdiagnosis will not suffer irreparable neurological or physical damage.

But the committee declined to publish its full report.

Samuel P. Bessman, a University of Maryland pediatrician and biochemist, was an exception to the rule that skeptical researchers did not take their case to the public. With political scientist Joseph Cooper, he lobbied tirelessly in opposition to legislated screening. Bessman and Cooper testified before state and federal legislative committees, gave speeches, wrote letters to the editor. Bessman also testified for the defense in a highly publicized malpractice suit, arguing that both the nature of the disease process and the efficacy of dietary therapy were open to question. He suggested that the improvement shown by children on the diet might be explained by the extra attention they received from parents, doctors, and other therapists (e.g. Bessman, 1966). A once highly respected PKU researcher, he is now something of a pariah in the field. But the association of Bessman with opposition to mass screening has obscured the fact that less confrontational colleagues shared many of his concerns. Thus in their summary of the proceedings of a 1966 Children's Bureau-supported conference on PKU, the editors criticized "legislation which has included public pressure for a specific therapy at a time when competent medical investigators are not in agreement on the criteria for diagnosis, natural course of the disease or optimal therapy or therapies" (Anderson and Swaiman, 1967, p. 238; see also Berry and Wright, 1967).

SOURCES OF SKEPTICISM

It was initially intended that the results of Guthrie blood testing would be compared with later more definitive blood or urine tests. In fact, little systematic comparison was ever carried out. Thus there was no direct comparison of this method with other screening tests at that time, no effort to define its reliability in the face of other compounds that might be present in newborn blood samples, no assessment of the test's reliability when administered very early in life, before the infant had been fed a sufficient protein load, and no agreement on what level of blood phenylalanine was optimal. Consequently, there was no clear "gold standard" for the diagnosis of PKU that would allow an accurate estimate of its sensitivity and specificity — characteristics crucial to determining the actual value the test would have. Moreover, in most cases, initial screening positives were confirmed by a second Guthrie test, with the test allowed to be its own criterion for accuracy (Edelson, 1994, p. 11). No detailed evaluation of testing was published in a peer-reviewed journal — a sensitive point with PKU researchers. (That Guthrie was an outsider to the human metabolic research community, who took his case directly to parents, press, and government, did not help his relations with scientists).

Moreover, newborn blood testing uncovered many more apparent cases of PKU than would have been predicted on the basis of studies of the mentally retarded. It was soon evident that elevated blood phenylalanine levels could result from conditions other than "classical" PKU. But it was unclear what proportion of these individuals were actually at risk of retardation.

Particularly worrisome was the prospect that some infants were being harmed through too-drastic treatment. Phenylalanine is an essential dietary amino acid, not synthesized by humans but necessary for survival and growth. Too little as well as too much may be harmful. Louis Woolf himself believed that as many as half of all newborn phenylketonurics might not require treatment, although he considered the financial costs and rigors and the dangers of unnecessary therapy "a small price to pay for preventing the mental deterioration otherwise inevitable" in those affected (Woolf, 1967, p. 58). While some researchers asserted that little harm would come from treating non-PKU infants, others worried that it could produce mental retardation. (For a range of views, see Anderson and Swaiman, 1967, pp. 59–61, 113–115). Problems in dietary management were compounded by uncertainty about the true phenylalanine content of foods and the unpalatability of the low-phenylalanine formula. In the first years of screening, some infants suffered severe malnutrition either as the result of too severe restriction of phenylalanine or of diet refusal (Kang, December 21, 1965).

Moreover, proponents of screening often asserted that only infants and young children required treatment. When gross brain development was complete — around the age of five — they thought it would be possible to resume a normal diet. But the skeptics maintained that no one knew if this were in fact true. A committee of the American Academy of Pediatrics noted: "Because an adequate and reliable diet first became available in this country in late 1958 and because early screening tests have only recently come into general use, only a few patients have been discovered within the first month of life. Even they have been treated for less than six years, and this period of time is inadequate for assessing child development and projecting eventual intellectual ability on optimum treatment" (Committee on the Handicapped Child, 1965, p. 502).

Some researchers also noted that intensive social and psychological support services would be required if dietary therapy were to be effective. Even after counseling, most parents understood little about the disease, and many could not cope adequately with the diet (Sibinga and Friedman, 1971). Skeptics predicted that these problems would only mount if it turned out that the restrictive and unpalatable diet had to be maintained beyond early childhood.

A few researchers were dubious that the abnormalities associated with PKU resulted from an excess of phenylalanine. Theirs was a minority view. But no one did understand why extra phenylalanine should be harmful or knew whether the damage was produced by the amino acid or one of its intermediary metabolities.

Bessman in particular argued that these and other scientific questions would be difficult or even impossible to answer once screening and treatment were universal. He protested that, "if all children with excess of some amino acid in their blood are placed on deprivation therapy, we shall never know the natural history of any new disease or what the actual risk of mental retardation may be" and pleaded for controlled clinical trials (Bessman, 1966, p. 337).

The lack of placebo-controlled trials disturbed members of the Children's Bureau technical advisory committee as well. They argued that experiments should be designed to show whether there was value in therapy for tyrosinemia "before medico-legal problems, which have arisen in PKU, prevent an objective and scientific evaluation of the treatment of this metabolic disease also" (Children's Bureau, March 11, 1966). Even a Children's Bureau official noted that she had become much more conservative as the result of defending the agency's stance regarding PKU and other metabolic disorders. "The implications of 'widespread distribution' of prematurely drawn conclusions have been criticized — and rightfully in my estimation," she wrote. "We have a responsibility not only to be 'purveyors of information relating to child health and welfare' but also to use judgment in what we 'purvey' and to foresee the results of such activity" (Kang, 1967). That view was echoed by the members of the Committee on the Handicapped Child, who complained that "exaggerated predictions for normal development regardless of the age of discovery and irrespective of the strictness of the diet or of the hereditary endowment have led to frustration and discouragement on the part of pediatricians and parents" (1965, p. 503).

CONCLUSION

Researchers in human metabolism were concerned that too little was known about the diagnosis and treatment of PKU to justify mandated screening. But while their concerns were wide-ranging, they were almost invariably technical. Only Joseph Cooper, the social scientist, asked whether screening programs would divert resources from more pressing needs of the mentally retarded.

> We must face the question: Would the same resources which are now put into rare conditions yield greater benefits if they were devoted to more pressing or more prevalent matters? ... less than 1% of the retarded population in hospitals consists of PKU-diagnosed patients. What of the other 99%? Other problems of more widespread significance ... are in need of social and legislative action. What are we doing, for example, about the home-situated retardees who awaken one day to find that their parents or relatives are gone or no longer able to care for them? What do we do about these people? They must certainly outnumber those with PKU (Cooper 1965, pp. 12–13).

Responding to Cooper's claim that undue emphasis given to PKU would result in a slighting of the needs of the vast majority of institutionalized retardates, Robert Guthrie charged that Cooper "missed the point that those of us who have been excited about PKU realize that PKU is a 'model' ... for future accomplishment with many of the other inborn errors of metabolism" and went on to note that "the specific prevention of mental retardation in PKU has had a marked effect in increasing the interest of medical investigators and practitioners in the general problem of mental retardation" (February 7, 1966). But Guthrie also missed Cooper's point that science could not solve the problem of individuals who needed social support in order to live outside an institution. Indeed, mass testing was followed by a turning-away from the service-oriented approach originally favored by both the Children's Bureau and the NARC. In the 1950s, their emphasis had been on putting services in place to permit children to stay in their homes. Screening programs promoted the attitude that the problem of childhood mental retardation was now in some sense "solved." Only with the emergence of de-institutionalization in the 1970s were movements for community-based services for the retarded revived.

But arguments that pit the interests of possible future beneficiaries against the interests of those with a clear and direct and immediate stake in the outcome are rarely effective. One constituency is already mobilized; the other is simply conjectured. Arguments about the need for controlled clinical trials seem equally abstract to legislators and the public. As in contemporary research on treatment for AIDS or high-dose chemotherapy for breast cancer, such arguments failed to persuade those whose interests were most immediate and who were thus most politically engaged. As David Hsia commented to a reporter, while he was personally unsure of the value of mass screening, "Yet the parents of such [PKU] children realize its worth" (Lyons, 1968). Then as now, the view that scientific explanations — especially in terms of molecules — can provide the solution to refractory social problems exerted a powerful appeal. Despite this appeal, the molecularization of treatment was vigorously contested. The issues that troubled the early history of metabolic screening continue to be among the thorniest and most intractable in contemporary biomedicine and society.

ACKNOWLEDGEMENTS

Diane Paul's research for this chapter is based on research supported by the National Science Foundation under Grant No. SBR-9511909. We are grateful to Marvin Natowicz of the Shriver Center for useful comments on a draft of this manuscript.

REFERENCES

Note on archives. Letters and memos cited are from the Records of the Children's Bureau, U.S. Department of Health, Education, and Welfare, Box 1000; Central Files, 1963–66; Record Group 102; National Archives, Washington, DC.

J.A. Anderson and K.F. Swaiman (eds), *Phenylketonuria and Allied Metabolic Diseases*. Proceedings of a Conference held at Washington, D.C., April 6–8, 1966 (Washington, D.C.: USGPO, 1967).

M.D. Armstrong and F.H. Tyler, "Studies on Phenylketonuria: Restricted Phenylalanine Intake" *Journal of Clinical Investigations* (1955), 34: 565–580.

M.L. Batshaw, D. Valle and S.P. Bessman, "Unsuccessful Treatment of Phenylketonuria with Tyrosine" *Journal of Pediatrics* (July 1981), 99: 159–160.

H.K. Berry and S. Wright, "Conference on Treatment of Phenylketonuria" *Journal of Pediatrics* (January 1967), 70: 142–147.

S.P. Bessman, "Legislation and Advances in Medical Knowledge — Acceleration or Inhibition?" *Journal of Pediatrics* (August 1966), 69: 334–338.

S.P. Bessman, M.L. Williamson and R. Koch, "Diet, Genetics, and Mental Retardation Interaction between Phenylketonuric Heterozygous Mother and Fetus to Produce Nonspecific Diminution of IQ: Evidence in support of the Justification Hypothesis" *Proceedings of the National Academy of Sciences* (March 1978), 75: 1562–1566.

S.P. Bessman and J.P. Swazey, "Phenylketonuria: A Study of Biomedical Legislation" in E. Mendelsohn, J. Swazey, and I. Tavis (eds), *Human Aspects of Biomedical Innovation* (Cambridge, MA: Harvard University Press, 1971), pp. 49–76.

H. Bickel, "The Effects of a Phenylalanine-Free and Phenylalanine-Poor Diet in Phenylpyruvic Oligophrenia" *Experimental Medicine and Surgery* (1954), 12: 114–118.

R. Brecher and E. Brecher, "Saving Children from Mental Retardation" *Saturday Evening Post* (November 21, 1959), 32–33, 109–111.

"Chemistry Causes Imbecility" *Science News Letter*, April 15, 1939.

Children's Bureau, U.S. Department of Health, Education, and Welfare, "An Inventory of Children with Phenylketonuria" November 1962.

Children's Bureau, U.S. Department of Health, Education, and Welfare, "Minutes of Children's Bureau Ad Hoc Committee on Medical Genetics" March 11, 1966. Records of the Children's Bureau, File 4-4-1.

Committee on the Handicapped Child, American Academy of Pediatrics, "Statement on Phenylketonuria" *Pediatrics* (March 1965), 35: 501–503.

R.E. Cooke, "The Postnatal 'Prevention' and Correction of Congenital Defects" in *Second International Conference on Congenital Malformations* (New York: International Medical Congress, Ltd., 1964), pp. 386–393.

J.D. Cooper, "Problems of Legislation in the Field of Mental Retardation" Scientific Seminars of the Rosewood State Hospital, Owings Mills, Maryland, November 21, 1965. Records of the Children's Bureau, File 4-5-11-5.

J.D. Cooper, "More Problems" *Saturday Review* (June 3, 1967), 50: 56–61.

P.J. Edelson, "History of Genetic Screening in the United States I: The Public Debate over Phenylketonuria (PKU) Testing" paper presented at the Annual Meeting of the American Association for the History of Medicine, New York, 1994.

P.J. Edelson, "Lessons from the History of Genetic Screening in the US: Policy Past, Present, and Future" in P. Boyle and K. Nolan (eds), *Setting Priorities for Genetic Services* (Washington, D.C.: Georgetown University Press, in press).

I. Følling, "The Discovery of Phenylketonuria" *Acta Paediatrica Supplement 407* (December 1994), 83: 4–10.

F. Clarke Fraser, "Genetics and Congenital Malformations," in A.G. Steinberg (ed), *Progress in Medical Genetics*, Vol. I (New York: Grune and Stratton, 1961), pp. 38–80.

G. Greer, "Unnecessary Menace to Young Brains: PKU" *Better Homes and Gardens* (February 1966), 25.

R. Guthrie, "Screening for 'Inborn Errors of Metabolism' in the Newborn Infant — A Multiple Test Program" *Birth Defects Original Article Series*, IV (November 1968), 6: 92–98.

R. Guthrie, "The Origin of Newborn Screening" *Screening* (1992), 1: 5–15.

R. Guthrie, "Mass Screening for Genetic Disease" in V. McKusick and R. Claiborne (eds), *Medical Genetics* (New York: HP Publishing, 1973), pp. 229–236.

J.B.S. Haldane, *New Paths in Genetics* (London: George Allen and Unwin, 1941).

M. Honeyman and I. Gabrielson, "Public Health Aspects of Genetics Screening" in D. Bergmsa (ed), *Human Genetics.* Birth Defects Original Article Series, IV (November 1968), 6: 99–104.

F.A. Horner and C.W. Streamer, "Effect of Phenylalanine-Restricted Diet on Patients with Phenylketonuria: Clinical Observations in Three Cases" *Journal of the American Medical Association* (1956), 161: 1628–1630.

D.Y.-Y. Hsia, "Recent Advances in the Detection of Heterozygous Carriers" in D. Bergmsa (ed.), *Human Genetics.* Birth Defects Original Article Series, IV (November 1968), 6: 75–87.

G.A. Jervis, "The Genetics of Phenylpyruvic Oligophrenia (a Contribution to the Study of the Influence of Heredity on Mental Defect)" *Journal of Mental Science* (1939), 85: 719–762.

G.A. Jervis, R.J. Block, D. Bolling and E. Kanze, "Chemical and Metabolic Studies on Phenylalanine. II. The Phenylalanine Content of the Blood and Spinal Fluid in Phenylpyruvic Oligophrenia" *Journal of Biological Chemistry* (1940), 134: 105–113.

G.A. Jervis, "Phenylpyruvic Oligophrenia: A Deficiency of the Phenylalanine-Oxidizing System" *Proceedings of the Society of Experimental Medicine and Biology* (1953), 82: 514–515.

G.A. Jervis and R.J. Block, "Chemical Investigations on Phenylpyruvic Oligophrenia" American Chemical Society, Division of Biological Chemistry, Abstracts of American Chemical Society Meeting (1939), pp. 21–22.

E.S. Kang, Consultant on Metabolic Diseases, Division of Health Services, to Mary Egan, Chief, Nutrition Section, Division of Health Services, January 4, 1967. Records of the Children's Bureau, File 4-5-11-5.

L.E. Kay, *The Molecular Vision of Life: Caltech, the Rockefeller Foundation, and the Rise of the New Biology* (New York: Oxford University Press, 1993).

J.F. Kennedy, "Statement" reprinted in *Journal of Pediatrics* (September 1962), 61: 494–498.

D.J. Kevles, *In the Name of Eugenics: Genetics and the Uses of Human Heredity* (New York: Alfred A. Knopf, 1985).

W.E. Knox, "An Evaluation of the Treatment of Phenylketonuria with Diets Low in Phenylalanine" *Pediatrics* (1960), 26: 1–11.

W.L. Laurence, "Find New Chemical to Sift Life Secrets" *New York Times*, April 4, 1939.

R.D. Lyons, "Brain Defect Risk is Eased by Diet" *New York Times*, May 3, 1968.

R. Masland, S. Sarason and T. Gladwin, *Mental Subnormality: Biological, Psychological, and Cultural Factors* (New York: Basic Books, 1958).

Massachusetts Department of Public Health, "Cost-Benefit Analysis of Newborn Screening for Metabolic Disorders" *New England Journal of Medicine* (1974), 291: 1414–1416.

T.A. Munro, "Phenylketonuria: Data on 47 British Families" *Annals of Eugenics* (1947), 14: 60.

New York Times, "A.M.A. Cites Gains by Medicine in '62" (January 2, 1963), p. 1.

D.B. Paul, "PKU Screening: Competing Agendas, Converging Stories," in E. Mendelsohn and M. Fortun (eds), *The Practices of Human Genetics* (Dordrecht: Kluwer, in press).

L. Pauling, "Sickle-Cell Anemia: A Molecular Disease" *Science* (1949), 110: 543–548.

L.S. Penrose, "Presidential Address. The Influence of the English Tradition in Human Genetics" in J.F. Crow and J.V. Neel (eds), *Proceedings of the Third International Congress of Human Genetics* (Baltimore: The Johns Hopkins Press, 1967), pp. 13–25.

L. Pompian, "Far-Reaching Fight Against Mental Retardation" *Today's Health* (May 1961), 39: 46, 60–63.

"Press Release" December 14, 1965. Children's Bureau. U.S. Department of Health, Education, and Welfare. Records of the Children's Bureau, File 4-5-11-5.

C.R. Scriver, "Science, Medicine, and Phenylketonuria" *Acta Paediatrica* (1994), 83: 11–18.

M.S. Sibinga and C.J. Friedman, "Complexities of Parental Understanding of Phenylketonuria" *Pediatrics* (August 1971), 48: 216–224.

"Suggest Premarital Test to Check Mental Defect" *Science News Letter*, August 25, 1956.

H.E. Sutton, *Genes, Enzymes, and Inherited Diseases* (New York: Holt, Rinehart and Winston, 1961).

J. Swazey, "Phenylketonuria: A Case Study in Biomedical Legislation" *Journal of Urban Law* (June 1971), 48: 883–931.

A. Wehrwein, "'Vast Medical Developments' Hailed" *Boston Herald*, January 3, 1963, p. 2.

L.I. Woolf, "Large-Scale Screening for Metabolic Disease in the Newborn in Great Britain" in J.A. Anderson and K.F. Swaiman (eds), *Phenylketonuria and Allied Metabolic Diseases*. Proceedings of a Conference held at Washington, D.C., April 6–8, 1966 (Washington, D.C.: USGPO, 1967), pp. 50–61.

L.I. Woolf and D.G. Vulliamy, "Phenylketonuria with a Study of the Effect upon it of Glutamic Acid" *Archives of the Diseases of Childhood* (1951), 26: 487.

CHAPTER 8

"A 'Cage' of Ovulating Females": The History of the Early Oral Contraceptive Pill Clinical Trials, 1950–1959

Lara Marks

In 1955 Katherine McCormick, the chief financier of the early development of the oral contraceptive pill, complained bitterly about the difficulties the investigators were experiencing in recruiting the right type of women to participate in trials. The real problem, as she put it, was how to get a "'cage' of ovulating females to experiment with".[1] What is most striking is McCormick's suggestion that women could be reduced to their reproductive physiology, i.e. that they could be seen merely as "ovulating females". Indeed, the notion that women could be reduced to their reproductive physiology was the very foundation on which the early trials of the pill were built. Reproductive physiology itself was also conceived merely as an interplay between various hormonal molecules which could be manipulated within the body by introducing new chemical hormones (Oudshoorn, 1994). In this sense, the pill was part and parcel of the molecularization of reproduction.

McCormick's words indicate, however, that this approach faced obstacles. It could be argued that in alluding to a "cage" of "ovulating females", McCormick believed that women could be treated merely as animals. Nonetheless, those undertaking the early clinical trials of the pill soon discovered that the practical application of a reductionist perspective of reproductive physiology was much more difficult in the case of the trials with women than with animals. As this chapter highlights, women, unlike animals, were free agents in their own right and could not be monitored 24 hours every day. McCormick herself admitted,

> Human females are not easy to investigate as are rabbits in cages. The latter can be intensively *controlled all the time*, whereas the human females leave town at

> unexpected times and so cannot be examined at a certain period; and they also forget to take the medicine sometimes; — in which case the whole experiment has to begin over again, — for scientific accuracy must be maintained or the resulting data are worthless.[2]

Despite this recognition, it was difficult for the scientists to break away from the parameters of the animal experiment. The need for accurate and constant observation was just as important in the trials conducted with women as it was with animals. Unlike animals, however, women could not be relied upon to be passive, and their active co-operation was needed if the scientists were to conduct their work.

Monitoring and testing was a problem common to other drug trials, but it was more difficult in the case of the pill because the drug was intended to be administered to healthy women out in the field and not, as was normally the case for other drugs, to sick patients under the close supervision of an outpatients' clinic or within a hospital. Aimed to be given to healthy women over long periods, the pill also raised important questions about injurious side-effects. While scientists and the medical community had been troubled about the hazardous impact of administering experimental drugs since the 19th century (Lederer, 1995; Rupke, 1987), the oral contraceptive pill, because it was aimed at healthy women of reproductive age, posed additional worries such as the possibility of causing sterility and foetus malformation.

Fears that the pill might result in detrimental side-effects were also heightened by the fact that contraceptive research was overall regarded as socially and politically unacceptable. As early as the 1920s and 1930s scientists in Britain, Germany and the U.S. had speculated about the possibility of developing an oral contraceptive on the basis of the growing knowledge about reproductive physiology and biochemistry, but few were willing to undertake such research for fear it would undermine their respectability (Vaughan, 1972; Borell, 1987, pp. 51–87; Oudshoorn, 1994, pp. 113–115). Strong legal obstacles also prevented such research. In the U.S. the Comstock Laws, passed in 1873 and not repealed until 1971, made contraception and contraceptive research illegal in a number of states, notably Massachusetts and Connecticut. Significantly, although the initial biological and laboratory testing for the oral contraceptive was undertaken within the U.S., the initial field studies with women were conducted in Puerto Rico, an American colony, which was outside the jurisdiction of the Comstock Laws.

The strength of the impediments facing researchers can be seen from the long and bitter struggle many campaigners, particularly women, had mounted since the late 19th century for the right to contraceptive knowledge and the means to control fertility. Many campaigners had experienced legal prosecution and imprisonment as a result of their activities. One of the most important figures in this battle was Margaret Sanger. Originally galvanized by the bad health and high rates of maternal mortality she witnessed while working as a nurse on the lower East side of New York at the turn of the century, Sanger succeeded in establishing a number of American birth control clinics and an international organisation to promote contraceptive research. One of her aims was to find a method of contraception for women that was divorced from coitus, and that ideally could be taken by mouth. Sanger was joined in her work by the suffragist Katherine McCormick, who for years had campaigned for birth control on the grounds that women's empowerment lay in their ability to control their fertility. Inheriting from her husband the fortunes of family who had founded the International Harvester Company, McCormick became the major sponsor of the development of the pill in the 1950s (McCann, 1994; Soloway, 1982; Gordon, 1977; Leathard, 1980; McLaren, 1990; Chesler, 1992).

This paper explores the early clinical trials of the pill in the 1950s, highlighting the complexities behind the application of a reductionist molecular model of medicine to the human body. As we shall see, while initially premised on a notion that all animals and women share a particular reproductive cycle that can be controlled by the manipulation of certain hormones, those developing the pill soon realised that this simple notion faced important challenges. Their success greatly depended not only on finding a large group of women who shared similar characteristics and who would co-operate for a long period of time, but also on having access to appropriate personnel who had the training and skill to collect uniformly accurate data and access to appropriate facilities for carrying out specific tests needed for measuring the effects of the pill. Nonetheless, the scientists succeeded in developing an effective contraceptive pill that was used by an increasing number of women despite on-going controversies about its safety.

Helping women to control their own fertility was not the only motivation in developing the pill. Indeed, many scientists and policy-makers saw the pill trials as a crucial step in the campaign to curb the problem of unprecedented population growth in the developing world and the resulting social, economic and political crises which they feared this would cause. Ideally aimed at the illiterate and poor population of developing countries, the success of the pill therefore rested on a formula that would be easy to use by women from a variety of cultures and educational backgrounds.

A VARIETY OF DISCIPLINES IN THE MAKING OF THE PILL

The development and testing of the oral contraceptive pill involved a number of scientists and medical practitioners from a variety of disciplines such as chemistry, endocrinology and gynecology. Foundations for developing the pill were built on the expertise accumulated during the early 20th century in steroid chemistry, particularly the experiments in the 1930s and 1940s of Max Ehrenstein which showed the possibilities of changing the structure of progesterone to make it more potent when swallowed, and those of Russell Marker which indicated that progesterone could be synthesized from plants such as the Mexican yam. The first chemical synthesis of an orally effective progestagen capable of being used as an oral contraceptive, norethisterone, was effected by Carl Djerassi and his student Luis Miramontes at the pharmaceutical company Syntex, in Mexico City in 1951 (Vaughan, 1971, pp. 10–19; Reed, 1978, 1983).

Initially seen as providing a potential cure for gynecological disorders, norethisterone was not automatically thought to be the means for devising an oral contraceptive. The actual inspiration for using synthetic hormones, such as norethisterone, as compounds for an oral contraceptive came from the biologist Gregory Pincus. As co-director of the Worcester Foundation for Experimental Biology in Shrewsbury, a private research institution based in Massachusetts, Pincus had greater freedom to pursue contraceptive research than researchers who were reliant on government funds elsewhere.[3] Highly dependent on donations and private sources for funding, Pincus saw the pill as potentially a lucrative research exercise for the Worcester Foundation as well as a means for his own professional advancement. Originally persuaded by Katherine McCormick and Margaret Sanger to develop an oral contraceptive in 1951, Pincus and a team of scientists and medical practitioners began a series of collaborative experiments and trials in the early 1950s (Vaughan, 1971, pp. 10–19; Reed, 1978). Funded by McCormick and directed by Pincus, these trials are the focus of this paper. Much of the work undertaken by Pincus and his team involved the screening of substances for their toxicity and ability to prevent pregnancy. The overall design of these trials was to establish an effective means of controlling fertility using various chemical hormonal preparations with the least side-effects to women. As we shall see, the techniques and methods used in these trials involved an intricate knowledge of reproductive physiology and endocrinology.

Pincus was well placed to undertake the development of the oral contraceptive pill. From his graduate years he had studied the mechanism of fertilization and by the early 1950s was an established expert on mammalian sexual physiology. Pincus's search to understand the developmental process of fertilization aimed to help medical experts to prevent spontaneous abortion and menstrual disorders. During the 1930s, however, he caused uproar among scientific and lay communities as a result of his experiments with the *in vitro* fertilization of rabbit eggs, and his claims to have produced fatherless (parthenogenetic) rabbits.[4] Nonetheless, by the 1950s he had become a key leader in the field of hormone and steroid research, and become a research consultant for the pharmaceutical company G.D. Searle. Through his contact with Searle and other pharmaceutical companies he not only accumulated funds to maintain the Worcester Foundation, but also had access to the latest discoveries in steroid chemistry, and by the 1950s the Worcester Foundation had become a key clearing station for many of the companies to test the physiological effects of their new drugs (Reed, 1978, pp. 343–344).

At the Worcester Foundation Pincus had at his disposal a team of scientists and technicians working on a variety of projects concerning reproductive physiology and endocrinology. Among these was the Chinese-born scientist Min-Chueh

Chang, whose early work at the Foundation involved investigating the process of fertilization.[5] It was to Chang that Pincus entrusted the initial work on animals for the development of the pill. While Chang undertook much of the thinking and planning of the initial trials, the crucial mechanics of the experiments were undertaken by technicians.[6] By 1954 the technician Anne Merrill was the chief organiser of the animal trials and the biological tests conducted among women recruited for the first human trials.

EXPERIMENTS WITH ANIMALS

Much of the development of the pill was inspired by the expertise in endocrinology and reproductive physiology built up from the early 20th century. First described by the physiologist Ernest Starling in 1905, hormones had become the subject of intense investigation among physiologists and chemists during the interwar years. By the 1930s scientists had not only isolated the female hormones, progesterone and oestrogen, but had understood their role in determining the menstrual cycle and were able to demonstrate the importance of progesterone in suppressing ovulation. At the same time the chronology of women's ovulation was recognised to be in the mid-cycle of the menstrual cycle and elaborate tests such as the vaginal smear and endometrial biopsies had been developed for detecting ovulation and endometrial changes (Pincus, 1965, pp. 200–201; Borell, 1977; Vaughan, 1971; and Reed, 1978, pp. 311–316).

The most direct stimulation for Chang's initial work on the pill came from the various animal experiments conducted by Makepeace, Weinstein, and Friedman from the 1930s onwards, who had demonstrated that progesterone could be used to prevent ovulation.[7] Between 1951 and 1952, Chang began investigations into the best method and appropriate dose for administering progesterone to prevent pregnancy. He conducted the tests with varying doses of natural progesterone and administered the substance in three different ways: by mouth, by injection and by vaginal pessary. Each animal was examined for the precise physiological effects of the progesterone, which involved examining its impact on spermatozoa, penetration of sperm into the egg, viability of the unfertilized egg, ovum cleavages and blastocyst growth, and lastly its interaction with the mechanism of implantation.

Chang used two types of animals for his experiments: rabbits and rats. Rabbits, because they only ovulate ten hours after mating, were particularly useful subjects for controlling and detecting ovulation. Having kept the rabbits in strict isolation from other animals for a number of weeks, progesterone could be administered and the animal then easily checked for ovulation suppression. Within 48 hours of mating, each rabbit was examined for ovulation through a laparotomy (opening the abdomen by surgical operation). Ovulation was determined by the detection of burst follicles in the rabbits' ovaries. Rabbits who ovulated were given a second laparotomy on the ninth day to examine the effect of progesterone on fertilization, cleavage and blastocyst development. All the rabbits were sewn up after each laparotomy and were reused for other experiments once the effects of the compound had worn off.[8]

Rats were also included in the investigations. Unlike rabbits, rats exhibit spontaneous ovulation in the course of their oestrous cycle and are thus more comparable with the human female (Pincus, 1965, pp. 57, 67–68). Ovulation, however, is less easily observed in rats, which necessitates more time-consuming and complicated procedures than carried out with rabbits. Before being given progesterone, the female rats had to be isolated from male rats, and then had to undergo daily vaginal smears to determine the exact stage of their reproductive cycle. Once given progesterone, the female rats were caged with male rats. During the following days the female rats were again subject to vaginal examinations, this time to detect sperm or certain vaginal secretions. Pregnancy could be determined by touching the animal's abdomen. However, unlike rabbits, which were expensive and therefore important to preserve, rats were usually sacrificed because they were relatively inexpensive and more delicate and thus more susceptible to damage. In each case the uterus and ovaries of the rats were scrutinized and weighed for signs of ovulation and the uterus examined for the presence of fertilised oocytes. This work was a time-consuming and laborious process.[9]

By January 1952 Pincus proclaimed that progesterone had effectively suppressed ovulation and prevented pregnancy, without destroying overall fertility of the animals in the long run. It was most effective and worked for long periods when given at high doses. One difficulty, however, was that while Pincus and others felt that an oral contraceptive was preferable to an injectable contraceptive or a vaginal suppository, progesterone appeared to be strongest when given by injection rather than by mouth.[10] Much of the work with animals was therefore guided towards finding a chemical substance which would be more active orally.[11] Between 1952 and 1953 a large variety of compounds, including synthetic progestins, were tried out on the animals. The explicit purpose of these tests was to find a substance that would be suitable for humans.[12] Until tested on animals the action of each compound was completely unknown. In each case a control group of animals was initially screened for determining the standard effect expected from different doses of pure hormones, and then matched against results observed in animals given synthetic compounds.[13] Animal experiments continued throughout the 1950s, the prime purpose being to screen different substances for their toxicity and effect on ovulation inhibition before being administered to women.

HUMAN TRIALS

The underlying logic of using animals for preliminary tests was that the process of reproduction was ultimately the same in animals as it was in humans. As Pincus argued in 1965,

> Fertility in women depends upon the normal production, fertilization, tubal transport and cleavage, uterine entry, blastocyst development, and implantation of ova and the

> maintenance of the fetus to parturition. Hartman has concluded: "the hazards of the human ovum from ovulation, fertilization and nidation to birth are the same fundamentally in the human species as in the rest of the Mammalia." Logically, then, the experimental studies concerned with female animals should find application in women and, indeed, most of them do. There are some variations in detail, but the basic controlling processes appear to be identical (Pincus, 1965, p. 197).

Theoretically the experiments that were conducted among the animals and the effects that were observed were therefore easily transferable to women.

Nonetheless, while the animal trials proved that progestational compounds had no deleterious side-effects and effectively controlled ovulation, no one could be certain of their effects on women.[14] Progesterone and its synthetic derivatives had been administered to women by medical practitioners, primarily for gynecological and reproductive disorders, since the 1930s. Such work had indicated that progesterone could be used in human subjects in very large doses and for long periods without ill-effects. Indeed, this was one of the reasons that Pincus chose to focus on progesterone and its derivatives rather than oestrogen and androgen substances which were known to have side effects.[15] Nonetheless, whether the progestational compounds could be used as an effective oral contraceptive in women remained a mystery. While progesterone had been tested a great deal during the luteal phase of the menstrual cycle of women, little was known about its impact on the pre-ovulatory stage of the cycle, which was critical for seeing whether progesterone was feasible as an oral contraceptive.[16]

As a biologist Pincus did not possess the authority or the clinical ability to conduct trials with women. In April 1953, John Rock, an obstetrician and gynecologist from Harvard University, agreed to join forces with Pincus to conduct tests on women. As a Roman Catholic, Rock had long been interested in finding a means to improve the efficacy of the rhythm method and had explored possibilities for developing easier means for ovulation detection. Based at the Free Hospital for Women in Boston, Massachusetts, and with the assistance of Miriam Menkin, and the doctors Herbert W. Horne, Angeliki Tsacona and Luigi Mastroiani, he had also treated women with gynecological disorders and problems of infertility for years. His treatment consisted of using progesterone and oestrogen to stimulate a pseudo-pregnancy, which he felt would correct dysfunctioning wombs and fallopian tubes and thereby lead to fertility.[17] Rock's assets were not only his expertise in the techniques for detecting ovulation and his use of hormone treatment for infertility, but also his access to a large number of infertile patients who were willing to undergo intricate and time-consuming tests with progesterone. In exchange for providing crucial evidence in developing an oral contraceptive, Rock hoped his involvement with Pincus's research would lead to advances in curing infertility.

SMALL-SCALE TRIALS

In May 1953 the first small-scale trials with women were launched at the Free Hospital for Women in Boston and at the Worcester State Hospital.[18] The aim of these trials was to find a "universally acceptable means to suppress temporarily and at will, human reproduction, without disturbing physical and mental well-being." As in the case of animals, the initial substance used was natural progesterone and it was administered to the women in three different ways: by pill, injection, and vaginal suppository.[19] Prior to running these trials, tests were performed in other places, such as Israel, on individual women to see what type of dosage should be used in the trials and how the substance should be administered and for how long.[20]

Pincus's original plan for the small-scale trials was to recruit between 30 to 40 women at the Worcester State Hospital with the help of the endocrinologist, Dr Harry Freeman, and between 10 and 15 for the one at the Free Hospital for Women, supervised by John Rock with the help of Miriam Menkin and Dr Angeliki Tsacona. Finding volunteers for these studies, however, they assumed would be difficult. As Pincus stated to McCormick, while the aim was to get 70 cases by the beginning of 1954, they would consider to have done well if 36 cases had been obtained by that stage. Nurses were sought as volunteers in the case of the Worcester State Hospital. By March 1954 only twelve women had been recruited at this hospital, reflecting the contending demands of summer vacations and job transfers. More success was achieved in recruiting women at the Free Hospital, where seventeen sterile women who had regular menstrual cycles joined the trial from the hospital's Fertility and Endocrine Clinic and private practice of the doctors involved in the clinic.[21]

The initial object of the two small-scale trials was to observe normal menstrual cycles (labelled "control cycles") among the women. This involved women taking their basal temperature every day, performing daily vaginal smears, collecting urine specimens, for analysis, over a 48-hour period on their seventh and eighth postovulatory days, and undergoing regular endometrial biopsies.[22] These observations were followed for two subsequent cycles with the oral administration of progesterone (supplied by Syntex) between the fifth and twenty-fifth day of each woman's cycle. Injections and intravaginal tests were also done with progesterone on those women recruited into the Free Hospital study. In order to check the impact of the progesterone on ovulation, each woman had to carry out the same tests as they had done when observing their normal menstrual cycles. Laparotomies were also performed on two women taking progesterone who were already undergoing abdominal operations for other reasons at the Free Hospital.

The evidence collected from these initial studies was analyzed statistically and suggested "a limited effect of progesterone on the first experimental cycle but a definite disturbance of ovulation during the second experimental cycle."[23] Further work, however, was needed to find out the exact doses that were needed to effectively suppress ovulation and to assess long-term effects. Some of the women encountered break-through bleeding while taking the progesterone.[24] By June 1954, 60 women had participated in these two small-scale trials, of whom only 30 had provided accurate information for Pincus and Rock to conclude that progesterone produced satisfactory contraceptive actions.[25]

Pincus felt it important to extend the small-scale trials in other places before launching larger studies. Various possibilities were discussed, including mounting trials at a clinic run by the Planned Parenthood Federation of America in New York, and the newly established Medical School at Puerto Rico. Japan, Hawaii, India, and Mexico were the other places considered.[26] There were, however, particular problems in extending the small studies. As Pincus admitted in 1954, there were few places where vaginal smears, endometrial biopsies and urinary pregnanediol tests could be carried out accurately and many institutions would "involve an extraordinary waste of time and energy." Similarly the tests were expensive to perform.[27] Pincus went on to say that it was also a question of the right personnel. What they needed was

> a physician who is completely familiar with the vaginal smear techniques, and who can check rather carefully on the work of the technicians. The technician takes everything that comes along and runs it through the routine. She may then read what are really rather poor slides either from the point of view of the way the smear has been collected or the way that it has been fixed and stained. This can lead to faulty diagnosis. Secondly, the physician in charge should have an accurate history of the patient, particularly from the point of view of her reproductive and menstrual history. This enables one to make a more intelligent appraisal of the immediate date [*sic*] obtained. This is a little hard to explain briefly, but may be perhaps best illustrated by the statement that the human menstrual cycle is not a clock-work affair and that characteristic individual variations in it are meaningful to the trained observer.[28]

Getting appropriate personnel who would co-operate in a small-scale trial was one of the reasons for the reluctance to undertake a trial in Japan.[29] Pincus was more confident about gaining accurate evidence from a trial in Puerto Rico, because, as he stated to McCormick, "those directing it have the American approach, and are American trained", and he would also be able to keep a closer eye on those running the trial.[30]

In addition to the difficulty of obtaining adequately trained staff, the investigators also faced the problem of finding suitable volunteers. While the women were not expected to be caged like animals, the trials demanded a high degree of co-operation on the part of each woman so that accurate results could be obtained. As McCormick wrote to Sanger in June 1954,

> The *headache* of the tests is the co-operation necessary from the women patients. — There is so much of it and it must be accurate. I really do not know how it is obtained at all — for it is onerous — it really is — and requires intelligent attention, persistent attention for weeks. Rock says that he can get it only from women wishing to become fertile: those who wish to be sterile are not ready to take so much trouble: I cannot see any way of overcoming this headache in the tests — unless one can furnish enough nurses to go around to their homes and see that the women patients do accomplish the tests regularly and correctly.[31]

As shown in Tables 8.1–8.3, the protocols laid down for selecting women for the trials run in Worcester and Boston (and used in later trials elsewhere) reveal that

Table 8.1 Procedures for women participating in the Worcester and Boston trials

Case procedure is

1. Observations are made of at least 4 menstrual cycles, the first of the 3 cycles is the "control" record during which no progesterone is administered. During the 2nd and 3rd cycle progesterone is administered in definite doses.
2. Each female is examined (before being accepted as an experimental subject) for the following qualifications
   a. Is she sufficiently intelligent to cooperate?
   b. Is she willing to cooperate?
   c. Are her home conditions such that she can cooperate?
   d. Would an unexpected pregnancy by a disaster for her?

Also necessary to ascertain:

a. Is her menstruation of the normal 4-week type?
b. Has she the classical bi-phasic oral temperature?
c. Has she a normal mucous lining of her uterus? (determined by biopsy)
d. Has she normal cell changes in the samples of the mucous lining of her vagina? (determined by smears)
e. Does her urine show normal excretion of pregnandiol and other steroids?

Source: "Procedure for progesterone tests", (n.d, *ca* 1954) Rock papers, Countway; and GP-LC, Box 17. See also document "Suitable Experimental Subjects", and "Objectives for Experimentation", November 1954, Rock's Papers, Countway.

Table 8.2 Procedures for accepting women for the trial run in Worcester and Boston

Women are asked:

Can you read a thermometer? If not, is there someone in your home who can and will? Is your domestic set-up such that you can take your temperature every morning as soon as you wake up, and at the same time every day? Can you record your temperature on the chart in the right place?

Can you, and are you willing to, take a vaginal smear every day on a special glass slide; can you put it in the jar of fixative for just the right time; can you put it in the right place in the slide-box?

Will you collect in the bottle we give you all the urine passed during a specified 48 hour period, and then bring it to the clinic?

Will you come to the clinic for a biopsy when requested?

During the second and third cycles we will give you tablets to take — 6 or 8 every day (about 1 every 6 or 8 hours) beginning with the 5th day of your menstrual cycle (5th day following onset of menstruation) and continuing through the 25th day. Will you take these and will you note on the chart we give you the time at which you take the tablets, and also any time you miss taking the tablets?

Can you and will you keep the charts accurately?

Will you follow this routine for a 3-month period exactly as we direct you?

Source: "Procedure for progesterone tests", (n.d, *ca* 1954) Rock papers, Countway; and GP-LC, Box 17. See also document "Suitable Experimental Subjects", and "Objectives for Experimentation", November 1954, Rock's Papers, Countway.

Table 8.3 Procedures once accepted for the trials to be run in Worcester and Boston

The following is the daily, weekly, and monthly routine in which women were to cooperate during the control cycles as well as during the cycles of treatment.

1. Temperature chart (*daily* oral temperature taken and recorded by the subject on awakening and at the same time every day.
2. Daily vaginal smear (taken by the subject on awakening every day, including those of the menstrual period. (The subject is taught how to do this and provided with the necessary equipment.) Using a pipette inserted into the vagina, she secures a sample of the vaginal contents; places it on a special glass slide numbered to correspond with numbered slots in slide box; drops the slide into a jar of fixative for a minimum period of twenty minutes; removes the slide and drops it into the correspondingly numbered slot box. These slides are kept for one entire cycle by the subject, then taken by her to the physician for laboratory study. (Proper interpretation of these slides to determine ovulation time requires special training.)
3. Endometrial biopsies are made by the physician on the 9th–10th "post ovulatory" day (estimated individually for each subject from the day the ovum, if released , is presumed to leave the ovary). The biopsied specimen is used for microscopical examination to determine whether or not a secretory endometrium (normal mucous lining is present.
4. Once in each cycle a 48-hour collection of all urine excreted is made by the subject on the 4th–6th post-ovulatory day. All the urine is collected by the subject in a large jar and this jar is then taken to the physician for a special laboratory examination to determine amounts of pregnandiol and other steroids by tests unobtainable in the ordinary urine-analysis laboratory.

   Suitable patients who are obliged to undergo abdominal surgery (for causes not connected with the progesterone experiments) may be given progesterone before the operation so that the surgeon may have an opportunity to observe direct from the body the effects of progesterone on ovulation, thus checking previously gathered clinical findings.

Source: "Procedure for progesterone tests", (n.d, *ca* 1954) Rock papers, Countway; and GP-LC, Box 17. See also document "Suitable Experimental Subjects", and "Objectives for Experimentation", November 1954, Rock's Papers, Countway.

women had to undergo more than just a thorough screening before joining the trial.[32] They were expected to stick to a set of complicated rules and undergo intense scrutiny, sometimes even surgery to investigate their ovaries.

In asking the women to undertake certain procedures it was clear that women were not merely passive subjects, but had an active role in the process of the trials and much depended on their co-operation. The investigators looked for women who were ideally "intelligent enough" so they could "be relied upon to carry out the procedure."[33] Not all women had the leisure, let alone the knowledge, to perform the intricate tests required for the trial. As Rock admitted to McCormick in July 1954, women with families could not be used in the trials as they did not have the time considered imperative for undertaking the elaborate tests on themselves. Making sure that the women performed the necessary tests was also time-consuming for the doctor. At the Free Hospital in Boston, 25 women occupied the full time of one doctor.[34]

Many of the difficulties that Pincus and Rock foresaw in extending the small-scale studies became a reality. The first extension of the small-scale trials was undertaken at the Margaret Sanger Research Bureau in New York, run by Dr Abraham Stone. The clinic was chosen because of the regular attendance of women at its birth control clinics and its well-coordinated personnel.[35] Between 1954 and 1955 thirteen women with normal menstrual cycles, twelve of whom came from the infertility clinic of the Bureau, participated in the study for sixteen cycles. Women underwent the same tests as those in Worcester and Boston, except that they received a higher amount of progesterone (1000 mg) and the pills were administered at a different time during their cycle, from the eighth to the twentieth day. Getting results from this investigation proved slow not only because inaccurate data were recorded for the vaginal smears, but the investigators found it hard to find "suitable patients who would meet the requirements for the study, and who would be willing to continue regularly and faithfully to take daily the prescribed medicine."[36]

Problems also occurred in Puerto Rico, where small-scale trials were initiated among female medical students at the School of Medicine at the University of Puerto Rico in San Juan in January 1955. One of the reasons for undertaking the trials at the medical school were that the personnel were reliable and sufficiently well-organized. Puerto Rico was also sufficiently near to the U.S. that Pincus and his team could visit the site easily and collect the vaginal smear slides and urine samples to analyze at the Worcester Foundation's laboratories.[37] Two of the Puerto Rican physicians were trained at the Worcester Foundation for two weeks in order to "understand the importance and the routine" of the tests in the trial.[38] Staff were also trained on site in Puerto Rico in the taking of temperatures, vaginal smears, endometrial biopsies and urine analysis. Such procedures were to be carried out a number of times on each subject before and after the administration of progesterone. All the vaginal smears were to be stained on location and then sent to the laboratory at the Worcester Foundation to ensure "uniformity in interpretation."[39]

Despite all these procedures to guarantee adequate training of staff and consistent collecting of accurate data, the greatest problem for the trial in Puerto Rico proved to be obtaining suitable volunteers. Prior to starting the project, the doctors in Puerto Rico feared that women would lose interest in the trial if they were expected to participate for too long.[40] Once started, the investigators faced other problems. In June 1955 Dr Celso-Ramon Garcia, who supervised the project, reported to Pincus considerable difficulty in keeping the women on the trial. Of the 23 women who started the trial only 13 completed both two control and two progesterone cycles. Some of the students dropped out on departing Puerto Rico to take up internships in the U.S., while one was forced to leave the school on account of failing her course and could not be traced to continue in the study.[41] Even students who did complete the pilot study often did not adhere strictly to the rules. Although they were volunteers, it has been suggested that students were threatened that their grades would suffer should they fail to comply with the regulations (Ramirez de Arellano and Seipp, 1983, p. 111). Further investigations were made to see if the trial in Puerto Rico could be extended to using student nurses from the San Juan City Hospital and female prisoners from the Women's Correctional Institute at Vega Baja, but these proved fruitless. Resentment on the part of the female prisoners towards the project proved enough to disrupt the discipline of the prison. In January 1956 investigators at the Medical School decided not to pursue any further work on the project because of the difficulties they had in obtaining suitable subjects.[42]

More success was achieved in retaining women on the trial among the fifteen schizophrenic women studied at the Worcester State Hospital from June 1956. These women were easier to monitor and control because they were within the confines of the hospital and the doctors were therefore less dependent on the women remembering to perform their own temperature readings, vaginal smears and other tests necessary for the project. As Pincus admitted, in working with psychotic women the scientists could undertake "fairly intensive observations", such as extra urine and blood tests that they could not do with other women. Much of the work that was done with these patients was to establish an effective dose and regime of hormones that would not result in the break-through bleeding that had been witnessed among women during the earlier trials run at the Free Hospital for Women in Brookline.[43] During the trial, psychiatric assessments were done on the women. Many experiments on animals and humans undertaken by scientists in previous years had indicated that progesterone might be used as a tranquillizer.[44] Sixteen men were also included in the trial to study the effects of progesterone on their testicular function and whether it would affect their psychiatric and sexual behaviour. Like the women, all the men underwent intensive medical tests, including testicular biopsies.[45]

It might be argued that in using psychotic patients the scientists had at long last found the cage of humans as envisioned by McCormick. While the investigators sought permission from the relatives of the patients before assigning them to the trial, the trials were nonetheless performed on people not choosing to undergo the treatment of their own free will (Vaughan, 1971, p. 39). It was also difficult to ascertain the contraceptive effects of the administered hormones in the women, for none of them were having sexual intercourse. Similarly, psychiatric problems can interfere with a woman's menstrual cycle, and thus it was difficult to establish whether the pattern of each woman's ovulation was affected by psychiatric causes or by the progesterone.[46] The men presented different difficulties. Investigators found it impossible to measure their potency and sexual arousal because of the difficulty in collecting sperm specimens and in monitoring the amount of time each man masturbated.[47]

LARGE-SCALE "FIELD" TRIALS

While the small-scale studies indicated that progesterone and its synthetic derivatives could suppress ovulation, Pincus and his team needed a much larger group of women if they were to prove to the outside scientific and lay communities the safety and effectiveness of such an oral contraceptive. The reservations scientists had about the research were articulated by Sir Solly Zuckerman (Professor of Anatomy at Birmingham University) when Pincus publicly announced his contraceptive research for the first time at the International Planned Parenthood Federation conference in Tokyo in October 1955. Zuckerman argued that Pincus' studies proved nothing new except in the form of administering the compounds, for scientists had known since the 1930s that oestrogen and progesterone could suppress ovulation. Similarly, he argued that the animal tests were inconclusive with respect to whether the oral contraceptive would be safe and effective for women. Even the small-scale trials conducted were considered unsatisfactory in proving whether the pill suppressed ovulation, because many of the women on trial were infertile. Similarly, many of the tests performed to detect ovulation were regarded as unreliable. Clearly larger and more prolonged trials were required before scientific experts would recognise the safety and efficacy of oral contraception and whether it could be used on a long-term basis (Vaughan, 1971, pp. 36–37).

Yet a large-scale trial was not easy to set up. In addition to the legal restrictions on conducting contraceptive research within Massachusetts, one of the biggest problems was finding a large number of women who would be motivated enough to co-operate over a long period. Up to this time only women within a hospital setting or who were primarily of a middle-class background had participated in the trials. Nonetheless, if the pill was to be acceptable and as universal as the investigators intended, they needed to find women from a wider range of educational, cultural, geographic and class backgrounds.

The first field studies were mounted in Puerto Rico. Puerto Rico was regarded as ideal for finding volunteers because of the desperation for birth control given its growing population and poverty, and its active family planning movement and well-established network of birth control clinics. Moreover, many of the women to be targeted were semi-literate or fully illiterate and had large families. They were therefore regarded as perfect models for testing whether the pill could be used for women around the world, particularly in places where population problems were seen to be at their worst.[48] Puerto Rico also had the advantage that it was an island and therefore had a relatively stationary population that could be easily monitored. As Dr Celso-Ramon Garcia pointed out,

> The basic aspect was trying to find an area where there were large numbers of fertile women that were in need of contraception; look, it was a question of the law of supply and demand and where you could carry out these studies under a close supervision of the individuals concerned ... The object was to set up a program in an area where you would have better control over the particular population that you are dealing with.[49]

By March 1956, a total of 225 women had been identified for the project who matched in age, parity, marital history, and prior contraceptive experience. Of these, 100 women were to take the pill and 125 were to form a control group who were to use conventional contraceptive devices or none at all.[50] Those entering the experimental group, i.e. taking the pill, were told they were participating in a birth control trial, while those in the control group were informed they were part of a survey on the size of the families residing in the housing project. In each case they were told that the trial was run by the Family Planning Association of Puerto Rico which was a private agency with no governmental links and whose objective was family planning.[51]

Launched in April 1956, the trial was under the medical supervision of Dr Edris Rice-Wray, a woman doctor who held a university appointment and was director of both the Puerto Rican Family Planning Association and the Public Health Service in San Juan. Permission for the project was granted by the Secretary of Agriculture and the Director of Social Relations. Initial volunteers for the trial were drawn from families who had been moved as a result of a slum clearance to a public housing project in Rio Piedras, a suburb of San Juan. Here the investigators effectively had what Garcia has called a "captive population", as many of the families living in the housing project prized their accommodation and were thus unlikely to move away during the course of the trial and were therefore easy to monitor.[52]

The women to be included in the trial were identified by the superintendent of the housing project who was supportive of contraception and knew the tenants well. Those first chosen were already visiting birth control clinics and had previously participated in short-term contraceptive trials (using conventional barrier methods) and were known to be desperate for an alternative means of contraception. Many of the women were having to cope with large families on a very small income and with little support from their husbands. The condition that many of these women were in was observed by the technician Anne Merrill who conducted regular tests on them,

> I would see these young women with many, many children, although I wouldn't know they were young women until I'd looked at their records. Some of them looked like what I would think of as a great-grandmotherly type ... All wizened up and gaunt to the point where you wanted to do something for them. I'd think it was due to old age, but then I'd look at the records and the women would be thirty-four ... with ten children (Vaughan, 1971, p. 49).

All those included in the study had to be under 40 years old and to already have had children and thus proven fertility. They also had to be prepared to have another child if oral contraception failed, and to guarantee long-term residence in the same area so that they could remain on the trial for a year.[53] Each woman was given a complete medical examination to check that they were in good health prior to participating in the trial. Data were collected on each women's menstrual and pregnancy history and stored on report forms.[54] Once on the trial, each woman was given a bottle of twenty pills supplied by Searle. Known by its trade name as Enovid, each pill contained progestin (synthetic progesterone made from Mexican yams) and oestrogen (Ethynol estradiol or "mestranol"). Each woman was expected to take one pill for twenty days on a daily basis from the fifth to the twenty-fourth day of her menstrual cycle.

> They were told that during the period when they have finished the 20 pills and are waiting for the menses, they may have intercourse without fear. They were made to understand that, two or three days after stopping they would have the menses. It was made clear to them that it is necessary to use the method properly if they want to succeed or do not want to have escape bleeding during the month. They were told that if they were to stop the tablet one day, the next day they would probably have bleeding. In such case, they were instructed to double the dose for two days and, if it did not stop then, to take three for one day.[55]

Adherence to the instructions was ensured by the regular visit of a social worker to each woman at the end of each month's supply of pills. The social worker checked whether the right number of pills had been consumed at the right time and asked detailed questions concerning side-effects such as break-through bleeding, length of menstrual cycle, and frequency of coitus.[56]

Despite social worker observation, the investigators soon discovered how problematic it was to set up an effective monitoring system. One of the major difficulties was the fact that the social workers did not always find the volunteers at home when they called. Where too much time had lapsed between visits, women often found it hard to recall crucial personal medical details.[57] As a result some of the data collected was distorted, causing dilemmas for those trying to analyze the evidence collected from the trial.[58] Problems also occurred in monitoring the consumption of the pills. In order to keep a close eye on the number of pills taken and to make it easier for the women to know when to stop taking the medication, the women were initially given only one month's supply of pills, which was then supplemented by the social worker at the end of each month. This arrangement however, soon proved impractical as the women were often absent when the social worker called. It thus became necessary to provide two bottles to ensure continuation of the regime. Even this system, however, was not foolproof. When supplied with two months of pills some volunteers continued to take the pills without stopping for five days to allow for bleeding, and some of them feared that any gap in taking the medication would result in an immediate pregnancy and thus became agitated when the pills ran out and the social worker had not replaced them immediately. In order to combat such confusion simple printed instructions were given out on the assumption that most women would be able to read or have a schoolchild read the instructions for them.[59]

By June 1957, 55 per cent of the total women enlisted on the project had left for a variety of reasons, such as objections on the part of husbands, desire for pregnancy or sterilization, or moving to a new area.[60] Other women left as a result of negative publicity. Within the first month of the trial 30 women stopped taking the pill or gave it away on hearing of an article printed in a local newspaper which claimed the trial to be a "neomalthusian campaign" that would result in sterilization and harm women.[61] Adverse reactions to the pill were another major contributing factor to the departure of women from the trial. While warned before taking the pill that they might experience side-effects, many women refused to continue in the trial because they experienced problems such as nausea, dizziness, headaches, and vomiting as a result of the medication.[62] As Rice-Wray admitted in the first few months of the trial, such side-effects threatened the viability of any long-term success of the pill.[63] Within the first six months of the study twelve of the original 100 chosen for the study had left on account of the side-effects.[64] Of the total 295 women enrolled by June 1957, 22 per cent had dropped out as a result of side-reactions.[65]

While many left the trial, securing new volunteers was easy. As the social worker Iris Rodriguez wrote, "we have more cases than what we can take for our study. Continuously they are ringing this office asking for the pill, going to see Dr Rice-Wray and calling on me when I make the visits." A waiting list had to be drawn up to cope with the demand.[66] Nonetheless, while it was easy to find new volunteers, the high turn-over of women posed a major dilemma given that the trial was intended to be monitoring a stable population of women over a long period, and the investigators had to gather consistent data in order to win approval from their scientific colleagues, and more importantly, the Food and Drug Administration on whom they were dependent for gaining market approval of the drug as an oral contraceptive.[67]

Part of the problem was resolved by initiating extra trials elsewhere. Trials were established in Humacao, a municipal district in a predominantly rural area in the eastern part of Puerto Rico, as well as in Port-au-Prince in Haiti, Mexico City, and Los Angeles.[68] Those conducting these trials, however, faced similar problems to those experienced on the Rio Piedras project. One of the biggest problems for the Haiti project was the women's inability to follow the instructions. Over 20 per cent of the cases in Haiti forgot to take the pill and some stopped the medication whenever their husbands went away.[69] Many of them were illiterate and innumerate and therefore found the instructions difficult to follow and could not remember when to take the pills even when supplied with a calendar as a form of reminder. Even when the women were given rosary beads which they were to move individually each day on taking a pill, a number wore the beads in the belief that they alone provided protection against pregnancy without the need to take the pill. Others ignored the instructions altogether and presumed that they should consume all the pills in one go.[70] Such difficulties, however, were not universal. In Mexico City and Humacao, for instance, even the poorest and illiterate women were faithful in taking the pill.[71] Clearly much depended not only on the educational background of women, but also on individual motivation in the success of following instructions, as well as the skill of the instructor.

A major problem for all the trials was the unpleasant side-effects caused by the pill, which resulted in many women leaving the trials prematurely. While fewer women dropped out of the trials in Haiti, Humacao, and Mexico City than in Rio Piedras, averaging between 10 and 30 per cent of the total cases recruited, many left on account of side-effects. In Los Angeles, where a higher number (66 per cent) departed than even in the Rio Piedras project, at least 37 per cent of those who departed did so because of side-effects (Vaughan, 1971, pp. 54, 56). Initially Pincus and some investigators presumed that leading questions had resulted in psychosomatic symptoms rather than real reactions.[72] This was somewhat ironic, for initially many of the questions were posed because women were shy of reporting any unpleasant reactions for fear of displeasing the investigators.[73] The undesirable reactions weakened the longer a woman stayed on the medication and decreased as the dose of the pill was lowered. Some of the symptoms were also relieved by distributing antacid tablets.[74] The number of subjects dropping out also diminished over time, and those participating for lengthy periods of time increased the longer a trial was conducted.[75]

Nonetheless, the departure of women from all the trials posed a difficult challenge for the investigators. The scientists overcame this problem, as Oudshoorn has shown, by publishing their results in terms of the number of menstrual cycles rather than in terms of individual women. Cycles therefore replaced individual women. In 1959 Pincus thus transformed the 830 subjects enlisted for the four Caribbean field trials into "8133 menstrual cycles, or 635 woman years" (Pincus, 1959, p. 81, cited in Oudshoorn, 1994, p. 132). As Oudshoorn points out,

> This representation strategy had important consequences. The very act of representing women as menstrual cycles resulted in a major increase of scale: the grand totals of the trials now included much more impressive numbers than a focus on the individual subject might have achieved. The trials were thus presented as having met their purpose: the testing of progestins on large numbers of women over longer periods, as a prerequisite for its approval as a safe and reliable contraceptive. (Oudshoorn, 1994, p. 132.)

DETECTING ADVERSE REACTIONS

It might be assumed that in representing women as menstrual cycles the investigators were masking the difficulties inherent in the large-scale clinical trials. Yet, the investigators themselves were hesitant about the long-term implications of administering the pill.[76] Early on Rock confessed to McCormick that he had reservations about using synthetic progestins, because they might, firstly, lead to permanent sterility by diminishing ova production, and, secondly, cause long-term damage to the endometrium.[77] In addition many feared that the pill could cause congenital or androgenous defects in future children and were relieved when women began to produce healthy babies once off the pill.[78]

Equally troubling was the question of whether the pill would cause long-term effects such as cancer and cardiovascular disease. Because of the statistical difficulties inherent in studying such diseases, these side-effects not only required a much larger sample size than involved in the original trials, but also a longer time-span.[79] Nonetheless, the long-term consequences of taking the pill were of enough concern to one of the first doctors involved in running the trials, Celso-Ramon Garcia, to insist that certain tests be performed on the patients during the trials, such as investigations for liver functioning, Pap smears, and tests for blood clotting. The investigative tests became more and more elaborate as time went on and the investigators learnt more about the possible side-effects they were looking for. By the end of the study, blood was being tested for red and white cell counts and clotting time, urine was being analyzed, and hepatic tests were being performed as were skin and cholesterol tests and endometrial smears.[80] Not everyone agreed with the need for such investigations. Garcia recalled that during the early small-scale trials conducted at San Juan Medical School, he fought with the Professor of Preventive Medicine who was unconcerned about the possibility of hazardous effects and saw such tests as pointless. The professor's attitude reflected a more general attitude of the period. Tight legislative regulations for testing drugs were not an established feature of clinical trials until the 1960s when the Thalidomide scandal prompted stricter protocols for the testing of drugs across the world.[81]

Garcia admitted that part of the reason that he advocated so strongly for tests and Pincus' acceptance that they form a routine procedure within the trials was that they felt a responsibility to the women they were working with. In addition to this, they felt that they had a particular duty to ensure good health care for the women in the trials in Puerto Rico. As he argued, many of these women did not have access to the good healthcare services that were available to the more affluent women included in the first trials of the pill in the U.S.[82] Regular medical check-ups were thus installed as part of the trials in Puerto Rico and Haiti.[83]

Undertaking regular medical check-ups and investigative examinations, however, was not an easy operation. The recollections of the technicians undertaking the bulk of the work for the investigative tests highlight the real difficulties of carrying out such work in the field as opposed to the enclosed environment of a small and well-equipped laboratory of the Worcester Foundation. One of the major problems was the lack of facilities in places like Puerto Rico and Haiti. As the technician Mary Ellen Johnson recalled, everything had to be taken from the Worcester Foundation to Puerto Rico and Haiti.

> … it was a major undertaking, taking all the stuff down there, I mean, you literally had to take the lab, a portable lab with you … We would box everything up ahead of time, we'd take an inventory list of what was in each box, and we'd have great big numbers on the box, and every time we'd get on a plane, we'd count … If we stopped anywhere, or they unloaded anything, someone would double-check to see if anything got off-loaded. And when we arrived there, we would take the inventory, make sure all the boxes were with us. It really was a major production, getting everything there.[84]

It was not uncommon for equipment to be lost on the way, which was a major problem as it could not be replaced once in Puerto Rico or Haiti. In addition to this, none of the equipment was disposable, which meant that everything (needles, syringes, and glass slides) had to be washed in order for it to be re-used. Even this was not a straightforward process as distilled water was unavailable. Because of the difficulties of conducting the tests out in the field, only the most basic of tests were performed, and blood and other samples were preserved so that more elaborate tests could be done once back in Worcester. Storage, however, proved difficult in this process.[85]

In addition to the technical problems, the tests had to rely on women's co-operation. The tests took place at certain pre-arranged times of the year when the technicians and medical practitioners flew out from Massachusetts. In many cases the tests were performed in a central place which was not always easy for the women to get to. In Haiti, for instance, some women had to walk all night in order to be tested and then had to wait for several hours before being seen. Women also did not know what to expect.[86] Even basic gynecological tests were often unfamiliar to them as they were unaccustomed to seeing a doctor except when giving birth. At the end of the day only a small number of those in the trials in fact underwent the rigorous physical examinations expected of them.[87]

CONCLUSION

The history of the clinical trial of the pill thus raises some interesting questions about the ways in which scientists conduct research and apply a reductionist molecular model of medicine to the human body. As we have seen, while premised on a notion that all animals and women have a particular reproductive cycle that could be controlled by the manipulation of certain hormones, the development of the pill challenges this simplistic notion. Unlike animals that could be caged and watched constantly once given certain compounds, women were free agents whose co-operation had to be constantly sought and won. Such co-operation was vital if they were to swallow the pill regularly and also submit to the investigative tests needed to check the action of the pill.

Women were not the only obstacle in transferring the molecular models of reproduction developed in the laboratory out into the field. Another problem in conducting the trials was the difficulty of finding an area where there were personnel who could carry out uniform testing and provide accurate data that lived up to the standards set by Gregory Pincus and his team. The trials being carried out in an international setting, uniformity was not always easy to achieve. Not only were there problems of finding appropriate staff, but basic equipment and facilities were often absent. The transfer of ideas and practices developed in the laboratory out into the field involved a complex set of factors including professional expertise, political and social attitudes towards contraception and research, and most of all women's active co-operation.

ACKNOWLEDGEMENTS

I would like to thank Dorothy Hunt, Anne Merrill, Mary-Anne Johnson, Adaline Satterthwaite and Celso-Ramon Garcia for recalling their experiences with the development of the pill. I am also grateful to Roger Short and John McCraken for their helpful comments on the development of endocrinology and to Harmke Kamminga, Soraya de Chadarevian, Colin Duncan, David Edgerton and Geoffrey Venning for their useful suggestions on earlier drafts of the paper. Research for this paper was sponsored by the Wellcome Trust.

NOTES

1 McCormick to Sanger, May 31 1955 (Margaret Sanger's papers, Sophia Smith Collection, henceforth MS-SS)

2 McCormick to Sanger, November 13 1953 (Margaret Sanger's papers, Library of Congress, henceforth MS-LC)

3 The Foundation was a joint enterprise set up by Pincus and Hudson Hoagland. Formerly a fellow student with Pincus at Harvard University, Hoagland had originally enabled Pincus to gain employment at Clark University when unable to get tenure at Harvard University. For more information on Pincus and the Worcester Foundation see Vaughan (1971) and Reed (1978).

4 According to Pincus, his involvement with the *in vitro* fertilization of rabbits' eggs and the scandal that they caused was a major factor leading to Harvard University's refusal to grant him tenure. For a more detailed description of Pincus's earlier work and the controversy it caused see Reed (1978, pp. 320–321).

5 In later years Chang became an expert on artificial insemination and sperm capacitation. For more information on Chang and his work see Reed (1978, pp. 349–351).

6 Interview with Dot Hunt by L. Marks, April 8 1995, transcript p. 12.

7 The following publications were also listed as the inspiration points for the experiments: E.B. Astwood, "Progesterone Inhibition of Ovulation in the Rat" *American Journal of Physiology* (1939), 127: 127; R.H. Dutt, and L.E. Casida, "Alteration of Oestrual Cycle in Sheep by Use of Progesterone and its Effects Upon Subsequent Ovulation and Fertility" *Endocrinology* (1943), 43: 208–217; A.W. Makepeace and C.L. Weinstein, and M.H. Friedman, "The Effect of Progestin and Progesterone on Ovulation in the Rabbit" *American Journal of Physiology* (1937), 119: 512–516; G. Pincus and R.E. Kirsch, "The Sterility in Rabbits Produced by Injections of Oestrone and Related Compounds" *Journal of Experimental Physiology* (1936), 115: 219–228; G. Pincus and N.T. Werthenssen, "The Comparative Behaviour of Mammalian Eggs *In Vivo* and *In Vitro*. III. Factors Controlling the Growth of the Rabbit Blastocyst" *Journal of Experimental Zoology* (1938), 78: 1–18; G. Pincus and N.T. Werthenssen, "An Analysis of the Mechanism of Oestrogenic Activity" *Proceedings of the Royal Society* (1939), 126: 506; G. Pincus, "Factors Controlling the Growth of Rabbit Blastocysts" *American Journal of Physiology* (1941),

133: 412–413; G. Pincus, "The Physiology of Ovarian Hormones" in *The Hormones*, Vol. II (New York, 1950), chapter 1; and G. Pincus, "Fertilization in Mammals" *Scientific American* (1951), 184: 3.

8 "Report of Progress to Planned Parenthood Federation of America (PPFA)", January 24 1952, GP-LC Box 12 and 14; Interview with A. Merrill and M.E. Johnson by L. Marks, April 8 1995, transcript, pp. 10, 11–12, 15–16; and interview with D. Hunt.

9 "Report of Progress to PPFA", January 24 1952 GP-LC Box 12 and 14; interview with A. Merrill and M.E. Johnson, pp. 10, 15, 41; and interview with D. Hunt, p. 34.

10 Worcester Foundation, "Report of Progress to PPFA", January 24 1952, Stone's papers, Francis Countway Library (henceforth FCL); same report is also in Pincus Papers, Library of Congress (henceforth GP-LC), Box 14; "Report of Progress to PPFA", January 23 1952, GP-LC, Box 14.

11 McCormick to Sanger, MS-LC, November 13 1953.

12 Worcester Foundation, "Report of Progress to PPFA", January 23 1953, GP-LC, Box 14; "Report to the Rockefeller Foundation of Research on the Physiology of Mammalian Eggs and Sperm", March 12 1953, GP-LC, Box 23.

13 Interview with A. Merrill and M. E. Johnson, p. 18; Pincus (1965, pp. 67–68).

14 Pincus to Henshaw, January 28 1953, GP-LC, Box 14; McCormick to Sanger, November 13 1953, MS-LC.

15 Pincus to Henshaw, January 28 1953, GP-LC, Box 14. By 1954 progesterone in oil was being sold in a capsule on the American market to women for preventing miscarriage. Sanger to McCormick, March 25 1954, MS-SS. See also Vaughan (1971, pp. 41–42).

16 Pincus to Henshaw, Febuary 19 1953, GP-LC, Box 14; Notes from McCormick's notes of conversation with Dr Rock, October 6 1955, MS-SS.

17 Searle, "Clinical Experience with and Practical Techniques for the Use of Enovid in Ovulation Control: A Closed Circuit Television Symposium", January 18 1961, p. 8; Clarence Gamble's papers, Francis Countway Library (henceforth CG-FCL) File 214, fo. 3326; Vaughan (1971, p. 31). For more information on Rock, see McLaughlin (1982); and Reed (1978, pp. 351–354).

18 Pincus to Rock, May 15 1953, Rock's papers, Countway Library. Henceforth JR-FCL).

19 "Procedure for Progesterone Tests", n.d, JR-FCL.

20 B. Zondek to Pincus, June 27 1953, GP-LC, Box 15.

21 Pincus to Rock, May 15 1953, JR-FCL; Worcester Foundation, "Report of Progress to PPFA", March 4 1954, MS-SS, and GP-LC, Box 17; Worcester Foundation, Application for a grant to the Robert L. Dickinson Memorial, PPFA, March 6 1954, GP-LC, Box 17; Pincus to McCormick, March 5 1954, GP-LC, Box 17, LC; Tsacona to Pincus, April 10 1954, GP-LC, Box 19.

22 Highly sophisticated tests for ovulation, such as vaginal smears developed by Papanicolaou in 1933, had been appearing since the 1930s. For more information on the historical development of these techniques see Pincus (1965, pp. 200–201).
23 Worcester Foundation, "Report of Progress to PPFA", March 4 1954, MS-SS and GP-LC, Box 17; "The Effect of Progesterone on Normal Ovulation", GP-LC, Box 17. For the statistical methods used to calculate the data of the trials see Pincus to Vogt, March 20 1954, JR-FCL.
24 Pincus to Sanger, March 31 1954, and Pincus to M. Matsuba, July 24 1954, MS-SS; Pincus to Sanger, March 23 1954, GP-LC, Box 17.
25 McCormick to Sanger, June 17 1954, MS-SS.
26 Pincus to McCormick, March 5 1954, GP-LC, Box 17; McCormick to Sanger, November 13 1953, MS-LC; McCormick to Sanger, June 17 1954, MS-SS; Sanger to Pincus May 29 1954, GP-LC, Box 17; McCormick to Sanger, July 21 1954, MS-SS; P. Henshaw to Pincus, August 6 1954, GP-LC, Box 16; Sanger to N. Larsen, September 10 1954, MS-SS; Pincus to McCormick, December 28 1955, GP-LC, Box 17; McCormick's notes on conversation with Dr Rock, January 9 1956, MS-SS; Pincus to McCormick, March 14 1956, GP-LC, Box 21.
27 Pincus to Vogt, March 30 1954, GP-LC, Box 17.
28 Pincus to Sanger, March 31 1954, MS-SS.
29 McCormick to Sanger, July 21 1954, MS-SS.
30 Notes on talk with Dr Pincus, February 1 1955, MS-SS.
31 McCormick to Sanger, July 21 1954, MS-SS.
32 "Suitable Experimental Subjects", and "Objectives for Experimentation", November 1954, JR-FCL.
33 McCormick to Sanger, October 10 1954, MS-SS.
34 McCormick to Sanger, July 19 1954, MS-SS.
35 Pincus to McCormick, March 26 1954, and Sanger to Pincus, March 29 1954, GP-LC, Box 17; McCormick to Pincus, June 10 1954, MS-SS.
36 A. Stone to McCormick, March 30 1956, GP-LC, Box 21. See also Rock to Pincus, April 18 1956, GP-LC, Box 23; H. S. Kupperman and A. Stone, "Preliminary Report of Progesterone Studies", March 1956, GP-LC, Box 21.
37 McCormick to Sanger July 19 1954, MS-SS.
38 Sanger to N. Larsen, September 10 1954, MS-SS; Pincus to McCormick, August 26 1954, GP-LC, Box 17.
39 M. F. Fuster to Pincus, October 12 1954, GP-LC, Box 17.
40 D. Tyler to Pincus, October 20 1954; M.F. Fuster to Pincus, November 1 1954, GP-LC, Box 17.
41 C. R. Garcia to Pincus, June 18 1955, GP-LC, Box 18; J. Diaz Carazo to Pincus, September 16 1955, GP-LC, Box 18.
42 J. Diaz Carazo to G. Pincus, September 16 1955 and Pincus to D. Tyler, September 30 1955, GP-LC, Box 18. J. D. Carazo to Pincus, January 31 1956, GP-LC, Box 22; McCormick's notes on conversation with G. Pincus, March 5 1956, MS-SS.

43 "Protocol on Progesterone Therapy in Psychotic Women", n.d, GP-LC, Box 17; "Progress Report: Studies of the Effects of Progestational Compounds Upon Psychotic Subjects", July 22 1957, GP-LC, Box 30.

44 G. Pincus, "Long Term Administration of Vilane to Human Subjects", *ca* 1957, GP-LC, Box 104.

45 Pincus, "Protocol: Effect of a New Steroid on Testicular Function", June 29 1956, GP-LC, Box 23; "Protocol: A Study in Castration Anxiety", to be presented to the Research Council Meeting, September 26 1956, GP-LC, Box 23. In the case of the women and the men, no marked behavioral changes were found among those taking the hormones, but the biopsies on the men disclosed a cessation of sperm production and a reduction of the interstitial cells among the men who took medication for the longest periods. "Progress Report, Studies of the Effects of Progestational Compounds Upon Psychotic Subjects", GP-LC, Box 30.

46 Interview with Ramon Celso Garcia by L. Marks, April 1995, transcript, p. 8; Vaughan (1971, p. 39).

47 Overall the investigators concluded that progesterone appeared to have no effect on the psychotic behaviour of either men or women. "Progress Report, Studies of the Effects of Progestational Compounds Upon Psychotic Subjects", GP-LC, Box 30; Vaughan (1971, p. 40).

48 McCormick to Sanger, Feb. 11 1956, MS-SS; Letter from E. Rice-Wray to W. Vogt, December 10 1953, and "A Proposal to the PPFA for a Program of Action in Fertility Control in Puerto Rico", in GP-LC, Box 17; Ramirez de Arellano and Seipp (1983, pp. 88, 109); Oudshoorn (1994, pp. 127, 128); Vaughan (1971, p. 41).

49 Garcia in Anonymous, "Historical Perspectives of the Scientific Study of Fertility," Transcript from the Conference on Historical Perspectives of the Scientific Study of Fertility, Session III. The Development of Endocrinology, American Academy of Arts and Sciences, Boston, Mass., May 5 1978, cited in Oudshoorn (1994, p. 126).

50 Pincus to A. Fishberg, February 28 1956, GP-LC, Box 22; Ramirez de Arellano and Seipp (1983, p. 114).

51 E. Rice-Wray, "Study Project of SC-4642", January 1957, CG-FCL, Box 50.

52 Garcia in Anonymous, "Historical Perspectives of the Scientific Study of Fertility," cited in Oudshoorn (1994, p. 126); E. Rice-Wray, "Study Project of SC-4642", January 1957, CG-FCL, Box 50.

53 McCormick's notes on conversation with G. Pincus, March 5 1956, MS-SS; E. Rice-Wray, "Study Project of SC-4642", January 1957, CG-FCL, Box 50; Interview with Merrill and Johnson, p. 20; Pincus (1958, p. 1335); Oudshoorn (1994, pp. 128–129).

54 Pincus to A. Fishberg, February 28 1956, GP-LC, Box 22.

55 E. Rice-Wray, "Study Project of SC-4642", January 1957, CG-FCL, Box 50.

56 Pincus to A. Fishberg, February 28 1956, GP-LC, Box 22; McCormick's notes on conversation with Pincus, March 5 1956, MS-SS; Pincus (1958, p. 1335).

57 E. Rice-Wray, "Study Project of SC-4642", January 1957, CG-FCL, Box 50; Pincus (1958, p. 1,335).

58 Interview with A. Merrill and M.E. Johnson, p. 35.

59 E. Rice-Wray, "Study Project of SC-4642", January 1957, CG-FCL, Box 50; Pincus (1958, p. 1335).

60 C. Tietze to M. Synder, August 15 1957, GP-LC, Box 28.

61 Pincus to McCormick, October 11 1956, JR-FCL; I. Rodriguez to Pincus, May 1956 and E. Rice-Wray to Pincus, May 10 1956, GP-LC, Box 22; E. Rice-Wray, "Study Project of SC-4642", January 1957, CG-FCL, Box 50.

62 Several patients also discovered that they were pregnant before taking the tablet. E. Rice-Wray to Pincus, 11 June 1956, GP-LC, Box 23.

63 Interview with Edris Rice-Wray by J. Reed, March 14 1974, Schlesinger Library; McLaughlin (1982, p. 134); Vaughan (1971, p. 50).

64 Pincus to McCormick, October 11 1956, JR-FCL.

65 E. Rice-Wray, "Study Project of SC-4642", January 1957, CG-FCL, Box 50; C. Tietze to M. Snyder, August 15 1957, GP-LC, Box 28.

66 I. Rodriguez to Pincus, May 1956 and E. Rice-Wray to Pincus, May 10 1956, GP-LC, Box 22; McCormick to Sanger, May 12 1956; E. Rice-Wray, "Study Project of SC-4642", January 1957, CG-FCL, Box 50.

67 Pincus to E. Rice-Wray, August 6 1956, GP-LC, Box 22. A letter written by a representative from Searle in 1960 highlights the importance attached by the FDA to having evidence from patients who had taken the pill for long periods. See J.W. Crosson to R. Nicholas, May 18 1960, GP-LC, Box 44 and Oudshoorn (1994, p. 130).

68 Interview with Adaline P. Satterthwaite by J. Reed, June 1974, transcript, p. 19, Family Oral History Project, Schlesinger Library; Reed (1978, p. 361); Williams (1978, pp. 321–329); Ramirez de Arellano and Seipp (1983, pp. 118–122); and Vaughan (1971, pp. 55–56).

69 F. Laraque to Pincus, July 7 1958, GP-LC, Box 32; L. Honorat to R. Crosier, August 27 1958, GP-LC, Box 32.

70 Interview with A. Merrill and M. E. Johnson, p. 35.

71 "Study of the Long-term Administration of 17 Ethinyl — 19 Nortestosterone in Fertility Control", CG-FCL, Box 109, fo. 1789; A.P. Satterthwaite and C.J. Gamble, "Control of Ovulation with Norethynodrel", 1962, CG-FCL, Box 53 fo. 849.

72 McCormick's conversation with R. Rock, February 22 1958, MS-SS. See also Oudshoorn (1994, p. 131).

73 Interview with A. Merrill and M.E. Johnson, p. 34.

74 E. Rice-Wray, "Study Project of SC-4642", January 1957, CG-FCL, Box 50; C. Tietze to M. Snyder, August 15 1957, GP-LC, Box 28; A.P. Satterthwaite and C.J. Gamble, "Control of Ovulation with Norethynodrel", 1962, CG-FCL, Box 53 fo.849.

75 McCormick's notes on conversation with Pincus, May 16 1957, MS-SS; Ramirez de Arellano and Seipp (1983, p. 116).

76 A.P. Satterthwaite and C.J. Gamble, "Control of Ovulation with Norethynodrel", 1962, CG-FCL, Box 53 fo.849; Interview with Satterthwaite by Reed, p. 24, interview with Adaline P. Satterthwaite by L. Marks, April 1995, transcript, p. 15. Tensions emerged between Pincus and Satterthwaite when Pincus prematurely announced in 1961 that the pill suppressed cervical cancer. Satterthwaite felt the evidence collected at that point was not convincing. For more on this, see Ramirez de Arellano and Seipp (1983, pp. 120–121).

77 McCormick's notes on conversation with R. Rock, October 10 1956, MS-SS. Similar anxieties were expressed by Garcia.

78 Vaughan (1971, p. 49); see also interview with Satterthwaite by L. Marks, pp. 23–24. Since the 1890s biological and medical literature had discussed the possibility of hazardous effects of drugs on foetuses, and this was a continual matter of debate throughout the early 20th century. For more information on this see Sjostrom and Nilsson (1972, pp. 168–172).

79 Interview with C.R. Garcia by L. Marks, p. 48; interview with Satterthwaite by Marks, pp. 27–28; and Ramirez de Arellano and Seipp (1983, pp. 120–121).

80 Interview with A. Merrill and M.E. Johnson, p. 21.

81 Interview with C.R. Garcia by L. Marks, pp. 29–32, 40–41. For more information on the history behind the establishment of strict controls in human experimentation, see McNeill (1993).

82 Interview with C.R. Garcia by L. Marks, pp. 29–32.

83 Interview with C.R. Garcia by L. Marks, p. 32; Pincus to McCormick, October 11 1956, JR-CL.

84 Interview with A. Merrill and M.E. Johnson, pp. 22, 29.

85 Interview with A. Merrill and M.E. Johnson, pp. 24, 30–31.

86 Interview with A. Merrill and M.E. Johnson, pp. 28, 43.

87 McLaughlin (1982, p. 131); Oudshoorn (1994, p. 130).

REFERENCES

M.E. Borell, "Biologists and Birth Control 1918–1938" *Journal of the History of Biology* (1987), 20: 51–87.

M.E. Borell, "The Origins of Modern Research on Fertility and Reproduction: Physiology and the Discovery of Hormones 1889–1930", unpublished paper prepared for Workshop on Historical Perspectives on the Scientific Study of Fertility at American Academy of Arts and Sciences, Boston, Mass., 9–10 December 1977.

E. Chesler, *Woman of Valor: Margaret Sanger and the Birth Control Movement in America* (New York and London: Simon and Schuster, 1992).

L. Gordon, *Woman's Body, Woman's Right: Birth Control in America* (New York: Grossman, 1977).

A. Leathard, *The Fight for Family Planning: The Development of Family Planning Services in Britain 1921–74* (London and Basingstoke: Macmillan, 1980).

S.E. Lederer, *Subjected to Science: Human Experimentation in America before the Second World War* (Baltimore and London: John Hopkins University Press, 1995).

C.R. McCann, *Birth Control Politics in the United States, 1916–1945* (Ithaca, New York: Cornell University Press, 1994).

A. McLaren, *A History of Contraception: From Antiquity to the Present Day* (Oxford: Blackwell, 1990).

L. McLaughlin, *The Pill, John Rock and the Church* (Boston, Toronto: Little Brown, 1982).

P.M. McNeill, *The Ethics and Politics of Human Experimentation* (Cambridge: Cambridge University Press, 1993).

N. Oudshoorn, *Beyond the Natural Body: An Archeology of Sex Hormones* (London: Routledge, 1994).

G. Pincus, "Fertility Control with Oral Medication" *American Journal of Obstetrics and Gynaecology* (1958), 75: 1335.

G. Pincus, "Progestational Agents and the Control of Fertility" *Vitamins and Hormones* (1959), 169: 81.

G. Pincus, *The Control of Fertility* (New York, London: Academic Press, 1965).

A.B. Ramirez de Arellano and C. Seipp, *Colonialism, Catholicism and Contraception: A History of Birth Control in Puerto Rico* (Chapel Hill, NC and London: University of North Carolina Press, 1983).

J. Reed, *The Birth Control Movement and American Society* (Princeton, NJ: Princeton University Press, 1978, rev. edn 1983).

R. Rupke (ed.), *Vivisection in Historical Perspective* (London: Croom Helm, 1987).

H. Sjostrom and R. Nilsson, *Thalidomide and the Power of the Drug Companies* (London: Penguin, 1972).

R.A. Soloway, *Birth Control and the Population Question in England, 1877–1930* (Chapel Hill, NC and London: University of North Carolina Press, 1982).

P. Vaughan, *The Pill on Trial* (London: Penguin, 1972).

D. Williams and G. Williams, *Every Child a Wanted Child: Clarence Gamble, M.D. and his Work in the Birth Control Movement* (Boston: Harvard University Press, 1978).

CHAPTER 9

Immunotherapy of Cancer from Coley's Toxins to Interferons: Molecularization of a Therapeutic Practice

Ilana Löwy

INTRODUCTION

In the early 1980s, the introduction of regulatory molecules which act upon the human immune system — the cytokines (interferons, and then interleukins) — into cancer therapy was presented as a major innovation in this domain. Thanks to genetic engineering methods, regulatory molecules produced by the body could be manufactured by industry and used as research materials and as drugs. [1] The availability of these purified regulatory substances, some experts proposed, finally brought the domain of immunotherapy of cancer into the realm of exact science. Thus in 1983, Robert Oldham, then the director of the Biological Response Modifiers Program of the U.S. National Cancer Institute (NCI), linked the increasing molecularization of this domain — or to be more precise the increasing availability of biological agents which modulate immune responses — with hopes of a therapeutic breakthrough in the cancer clinics: "The revolution in molecular biology, the advent of computers and the improvement in hybridoma technology have combined to enable the scientists to produce the highly purified reagents necessary to begin to ask rather specific questions about immunological mechanisms involved in cancer growth, metastasis and therapy" (Oldham and Smalley, 1983, p. 3).

The efforts to unravel the molecular basis of the control of malignant cells by immune mechanisms started in the 1930s. They originated in early attempts to cure cancer through the stimulation of unspecified "defense mechanisms" by bacterial products. This therapy, first introduced in the late 19th century, was occasionally reported to be successful in the treatment of inoperable cancer, but it was perceived as unreliable and dangerous, and disappeared with the advent of radiotherapy of malignant tumors in the 1910s and 1920s. From the 1930s on, the growing importance of biochemical explanations of physiological phenomena stimulated a search for the "active fraction" of bacterial toxins which, scientists believed, might become a more efficient and more reproducible anti-cancer treatment. These attempts did not, however, lead to the development of efficient anti-cancer therapies. In the 1960s clinicians returned to the application of substances credited with the property of stimulating the "natural defenses of the body" — bacterial vaccines and bacterial products — to cancer therapy. At the same time, the focus of immunological investigations gradually switched from studies of the chemical structure of antigens, that is, of extraneous substances which induce immune reactions, to the investigation of biochemical mechanisms underlying immune responses. Consequently, studies in the immunotherapy of cancer shifted from the search for chemical substances (bacterial vaccines, purified fractions of these vaccines, and other "immunostimulatory" molecules) credited with the ability to stimulate anti-tumor responses of the body, to the investigation, and then the clinical application, of regulatory molecules (such as interferon and interleukins) produced by the body in response to external stimulation. The advent of genetic engineering techniques and its consequence, the availability of industrially produced recombinant cytokines, greatly accelerated these studies.

Oldham's faith in the advantages of the introduction of purified cytokines into cancer studies was justified insofar as the use of these molecules in the immunology and the experimental oncology laboratory indeed supplied new insights into the interaction between immune mechanisms and malignant cells in the test tube and in the laboratory animal. The introduction of cytokines into the cancer clinics was not, however, followed by a rapid elaboration of efficient anti-cancer treatments. In the years 1975–1995, interferons, interleukins and colony growth factors were integrated into routine and semi-routine treatments of selected human malignancies, but these substances are usually perceived (at least for the moment) as minor improvements rather than major therapeutic breakthroughs (Keidar and Klein, 1992; Balkwill, 1994). The translation of the "non-specific resistance to cancer" into molecular terms and its integration into a broader framework of "regulatory approaches to the cancer problem" nevertheless helped to bridge the gap between the reductionism of the "molecular vision of life" and the complexity and non-predictability of human cancers, that is, between the practices of fundamental researchers and of oncologists. The introduction of these compounds changed practices in the experimental oncology laboratory and in the cancer ward, and from the 1980s on, linked these sites with the biotechnology plant. In addition, this chapter proposes, the trajectories of individual compounds mirrored — and occasionally affected — the trajectories of scientists and physicians who produced and employed these compounds. The story of immunotherapy of cancer told in this chapter is therefore not about an ascending, linear pathway which culminated in an efficient molecularization of a clinical domain, but rather about ways in which the ideal of molecularization-tion shaped practices of scientists and doctors, and about the introduction of a "molecular culture" into the cancer clinics.

COLEY'S TOXINS: IMMUNOTHERAPY OF CANCER BEFORE MOLECULES

In 1909 an American surgeon, William B. Coley (1862–1936), Professor of Surgery at Cornell University, presented his method of curing sarcoma through the injection of bacterial toxins at a meeting of the Royal Society of Medicine, London. The comments of the physicians present at the meeting were favorable, until the intervention of Sir Almroth Wright, the director of the Inoculation Department at St. Mary's Hospital, London. Wright attacked Coley and expressed strong doubts about the validity of his therapy. He ended his intervention by accusing Coley of using his toxins "after the manner of quack medicine" (Coley, 1909). The sharp exchange between Coley and Wright revealed divergent ideas not only on the way to treat advanced cancer, but on what counts as "scientific" in the clinics.

The cancer therapy which provoked this debate was developed by Coley, then surgeon at the Orthopedic Surgery Ward of Memorial Hospital, New York, in the 1890s. In 1891 Coley observed a spontaneous regression of advanced inoperable sarcoma in a patient suffering from erysipelas — a severe streptococcal infection. Following this observation, he screened the literature for similar cases and found 38 descriptions of regression of advanced malignant tumors following an erysipelas attack, ten of which were permanent. Coley decided therefore to attempt to cure patients suffering from advanced, inoperable cancer by an artificial introduction of an erysipelas attack (Coley, 1891; Coley, 1893; Nauts, Swift and Coley, 1946). The therapy was not always successful, and several patients succumbed to the streptococcal infection.[2] Coley then turned to attempting to mimic the results of a streptococcal infection through the injection of heat-killed bacilli or their culture filtrate ("toxin"). The streptococcal toxins alone were not effective, but when the Streptococcus was co-cultured with another bacterium, *Bacillus prodigiosus* (*Serratia marcescens*)[3], the toxin-containing culture supernatant did induce occasionally (and in non-predictable ways) the regression of advanced malignant tumors, in particular of sarcoma. It was, however, a very harsh therapy. Repeated injection of bacterial toxins induced severe iatrogenic effects such as high fever, bone pain and fatigue (Coley, 1894; Coley, 1896; Coley, 1898).

In the absence of therapy for inoperable tumors, Coley's method evoked the clinicians' interest, its multiple secondary effects and erratic clinical results notwithstanding. As early as 1895 the Lister Institute, London, undertook the commercial preparation of Coley's toxins. Commercial preparations of Coley's toxins were distributed in the U.S. by Parke, Davis & Company. The new method had at least moderate success. Between 1894 and 1916 there were 132 reports — besides Coley's own work — on clinical results of the application of Coley's toxins to the cure of cancer (again, the majority of long-term regressions of tumors were observed in sarcoma, a relatively rare form of malignancy) (Nauts, 1991). In 1907, Dr Silas P. Beebe and Dr Martha Tracy from the Huntington Cancer Research Fund at Cornell University developed an improved method of preparation of toxins. They then showed that their preparation of toxins induced the regression of lymphosarcoma in dogs (Beebe, 1907; Beebe and Tracey, 1907).[4]

Coley's talk at the 1909 meeting of the Royal Society of Medicine was made at the height of the fame of his therapy. Wright's opposition to Coley's method was not based on claims of its lack of efficacy, but on lack of an adequate scientific basis. Coley's method, Wright explained, "rested on a purely empirical basis and was applied without any kind of scientific control". Coley was unable to propose a specific mechanism for the anti-tumor activity of the toxins, and did not attempt to quantify the "modifications of the blood" induced by his treatment. Such quantification, Wright affirmed, was necessary in order to control and guide the treatment, and in order to transfer it successfully to different clinical settings. (Coley, 1909, p. 46). Speaking about quantification of changes in the blood, Wright made an obvious allusion to his "opsonin theory", which, he believed, gave a scientific explanation of "resistance phenomena" in terms of the coordinated activity of specific antibodies in the serum ("opsonins") and of phagocytic cells.[5] It was possible, Wright explained, to quantify "resistance phenomena" through the study of the "opsonin index" of the serum of a given individual (that is, its ability to enhance phagocytosis). It is difficult to classify Wright's "quantitative method" as "molecular", since circa 1909 it was a non-proven assumption that the entities responsible for the resistance reaction — "opsonins" or "antibodies" — were well-defined chemical entities.[6] The measure of "opsonin index" proposed by Wright , however, reduced complex physiological phenomena to a set of carefully codified measures in the laboratory. The proposal to introduce it to the study of interaction between "resistance mechanisms" and cancer may thus be related to later efforts to understand the molecular basis of tumor immunology.[7]

Coley strongly rejected Wright's accusation and asserted that his therapeutic method was "an absolutely scientific one" because the results obtained in patients were later replicated in dogs (by Beebe and Tracy) and, above all, because "the method worked" and could be documented with detailed description of clinical cases, complete with "before" and "after" photographs. The fact that experimental studies in animals were made in order to confirm clinical findings highlights the short and well-integrated trajectory of Coley's toxins. The bacteria were grown in the laboratory, their toxins extracted through a simplified procedure and then injected directly into patients. The whole process could have been (and often was) controlled exclusively by clinicians — hence the relative lack of interest in mechanism(s) of anti-tumoral activity of the toxins. One may add that Coley did have a (rather vague) supposition how bacterial products affected malignant growths. He believed that microorganisms induce and/or stimulate tumors in humans, hence the possibility of a protective role of the stimulation of "defense mechanisms of the body" by bacterial toxins. Such speculations about the role of "resistance mechanisms of the body" in fighting cancer were not unusual in the early 20th century. Dr James B. Murphy, the head of the cancer laboratory at the Rockefeller Institute, was one of the zealous defenders of such a view. Dr Beebe, the same researcher who had studied Coley's toxins in dogs, was another. When the U.S. press announced in 1915 that Murphy had found that lymphocytes play an important role in "resistance to tumors", Beebe's associates published an angry letter in which they claimed that Beebe had worked for a long time on the role of white blood cells in resistance, and had made the same discovery several years earlier.[8] However, Coley, unlike Beebe, never attempted to prove his "resistance" hypothesis. He was a practicing surgeon, and was interested above all in patient response rates, in the disappearance of patients' symptoms, in "before" and "after" X-ray plates, in the survival of otherwise terminal patients. For him, a "scientific method" in medicine was not based on studies of physiological mechanisms of disease or cure, but on accurate bedside observations confirmed by histological diagnosis, that is, on appropriate establishment of "clinical facts". The central point was *whether* the treatment worked, not *how* and *why* it worked (Coley, 1912).

Coley's practice-based approach to cancer therapy was seen as increasingly outmoded in the 1920s and 1930s. The main reason was the advent of a new method to treat inoperable malignant tumors – radiotherapy. Radiotherapy was based on established physical principles and enjoyed the prestige of a method on the cutting edge of technology. An additional reason for the abandonment of Coley's toxins was the low efficacy of this method and the great variability and lack of predictability of its clinical results. Coley, and later his daughter Helen Nauts, an enthusiastic advocate of his therapeutic method, attributed this variability to the low quality of commercial preparations of toxins. Only a strict adherence to the empirical methods of preparation of toxins developed by Tracy and their inoculation in the precise manner recommended by Coley (which implied repeated injection of non-purified supernatants of bacterial cultures into patients) might lead to cures (Nauts *et al.*, 1946; Nauts, 1991). The "chemical turn" of immunology in the 1920s, which was parallel to a rapid development of biochemistry in the interwar period, did not, however, favor the therapeutic use of a mixture containing unknown amounts of unidentified chemical compounds secreted by bacteria (Silverstein, 1988). A renewal of interest in Coley's toxins in the 1930s has its roots in the hope to attribute their anti-tumor activity to a well-defined molecule(s). The identification of such molecules, it was expected, would open new areas of fundamental investigations in oncology, and would lead to the development of a more efficient therapy for cancer.

BACTERIAL TOXINS IN THE 1930S AND 1940s: THE CHEMICAL PATHWAY

In the mid-1930s, Dr Murray Shear from the U.S. Public Office of Field Investigations of Cancer at Harvard University (consolidated in 1937 with the

Pharmacology Laboratory of the old National Institute of Health to form the nucleus of the National Cancer Institute) became interested in Coley's toxins. The toxins, he explained, worked sometimes, although in an entirely unpredictable way. The erratic therapeutic effect of the toxins could perhaps be explained if one assumed that different preparations of toxins contained different quantities of an unknown anti-tumoral substance. The identification and purification of such a substance might transform Coley's toxins from a marginal therapy and a medical curiosity into a reliable anti-cancer treatment. Shear's optimism to be able to find specific anti-tumoral substances in bacterial filtrates was probably partly based on the observation that the so-called "Schwartzman phenomenon", local skin reactivity to bacterial filtrates, could be extended to the induction of hemorrhagic necrosis of malignant tumors. Injection of bacterial filtrates induced rapid and selective destruction of well-accepted transplanted tumors in laboratory animals (Gartia and Linz, 1931; Schwartzman and Michailovsky, 1932). In 1933 Francisco Duran Reynals, who studied the Schwartzman phenomenon at the Cancer Laboratory of the Rockefeller Institute (then headed by Murphy) linked this phenomenon explicitly to the regression of lymphosarcoma in dogs after the injection of Coley's toxins (Duran Reynals, 1933; Beebe and Tracey, 1907).

Shear decided to test all the purified fractions of Coley's toxins in a single screen system, sarcoma S37 transplanted in mice. He thus proposed a different pathway for a putative anti-cancer compound(s) extracted from bacteria: long passage in the laboratory which included the purification of specific bacterial compounds and screening of their anti-tumoral effects in animals, followed by a relatively short period of clinical tests of promising molecules, conducted by physicians. At first he and his collaborators were successful. They isolated and purified a polysaccharide of *Serratia marcescens* which induced hemorrhage in S37 tumors in mice, probably through the production of a "shock-like" state and the decrease of blood supply for the vulnerable tumor (Shear, 1943; Hartwell *et al.*, 1943; Kanler *et al.*, 1943).

Shear and his colleagues checked the purity of their polysaccharide through electrophoresis and ultracentrifugation, then organized clinical trials of this polysaccharide in cancer patients. Some clinical experiments were conducted at Lankenau Hospital in Philadelphia while additional assays were conducted at Harvard Medical School. Twenty one patients were treated with the polysaccharide in the years 1944–46. Some reacted to the new therapy by hemorrhage into the patient's tumor and occasionally its regression, but the affected tumors were not totally destroyed, and they regenerated later. Patients also rapidly developed anti-polysaccharide immunity, making prolonged therapy difficult. In addition, injections of polysaccharide induced severe side effects such as high fever and a sharp fall in blood pressure, frequently to shock levels. The polysaccharide was therefore declared to be too toxic for routine clinical use (Brues and Shear, 1944; Hollman, 1947; Oakley, 1947). For Shear and his collaborators, the search for an active substance in Coley's toxins was an attempt to uncover an efficient anti-cancer drug, not the investigation of substances able to activate defense mechanisms of the body.[9] The articles on the anti-tumor activity of *Serratia* polysaccharide were published as a series on "chemical treatment of tumors". It is not surprising that following the initial failure of therapy with *Serratia* polysaccharide, Shear and his colleagues turned to screening of natural and synthetic compounds for anti-tumor activity (Shear, 1951). Coley's toxins became the "precursors" of cancer chemotherapy, not of immunotherapy.

Shear's relatively modest "screening service" at the NCI, practically devoid of clinical facilities, was replaced, from 1955, by the "screening empire" of the Cancer Chemotherapy National Service Center (CCNSC) supervised by the NCI. The CCNSC, active between 1955 and 1966, did not contribute directly to studies of the relationship between immunity and cancer. It did, however, change the pattern of clinical research in oncology, a change which affected studies of immunotherapies of cancer too. One of the CCNSC's main tasks was the organization of large-scale clinical trials of new drug therapies of cancer. The tradition of organization of collaborative clinical trials stemmed directly from wartime and postwar research. War research "acclimatized" physicians with large-scale projects, coordinated actions and the need to suspend part of their expertise-based authority to achieve collective goals. It thus favored the development of cooperative enterprises such as the clinical trials of penicillin and streptomycin (Marks, 1987).

Multicenter trials, which included large numbers of patients, were particularly important when testing anti-cancer drugs which, as a rule, have low efficacy and high toxicity. They allowed, in addition, an efficient linking of the laboratory and the clinics through collaborative structures (Löwy, 1995). From the 1960s, multicenter clinical trials became the "obligatory passage point" for the introduction of new therapies of cancer. Treatments such as bacterial products and later interferons and other "biological response modifiers" were no exception (Zubrod *et al.*, 1966; Bud, 1978; Zubrod, 1984). Shear, the pioneer of chemical studies of Coley's toxins, and later the director of the first screening program of the NCI, stressed as early as 1951 the importance of large-scale clinical trials and of close collaboration between the laboratory and the clinic, adding that "it is already apparent that much careful thought will need to be devoted to the relationships between organizational units and among the human elements if such large ventures are to be productive enough to justify the large expenditures" (Shear, 1951, p. 571). The inspiration for the organization of such large-scale scientific-medical ventures, Shears proposed, should come from industry: "many large industrial organizations ... have taken intelligent steps to foster good morale and a cooperative spirit" (Shear, 1951, p. 571).

FROM BACTERIAL VACCINES TO REGULATORY MOLECULES: THE RETURN OF PHYSIOLOGICAL APPROACHES

In 1959, the immunologists Old, Clarke and Benacerraf noted that mice infected with attenuated strain of the tuberculosis bacillus, the BCG (*Bacillus Calmette-Guérin*), employed from the 1920s to vaccinate against tuberculosis, were more resistant to a graft of transplantable murine tumor (Old, Clarke and Benacerraf, 1959). Later, investigators at the same laboratory demonstrated that the BCG-infected mice rejected skin grafts more rapidly. These observations were made in the framework of studies of transplantation immunology, a subject which came to the fore in the late 1950s, following the highly publicized success of kidney grafts in identical twins. Medawar's studies on skin grafts in rabbits confirmed that the phenomenon of "second set rejection" (a more rapid rejection of a second graft from the same donor), first observed in patients, is highly specific. It should therefore be classified, Medawar proposed, among immune phenomena (Gibson and Medawar, 1943; Medawar, 1944; Medawar, 1945). In the 1940s, immunity was invariably linked to the presence of specific antibodies in the serum. However, when new, sensitive methods for detecting minute amounts of proteins failed to uncover the presence of specific antibodies in sera of animals which rejected skin grafts, Medawar and his collaborators became interested in "cell mediated immunity" and in physiological mechanisms in graft rejection. These studies were practice-oriented, and they were directly connected to the prestigious new field of organ and tissue transplantation (Billingham, Brent and Medawar, 1953; Mitchison, 1953; Alfred and Medawar, 1959).

Investigators who studied transplantation immunology in the 1950s made an explicit parallel between the physiological manifestations of graft rejection and other immunopathological phenomena, such as "delayed hypersensitivity reaction" (Merrill, 1961). Immunopathology was a relatively neglected domain during the "chemical period" of immunology. Before the advent of antibiotics, clinical immunologists were mainly interested in the study of infectious diseases, and thus in bacterial antigens and in specific antibodies against bacteria (serodiagnosis, serotherapy). Sulfa drugs, and then antibiotics, replaced serotherapy, however, and diminished the importance of serodiagnosis for the clinician. Penicillin, it was rapidly found, killed with the same efficacy all the serological strains of pneumococci. In the 1950s the focus of medical activities in Western countries shifted to the study of chronic diseases, some of which had an important immunological component. The growing importance of immunopathology, together with interest in mechanisms of graft rejection, contributed to the "physiological turn" of immunology in the 1950s and 1960s.

The "physiological turn" of immunology was not merely a return to pre-molecular ideas of late 19th and early 20th centuries. It was influenced by recent conceptual, technical and institutional changes in biology. Immunologists adopted new techniques to study proteins and to observe cell-bound molecules (ultracentrifugation, chromatography, radiolabelling, fluorescent labeling, electron microscopy). Some of these new techniques were developed through a close collaboration of protein chemists and immunologists, a collaboration favored by the expansion of fundamental biological research after World War II. Radiolabeled or fluorescein-labeled specific antibodies became a powerful tool for tracing other molecules, including cell-bound molecules. In addition, in the late 1950s and the 1960s the function of immune mechanisms was redefined as a distinction between "self" and "non-self" and a reaction to the latter (Tauber, 1994). Immunologists who previously had focused their investigations on the antigenic structure of microorganisms, that is, on extraneous chemical structures which elicit the formation of antibodies, now became increasingly interested in the "immune system", interpreted as a network of cells and specific or non-specific mediators which together were responsible for immune reactions (Silverstein, 1989; Moulin, 1991). This redefinition rapidly affected the orientation in cancer research: in 1966 a programmatic statement on the activities of the NCI-sponsored Cancer Chemotherapy National Service Center defined among priority goals the study of "immune defense modifiers" (Zubrod *et al.*, 1966).

The observation that BCG — a substance readily available on the market — accelerated the rejection of transplanted grafts, led to investigations of the possible role of BCG and other bacterial products in the cure of cancer. These inquiries were legitimated by three assumptions which were partly supported by experimental findings: (a) BCG stimulates cells of the immune system, in particular cells involved in "cell-mediated immunity", (b) cellular immune mechanisms responsible for the rejection of grafted tumors and tissues are akin to mechanisms able to destroy malignant cells, and (c) such immune mechanisms are compromised in cancer patients. The investigations of the influence of BCG on tumors may seem similar to the studies of the effect of Coley's toxins on lymphosarcoma in dogs (Beebe and Tracey, 1907). Indeed, on the practical level the experiments were not very different: tumor-bearing mice or guinea pigs were injected with a bacterium or with a bacterial product, and researchers observed the possible regression of the tumor. The background of the new experiments was, however, distinct. The rationale behind these studies was an assumption that a better understanding of mechanisms of immune reactions in the cancer patient would lead to the development of a better immunotherapy. Consequently, different parameters of the immune response of tumor-bearing animals treated with bacterial substances, and later of patients who had undergone immunotherapy were carefully measured and evaluated. This approach thus connected the immunology laboratory, the experimental oncology laboratory and the clinic, through the circulation of objects (the patient's serum and blood cells), of techniques (measurement of immune parameters, often first elaborated in laboratory animals, then adapted to use in humans), and of immunological concepts which supplied a shared framework for the interpretation of pre-clinical and clinical data.[10] The focus in BCG studies was the investigation of mechanisms of cellular immunity in laboratory animals. The comparison between the trajectories of *Serratia* polysaccharide and BCG points to the absence of linearity in attempts to molecularize cancer immunotherapy. In the 1930s and 1940s studies of relationships between tumors and immune mechanisms were limited to an investigation of therapeutic effects of a well-defined chemical substance. In the 1950s and 1960s, these studies were extended to a broader domain of cellular interactions, but they were focused on physiological reactions which were defined in functional, not in molecular terms.

In the early 1960s, studies with laboratory animals confirmed that the white blood cells of BCG-treated animals destroy tumor cells more rapidly in a test-tube. A group of French investigators who studied this phenomenon in the laboratory rapidly turned to clinical experimentation and attempted to prolong remissions in leukemia patients through repeated injection of BCG. This therapy, they maintained, was successful. Patients treated with BCG lived longer (Mathé *et al.*, 1969). Other researchers affirmed that BCG therapy was also efficient in malignant melanoma (Gutterman *et al.*, 1974; Morton *et al.*, 1974). Following the first optimistic reports, attempts at immunotherapy of cancer were made in numerous specialized cancer treatment centers. In parallel, researchers (often the same ones who conducted experimentation in patients) intensified animal studies of immunotherapies of cancer. In 1972, a group of investigators met at the NCI to discuss the need to monitor and coordinate clinical experiments with immunotherapy (attempts to treat patients with BCG, with other bacterial vaccines such as *Corynebacterium parvum*, and with BCG extracts). They decided to found an NCI-sponsored international registry of trials of cancer immunotherapy. The number of protocols recorded by this registry increased from 50 in 1973 to 347 in 1976, reflecting the rapid diffusion of the new therapeutic approach (Windhorst, 1978).

The rapid spread of the new therapy did not, however, follow positive clinical results. Just the contrary was true. While early trials of immunotherapies (except for the French trial of BCG in leukemia) were made with small groups of patients, the first large-scale multicenter trial of effects of BCG therapy in the treatment of acute leukemia failed to confirm initial optimistic reports (Concord trial, 1971). Other coordinated clinical trials did not fare better (Mastrangelo *et al.*, 1978). Some of their advocates attempted to contest the very principle of multicenter trials which eliminated the advantages of local expertise. Positive results, one of the French pioneers of immunotherapy explained, were obtained in a trial which was above criticism "because it was conducted on a homogenous population of patients", while "negative trials were conducted by cooperative groups specialized in negative results" (Mathé, 1976, p. 383). Such an argument was, however, difficult to sustain in the oncological milieu in the 1970s, organized, in particular in the U.S., as a trial-oriented professional segment. Medical oncologists were socialized in the "trialist ethos", participated in cooperative groups which coordinated clinical experimentation, and legitimated their clinical activities by the organized search for the cure of cancer. Some even used arguments which were not very different from those advanced by Coley in 1910: "The protagonists argue that even though a clear scientific basis for such immunotherapy may not exist, it nevertheless works and can act as a valuable adjunct to the more routine chemotherapeutic regimens" (Davies *et al.*, 1980, p. 3). The momentum of diffusion of clinical experiments with bacterial vaccines continued in the 1970s, but the accumulation of negative clinical results of multicenter trials became increasingly problematic for the promoters of these experiments.

Two solutions were proposed to overcome the difficulties of the new domain of cancer immunology. The first — intensification of fundamental, pre-clinical and clinical research in tumor immunology and immunotherapy — was a direct continuation of previous trends.[11] Oncologists and immunologists interested in immunotherapy of cancer agreed in the mid-1970s that the poor clinical results of immunotherapies of cancer reflected insufficient understanding of cellular and molecular events underlying immune reactions to cancer cells, or, as William Terry from the Division of Cancer Biology and Diagnosis of the NCI put it: "Negative results are not only predictable but they are absolutely anticipated when you initiate clinical applications in a field of this complexity" (Terry, 1978, p. 671). The solution for clinical difficulties, they hoped, would be found in a test tube. Some specialists explained that in the not-too-distant past, "tumor immunology, which was a generally primitive undertaking characterized by a phletora of theories and a paucity of data, has progressively evolved into a subdiscipline of immunology, using sophisticated experimental approaches and generating interesting results". This development, however, led to new difficulties, because "the functioning of the immune system *in vivo* has been shown to be extremely complex, characterized by numerous types of cells and cell-products involved in intricate reactions" (Terry and Yamamura, 1979, p. ix). The proposed solution was to turn to *in vitro* studies of tumor immunity, "in order to reduce the complexity and to achieve interpretable data" (Terry and Yamamura, 1979, p. x). This approach was indeed successful: "During the last decade", authors of a textbook on immunotherapy of human cancer explained in 1979, "the development of techniques through which human immune response could be critically evaluated ... has added strong, rational, scientific basis to this field" (Clark *et al.*, 1978, p. ix). Another book, centered on clinical trials of immunotherapies of cancer, affirmed that "there is new information every day about the nature of cell surfaces, about the nature of weak antigens, about what constitutes the ground for establishing stronger immune response against weaker antigens, information about specific and non-specific suppressing cells and factors. And all this information is being very rapidly applied to tumor immunology and immunotherapy" (Terry, 1978, p. 669). Major new developments in the immunotherapy of cancer came, however, not from the continuation of previous researches but from a new direction: studies of cytokines, starting with interferon. A new actor, the biotechnology industry, played a central role in this development: first through the promise of mass production of biologically active molecules, then through the fulfilment of this promise.[12]

Interferon was first described (in 1957) as an anti-viral substance produced by the body. In the 1960s clinical trials made with this molecule concluded that it was not cost-effective in the treatment of viral infections. In the meantime, the growing importance of the search for cancer-inducing viruses (as exemplified by the "Virus-Cancer" program of the NCI) (Gaudillière, 1993, and this volume), led to an interest in interferon as a potential anti-tumor agent, able to eliminate tumors of viral origin. Scientists who studied interferon then found that this substance also had inhibitory effects on tumors which do not have an obvious relation to viruses (e.g., chemically-induced tumors) (Sikora, 1983). This unexpected finding was attributed to the indirect effects of interferon, in particular the stimulation of a specific category of lymphocytes, the natural killer cells (NK), thought to be implicated in the elimination of tumor cells by the body. Promoters of cancer immunotherapy added interferon and interferon inducers to the (very long) list of immunostimulatory substances which might find an application in oncology clinics (Hersh *et al.*, 1973, p. 93). In the early 1970s results obtained in interferon-treated laboratory animals looked promising enough to start clinical experimentation with this substance. The major difficulty, however, was interferon's scarcity. Interferon is species-specific (that, is only human interferon can be used in humans), and obtaining it from its major natural source, white blood cells, was labor intensive and expensive. Nevertheless, interferon obtained in this way by Dr Kari Cantell of the Finnish Red Cross was tested on a small number of cancer patients suffering from malignant osteosarcoma. The results, although not statistically significant, were presented as encouraging (Strander *et al.*, 1974).

The next step was the organization, in the U.S., of an active "interferon lobby" led by Dr Mathilde Krim from New York. The lobby advocated allocation of funds for interferon research, and the stimulation of mass-production of this molecule. Between 1975 and 1978, this lobby was successful in enlisting some of the leading U.S. experts in cancer research. Its activity at first met with resistance from other experts, who were less enthusiastic about interferon's promise, but it ended by being successful. In 1979, a subcommittee of the NCI recommended to establish a program with a strong emphasis on interferon-related studies — the Biological Response Modifiers Program (Mihich and Fefer, 1983). In parallel, the American Cancer Society decided to spend two million dollars on the purchase and testing of interferon. The flux of funds led to public interest in interferon, presented as a "miracle drug" for cancer. It also enhanced the biotechnology industry's interest in this molecule. Indeed, the cloning of interferon in 1979 opened the way for its mass production by the industry.

Circa 1980, interferon became the "demonstrative substance" of this industry. It displayed at the same time biotechnology's promise to increase general well-being and its potential to generate substantial profits (Kenney, 1986, p. 198). Rapidly the demonstration became problematic, however. The first large-scale clinical trials of interferon, made with interferon extracted from white blood cells, did not demonstrate impressive therapeutic effects of this molecule. Moreover, they indicated that, contrary to earlier assumptions, this molecule was toxic in patients. At the same time, fundamental and pre-clinical studies were unable to show ways to increase interferon's therapeutic efficacy. Interferon promoters hoped that purified recombinant interferon produced through genetic engineering methods would be more efficient and less toxic. When large quantities of recombinant interferon became available, they found, however, that interferon's toxicity was due to the molecule itself, not to accidental contaminants, and that the purified molecule was not a more efficient anti-cancer treatment (Powledge, 1984; Fent and Zbiden, 1987). Consequently, the promised profits from interferon production became more problematic, and in the mid-1980s some experts even believed that interferon was a failure. The demonstration of the industry's ability to produce complex biological molecules, they explained, should not be confused with its ability to produce marketable drugs.[13] These pessimistic predictions did not prevail, however. Interferon did not become a "miracle drug" for cancer, but the molecule did show a limited, but uncontested efficacy in the treatment of a certain number of malignancies. The persistent efforts of oncologists and industrialists therefore led to the integration of interferon into routine therapies for these malignancies, despite difficulties in correlating its anti-tumor effects in patients with the activation of known biological pathways (Jahiel and Krim, 1988).[14]

In the 1970s, with the introduction of interferon (and later other cytokines) into cancer therapy, some experts hoped that the use of a well-defined molecule to stimulate the regression of malignant tumors would lead to a symmetrical "molecularization" of views of the mechanisms by which this molecule affects selected malignant growths. However, investigators rapidly found that interferon, like numerous other biologically active substances, acted on multiple levels, and the untangling of the anti-tumor effects from other physiological effects of this molecule was not an easy task. Data accumulated after the introduction of interferon into the clinic did not allow determination of what the (major) mechanism of the anti-tumor activity of interferon in humans was.[15] The clinicians' failure (1975–1995) to uncover a correlation between the clinical status of interferon-treated cancer patients and the activation of putative anti-tumor immune mechanisms in these patients, left open the debate on the nature of the anti-tumor activity of this molecule: activation of immune mechanisms (or, possibly, other complex regulatory circuits) versus direct toxicity to malignant cells (Gresser, 1977; Foon, 1989, pp. 1621–1622). This debate had practical consequences in the clinic. Interferon could be presented as the "fourth modality" of cancer therapy — immunotherapy, or, for some "biotherapy" — which should be added to the other three modalities of treatment of malignant growths: surgery, radiotherapy and chemotherapy. This innovative "fourth modality", its advocates proposed, connected the oncology clinic with the newest developments in the immunology, cell biology and molecular biology laboratory, and with the biotechnology plant. At the same time, interferon could be presented (mainly by clinicians) as but another molecule which destroys malignant cells, the use of which may be assimilated to other drug therapies of tumors. Interferons may thus be seen as "boundary objects" that are viewed by scientists as molecules which have specific effects in well-defined experimental systems, by doctors as anti-tumor drugs in patients, and in their shared uses by scientists and doctors as loosely defined "biological response modifiers". This substance, one may conclude, has a particularly complicated pathway, which loosely but efficiently associated the clinics, the laboratory and the production plant.

IMMUNOTHERAPY OF CANCER BETWEEN LABORATORY AND CLINIC

In the early 1970s the historian of science June Goodfield interviewed a large number of cancer specialists and questioned them on the current status of cancer studies. Her conclusion was that, "of all scientists working on cancer, immunologists gave me the stronger impression, not so much of success, as being able to focus precisely on a series of intellectual and therapeutic targets. Here I finally sensed, for the first time, just where T.H. Huxley's physician, the 'blind man hitting out with a club', will metamorphose into the sniper picking off intruders with lethal accuracy" (Goodfield, 1975, p. 113). In one sense, June Goodfield's prediction was not fulfilled. Immunological studies done in the 1960s and early 1970s did not lead to the rapid development of efficient anti-cancer therapies. Other targets were hit, however. The introduction of immunotherapy (and more broadly of "regulatory approaches to cancer") helped to reduce the gap between the research laboratory and the oncology clinic. Studies in experimental oncology centered on the mechanism of malignant transformation aimed mainly at the development of prevention strategies and of methods of early diagnosis. Early and/or accurate diagnosis was presented as the key issue in cancer therapy, because, cancer specialists explained, it can transform cancer from an incurable disease into a treatable one. In the meantime, however, that is until an efficient method of prevention or early diagnosis was to be found, the bulk of the work on cancer wards was (and still is) dedicated to alleviation of the suffering of cancer patients, often with an advanced disease. The research laboratory had relatively little to offer these patients. Immunotherapy of cancer was one of the vehicles which made possible an articulation between the rapidly changing domain of cancer research and the cancer clinic. From the mid-1970s on, immunotherapy of cancer (interferons, then interleukins and other cytokines) became one of the means of extension of molecularization of cancer to pre-clinical and clinical investigations, and of consolidation of the interactions between the research laboratory and the cancer ward. Such interactions were already present during the chemotherapy trials of the 1950s and 1960s, but their scope was often limited. When a given molecule reached the stage of clinical experimentation, the contacts between researchers who elaborated anti-tumor products (chemists who synthesized new compounds, pharmacologists and toxicologists who tested their physiological properties, biologists who screened their efficacy in cancer-carrying animals), and clinicians who investigated their effects in patients, were usually restricted to practical questions such as inquiries into the toxicity and tolerance of these products. Immunotherapies of cancer consolidated the links between the immunology laboratory and the oncology ward. The promoters of these therapies stressed the key role of the laboratory in their development and strongly advocated the introduction of advanced immunological techniques into the clinics in order to monitor patients who underwent immunotherapy.

In parallel, the shift in perception of the nature of "resistance to tumors" favored the tightening of the links between the immunology laboratory and the oncology clinic. In the 1960s and 1970s, "resistance to tumors" became gradually perceived as complex interactions within a complicated network of molecules, some of which are cell bound and some of which are inter-cellular messengers. According to this new explanatory model, molecules produced outside the body (bacterial antigens, interferon inducers, tumor necrotic factor) or inside it (tumor antigens, other products of tumor cells) start a complicated cascade of events such as secretion of cytokines, hormones, and growth factors and stimulation of selected categories of immunocompetent cells (lymphocytes, macrophages, and later sub-groups of lymphocytes). The term "resistance to tumors" was translated into a new language — that of immune reactions — and into new practices at the bench and the bedside — studies of the patients' immune parameters. The redefinition of the "resistance" problem in terms of "networks of cells" and then of "networks of molecules", paralleled the development of networks of specialists which linked the immunology laboratory, the experimental oncology laboratory, the cancer ward, and later also the molecular biology laboratory and the biotechnology plant.[16] The clinical trials of bacterial immunostimulants such as BCG were early examples of such coordinated enterprises. The "interferon crusade" which combined intensive effort to purify (and later produce through genetic engineering methods) a regulatory molecule, large-scale clinical trials of this molecule, and extensive studies of its physiological and immunological effects in patients, may be seen at the same time as an extension of existing trends and as their qualitative transformation through the addition of a new and powerful actor: the biotechnology industry.[17]

The parallel between the development of networks of cells and molecules and networks of experts who study these cells and molecules may be extended, I argue, to a study of the interdependence of trajectories (or "careers") of reagents (but also of tools, instruments, and scientific concepts) and of trajectories of the professionals who produce them and/or who define their properties and uses. A close look at the trajectories of substances used in the immunotherapy of cancer may thus illuminate selected aspects of the development of this domain. The production of Coley's toxins in the bacteriology laboratory, and later in commercial plants, was a low-skill endeavor while the toxins' pathway was very short. Bacterial subtances to be used in humans were neither tested nor standardized in the laboratory, and were directly transferred to the doctor's office or to the hospital ward. Toxins were perceived exclusively as anti-tumor treatment, not as biological agents or regulatory substances. This view reflected the closed circuit of therapy-oriented clinical practice before World War I, which included the hospital ward and the laboratory affiliated with this ward, but excluded the research laboratory. The end of this trajectory — the near complete demise of the toxins in the 1920s — is related to their erratic clinical record and their absence of an alternative career in the research laboratory.

The trajectory of the polysaccharides of *Serratia marcescens*, which included the research laboratory — with its strong chemical component — and the oncology clinic testifies to the growing importance of biochemical methods (and thus of molecularization) in the medical sciences in the 1930s and 1940s, and the relative eclipse of physiological approaches in immunology. This trajectory was also short, and was limited to the research laboratory, with a brief and unsuccessful excursion to the clinics. The perception of this (potential) therapy was therefore, one may assume, dominated by the view of the researchers who studied it: the toxin was seen as a chemical compound endowed with the capacity to eliminate tumor cells. Consequently, the failure to transform polysaccharides into efficient anti-cancer drugs did not lead to the intensification of study of the mechanisms by which these molecules harm malignant cells or the mechanisms of their toxicity to the organism, but to a search for different molecules endowed with anti-tumor properties, and *in fine,* to the development of NCI's "screening enterprise", which aimed at the uncovering of the "antibiotics of cancer".

The trajectory of BCG (a commercially-available product) as an anti-cancer therapy started in the cellular immunology laboratory, and later oscillated between physiology-oriented laboratory studies and large-scale testing in the clinics. This trajectory (which, one should not forget, was but a small appendix to the large body of studies of chemotherapy of cancer, grounded in the skills of chemists, biochemists, pharmacologists, toxicologists, oncologists, and statisticians) reflected the consolidation of the alliance between laboratory and clinic after World War II, and the importance of clinical tests of new cancer therapies in the development of oncology in the 1960s and 1970s.[18] At first these tests were focused on physiological rather than molecular data. In the 1960s and early 1970s, cancer immunotherapy was defined as the action of complicated biological agents, such as live or killed bacteria, on complicated biological mechanisms — the control of malignant growths. The basic and pre-clinical research in this area was thus usually centered on studies of cell-mediated and serum-mediated anti-tumor immunity in laboratory animals, while clinical studies followed the stimulation of immune responses in patients. The conviction that BCG's (putative) anti-tumor effects should be attributed to the activation of numerous cellular interactions and biochemical pathways had a double effect. It stimulated fundamental and preclinical investigations, but also discouraged clinicians from studying in detail the physiological effects of this molecule, limiting their other than purely utilitarian interest in BCG therapy. It may be proposed that the absence of visible clinical efficacy, coupled with the low level of investment in basic research on this substance (hampered by the difficulty of studing complicated biological systems) and with the absence of significant commercial and industrial interest in this agent, sealed BCG's fate as an anti-cancer therapy.

Interferon, in the 1960s mainly a substance investigated in the virology laboratory, started its career as an anti-tumor therapy in the mid-1970s thanks to increased funding for cancer research, to the development of a method of obtaining large quantities of interferon from blood, and above all to the hope, rapidly materialized, that the newly developed techniques of genetic engineering would soon allow industrial production of this molecule. The initial hopes of high therapeutic efficacy of interferon as an anti-cancer therapy were not materialized and, moreover, experimental data concerning interferon could not be correlated with clinical ones. Nevertheless, after an initial period of uncertainty, interferon found a stable place in the oncology clinic, and secured its share of the market of pharmaceutical compounds. The trajectory of interferon, from the virology and immunology laboratory to the clinic via the biotechnology industry — and the importance of commercial aspects of this career — reflect the recent consolidation of the "technoscience" aspect of biomedicine (Pickstone, 1993), mediated through the diffusion of genetic engineering techniques. Thus the fact that interferon, unlike other physiologically active substances described in this paper, has "resisted" a first wave of near-negative results of clinical trials and, thanks to the persistence of its promoters, has found clinical applications in oncology, may be attributed to the desire to bring to fruition the important scientific and commercial investments in this molecule. On the other hand, the incomplete molecularization of immunotherapy of cancer has prevented an efficient translation of phenomena observed in interferon-treated patients into biological terms. Interferon's success shows the robustness of the ideal of molecularization of cancer therapy, while interferon's double image in the cancer clinic, as a "biological response modifier" and an "anti-tumor drug",[19] illustrate the strength — but also the fragility — of recent ties between the research laboratory, the oncology ward and the production plant.

NOTES

1 Molecules which affect physiological activities may be at the same time drugs, integrated into commercial circuits, and a starting point for basic and pre-clinical investigations. On the role of such industrially-produced substances in fundamental research, see Goodman (in press).

2 Twenty four cancer patients were infected with the erysipelas agent in the early 1890s. Eighteen developed a virulent form of the disease, two had a permanent regression of the tumor, six a temporary regression, four died as a result of erysipelas infection (Moullin, 1898).

3 This bacterium was initially selected because it was reported to increase the virulence of Streptococcus. Later it was found that the two microorganisms may be cultivated separately and their toxins mixed before use.

4 Beebe and Tracy believed that bacterial substances were directly toxic for malignant cells.

5 On Wright, see also Colebrook (1949) and MacFarlane (1984). Wright's theories were immortalized in G.B. Shaw's play *The Doctor's Dilemma* of 1906.

6 In the early 20th century, some experts (e.g. Paul Ehrlich) assumed that this was indeed the case and viewed antigen-antibody reaction in terms borrowed from chemistry, as an interaction of two complementary chemical structures (lock and key theory). Other specialists (e.g., Jules Bordet) disagreed, however, and proposed to view antibodies as "states" of the

serum, and the antigen-antibody reaction as the expression of dynamic interactions of colloids (Moulin, 1991, pp. 99–104).

7 Wright's opponents affirmed that just the opposite was true: the "opsonin index " was extremely difficult to reproduce outside the "Innoculation Department" (Keating, 1988; Worboys, 1992). Wright was perhaps more interested in individually-tailored therapies which obliged patients to seek help in his Innoculation Department at St. Mary's Hospital (and to pay for it) than in the large-scale diffusion of standardized techniques (Chen, 1992).

8 *New York American,* Tuesday, August 16, 1915; Press cuttings from 1915, Murphy's Papers BM 956, American Philosophical Society, Philadelphia.

9 Shear started his investigations on malignant tumors by a study of cancerogenic compounds. His turn to the search for cancer-destroying chemical substances may be seen as a continuation of the previous line of inquiry. In the 1930s cancer researchers (such as Alexander Haddow in England, or Antoine Lacassagne in France) believed that chemical compounds akin to the substances which induced cancers (steroid hormones, aromatic compounds) may become efficient anti-cancer drugs.

10 Thus, a textbook of clinical immunotherapy affirmed that while early attempts at immunotherapy of cancer lacked a scientific basis, "a second phase in the history of tumor immunotherapy was intitiated in the mid-1950's with the development of inbred strains of laboratory animals. These strains allowed the development of modern tumor immunology based on animal tumor models... Observations made in animal models ... led to a resurgence in clinical trials of tumor immunotherapy" (Lobuglio, 1980, p. 1).

11 A dissenting opinion was, however, expressed by the British cancer specialist Graham Currie, who affirmed in 1974 that available experimental and clinical data do not support the hypothesis that an activation of immune mechanims may lead to a cure of human cancer (Currie, 1974).

12 The introduction of interferon into cancer therapy was studied by Sandra Panem and Toine Pieters (Panem, 1984; Pieters, 1993a; Pieters, 1993b).

13 This was, for example, the opinion of Stephen K. Carter, from the Cancer Research Division of Bristol-Myers Company, as quoted by E. Shorter (Shorter, 1987, p. 250).

14 Interferon also became a highly profitable product, with a market which expanded from 13.6 million dollars in 1986 to an estimated one billion dollars in 1993 (Dodet, 1994).

15 Thus the physiological mechanisms of activity of numerous drugs were elucidated long after the introduction of these drugs to the clinics. The classic example is aspirin (salicylic acid). Scientists first proposed a mechanism to explain its anti-pyretic and anti-inflamatory effects a hundred years after its introduction as a therapy.

16 The immunologist Niels Jerne first proposed to link immunological concepts developed in the 1950s and 60s with the "proliferation of immunologists" (Jerne, 1967, p. 601; see also Moulin, 1989, p. 294).

17 The expression "interferon crusade" was coined by Sandra Panem (Panem, 1984). The possibility to isolate from human blood "non specific factors" for treatment of neoplastic disease was first explicitly evoked in 1960, following a discussion of the anti-tumor activity of Coley's toxins, but it was then rejected as impractical (Southam, 1960, p. 286).

18 The pharmaceutical industry was an important partner in the organized search for new anti-cancer drugs which started after World War II and culminated in NCI's screening empire. The industrial investment in individual chemical compounds tested for their potential anti-tumor properties was low, however (Endicott, 1957).

19 This "double image" may contrast with the relatively straightforward view of interferon as an anti-viral drug (one of the most important clinical applications of interferon in the 1990s is the treatment of chronic viral hepatitis). Medical virologists perceive interferon above all as a molecule which directly neutralizes pathogenic viruses.

REFERENCES

F. Alfred and P.B. Medawar (eds), *Biological Problems of Grafting* (Oxford: Blackwell Scientific Publications, 1959).

F.R. Balkwill, "Cytokine Therapy of Cancer. The Importance of Knowing the Context" *European Cytokine Network* (1994), 5(4): 379–385.

S.P. Beebe, "The Growth of Lymphosarcoma in Dogs" *Journal of the American Medical Association* (1907), 49: 1492–1493.

S.P. Beebe and M. Tracy, "The Treatment of Experimental Tumors with Bacterial Toxins" *Journal of the American Medical Association* (1907), 49: 1493–1498.

R. Billingham, L. Brent and P.B. Medawar, "Actively Acquired 'Tolerance' to Foreign Cells" *Nature* (1953), 172: 603–606.

A.M. Brues and M.J. Shear, "Chemical Treatment of Tumors: Reactions of Four Patients with Advanced Malignant Tumors to Injection of Polyscaccharide from *Serratia marcescens* Culture Filtrate" *Journal of the National Cancer Institute* (1944), 5: 195–208.

R.F. Bud, "Strategy in American Cancer Research after World War II: A Case Study" *Social Studies of Science* (1978), 8: 425–459.

W. Chen, "The Laboratory as Business. Sir Almroth Wright's Vaccine Programme and the Construction of Penicillin", in A. Cunningham and P. Williams (eds), *The Laboratory Revolution in Medicine* (Cambridge: Cambridge University Press, 1992), pp. 245–293.

R.L. Clark, R.L. Hickey and E.M. Hersh (eds), *Immunotherapy of Cancer* (New York: Raven Press, 1978).

L. Colebrook, *Almroth Wright, Provocative Doctor and Thinker* (London: William Heinemann Medical Books, 1949).

W.B. Coley, "Contribution to the Knowledge of Sarcoma" *Annals of Surgery* (1891), 14: 199–200.

W.B. Coley, "The Treatment of Malignant Tumors by Repeated Inoculation of Erysipelas, with a Report of Original Cases" *American Journal of Medical Science* (1893), 105: 487–511.

W.B. Coley, "Treatment of Inoperable Malignant Tumors with the Toxines of Erysipelas and *Bacillus prodigious*" *Transactions of American Surgical Association* (1894), 12: 183–212.

W.B. Coley, "Further Observations Upon the Treatment of Malignant Tumors with the Toxines of Erysipelas and *Bacillus prodigiosus*, with a Report of 160 Cases" *Bulletin of the Johns Hopkins Hospital* (1896), 65: 157–162.

W.B. Coley, "The Treatment of Inoperable Sarcoma with the Mixed Toxines of Erysipelas and *Bacillus prodigiosus*: Immediate and Final Results in 140 Cases" *Journal of the American Medical Association* (1898), 31: 389–395; 456–465.

W.B. Coley, "Treatment of Inoperable Sarcoma by Bacterial Toxins (the Mixed Toxins of the *Streptococcus erysipelas* and the *Bacillus prodigiosus)*" *Proceedings of the Royal Society of Medicine,* section 3 (London: John Bale & Sons, 1909), pp. 1–44. Discussion pp. 44–48.

W.B. Coley, "Some Unsolved Problems in Medicine and Surgery" *The Therapeutic Gazette* (Detroit) (1912), 26: 73–82.

"Concord Trial: A Preliminary Report" *British Medical Journal* (1971), 4: 189–194.

G.C. Currie, "Immunotherapy of Human Cancer: Prospect and Retrospect" in G.C. Currie (ed.), *Cancer and the Immune Response* (Baltimore: The Williams and Wilkins Company, 1974), pp. 74–96.

A.J.S. Davies, D.C. Edwards and P.E. Thrope, "An Introduction" in B. Serrou and C. Rosenfeld (eds), *International Symposium on New Trends in Human Immunology and Cancer Immmunotherapy* (Paris: Doin Editeurs, 1980), pp. 3–4.

B. Dodet, "Cytokines in the Clinics. Choose Your Weapon" *European Cytokine Network* (1994), 5: 369–376.

F. Duran Reynals, "Reaction of Transplantable and Spontaneous Tumors to Blood-Carried Bacterial Toxins in Animals Unsusceptible to the Schwartzman Phenomenon" *Proceedings of the Society of Experimental Biology and Medicine* (1933) 31: 341–344.

K.M. Endicott, "The Chemotherapy Program" *Journal of the National Cancer Institute* (1957), 19: 275–293.

K. Fent and G. Zbiden, "The Toxicity of Interferon and Interleukin" *Trends in Pharmacological Sciences* (1987), 8: 100–108.

K.A. Foon, "Biological Response Modifiers: The New Immunotherapy" *Cancer Research* (1989), 49: 1621–1639.

A. Gartia and R. Linz, "Le Phénomène de Schwartzman dans le Sarcome du Cobaye" *Comptes Rendus de la Société de Biologie* (1931), 108: 427–428.

J.P. Gaudillière, "NCI and the Spreading Genes: About the Production of Mice, Viruses and Cancer", in E. Mendelsohn (ed.), *Human Genetics*, Sociology of Science Yearbook (Dordrecht: Kluwer, 1993).

T. Gibson and P.B. Medawar, "The Fate of Skin Homograft in Man" *Journal of Anatomy* (1943), 77: 299–300.

J. Goodfield, *The Siege of Cancer* (New York: Random House, 1975).

J. Goodman, "Can it Ever be Pure Science: Pharmaceuticals, the Pharmaceutical Industry and Biomedical Research in the Twentieth Century", in J.P. Gaudillière and I. Löwy (eds), *The Invisible Industrialist: Manufactures and the Production of Scientific Knowledge* (London: MacMillan, in press).

I. Gresser, "Commentary: On the Varied Biological Effects of Interferons" *Cellular Immunology* (1977), 35: 406–415.

J.U. Gutterman, G. Mavligit, C. Macbridge, E. Hersh and R. Reed, "Active Immunotherapy with BCG for Recurrent Malignant Melanoma" *Lancet* (1974), ii: 128–135.

L. Hartwell, M.J. Shear and J.R. Adams, "Chemical Treatment of Tumors: VII. Nature of the Hemorrhage-Producing Fraction for *Serratia marcescens (Bacillus prodigiosus)* Culture Filtrate" *Journal of the National Cancer Institute* (1943), 4: 107–122.

E.M. Hersh, J.U. Gutterman and G. Mavligit, *Immunotherapy of Cancer in Man: Scientific Basis and Current Status* (Springfield, Ill.: Charles C. Thomas, 1973).

A.L. Hollman, "Reactions of Patients and of Tumors to the Injection of S. Marcescens Polysaccharide in Eight Cases of Malignant Disease", in F.M. Moulton (ed.), *Approaches to Tumor Chemotherapy* (Washington D.C.: *American Association for the Advancement of Science,* 1947), pp. 273–276.

R.I. Jahiel and M. Krim, "Interferons, Clinical Trials and Effects on Hematologic Neoplasms", in J.W. Chiao (ed.), *Biological Response and Cancer Therapy* (New York and Basel: Marcel Dekker Inc., 1988), pp. 197–266.

N.K. Jerne, "Summary: Waiting for the End" *Cold Spring Harbour Symposia for Quantitative Biology* (1967), 32: 591–603.

H. Kanler, M.J. Shear and J.L. Hartwell, "Chemical Treatment of Tumors: Ultracentrifugal and Electrophoretic Analysis of the Hemorrhage-Producing Fraction for *Serratia marcescens* (*Bacillus prodigiosus*) Culture Filtrate" *Journal of the National Cancer Institute* (1943), 4: 123–129.

P. Keating,"Vaccine Therapy and the Problem of Opsonins" *Journal of the History of Medicine and Allied Sciences* (1988), 43: 275–296.

E. Keidar and E. Klein, "Cancer Immunotherapy: Are the Results Discouraging? Can They Be Improved?" *Advances in Cancer Research* (1992), 59: 245–322.

M. Kenney, *Biotechnology: The University-Industrial Complex* (New Haven and London: Yale University Press, 1986).

A.F. Lobuglio, "Approaches to Human Immunotherapy", in A.F. Lobuglio (ed.), *Clinical Immunotherapy* (New York and Basel: Marcel Dekker Inc., 1980), pp. 1–5.

I. Löwy, "Nothing More to Be Done: Palliative Care Versus Experimental Therapy", *Science in Context,* (1995), 8: 209–229.

G. MacFarlane, *Fleming: The Man and the Myth* (Oxford: Oxford University Press, 1984).

H.M. Marks, "Notes From the Underground: The Social Organization of Therapeutic Research", in R.C. Maulitz and D. Long (eds), *Grand Rounds. One Hundred Years of Internal Medicine*, (Philadelphia: University of Pennsylvania Press, 1987), pp. 297–334.

M.J. Mastrangelo, D. Berd and R.E. Bellet, "Limitations, Obstacles and Controversies in the Optimal Development of Immunotherapy", in R.L. Clark, R.L. Hickey and E.M. Hersh (eds), *Immunotherapy of Cancer* (New York: Raven Press, 1978), pp. 375–394.

G. Mathé, J.A. Amiel, L. Schwartzenberg *et al.*, "Attempts at Immunotherapy for Acute Immunoblastic Leukemia" *Lancet* (1969), i: 697–699.

G. Mathé, "Round Table", in G. Lamoureux, R. Turcotte and V. Portelance (eds), *BCG in Cancer Immunotherapy* (New York, San Francisco and London: Grune & Stratton, 1976).

P.B. Medawar, "The Behaviour and Fate of Skin Homografts in Rabbits" *Journal of Anatomy* (1944), 78: 176–179.

P.B. Medawar, "A Second Study on the Behaviour and Fate of Skin Homografts in Rabbits" *Journal of Anatomy* (1945), 79: 157–160.

J.P. Merrill, "The Production of 'Delayed Type' Cutaneous Hypersensitivity to Human Donor Leucocytes as a Result of the Rejection of Skin Homografts" *Journal of Clinical Investigations* (1961), 51: 631–643.

E. Mihich and A. Fefer, *Biological Response Modifiers: Subcommitteee Report*, NCI Monograph n° 63, (Bethesda, MD: NIH Publication n° 83–2606, 1983).

N.A. Mitchison, "Passive Transfer of Transplantation Immunity" *Nature* (1953), 171: 267–268.

D.L. Morton, F.R. Eiber, E.C. Holmes, F.A. Malmargen and W.C. Woods, "BCG Therapy of Malignant Melanoma: Summary of Seven Years of Experience" *Annals of Surgery* (1974), 10: 635–642.

A.M. Moulin, "Immunology Old and New: The Beginning and the End", in P. Mazumdar (ed.), *Immunology 1930–1980: Essays on the History of Immunology* (Toronto: Wall & Thompson, 1989), pp. 291–298.

A.M. Moulin, *L'Histoire de l'immunologie de Pasteur au Sida* (Paris: PUF, 1991).

C.M. Moullin, *The Treatment of Sarcoma and Carcinoma by Injections of Mixed Toxins* (London: John Bale Sons & Danielson, 1898).

H.C. Nauts, W.E. Swift and B.L. Coley, "The Treatment of Malignant Tumors by Bacterial Toxins as Developed by the Late William B. Coley, Reviewed in the Light of Modern Medical Research" *Cancer Research* (1946), 6: 205–216.

H.C. Nauts, *Bibliography of Reports Concerrning the Clinical or Experimental Use of Coley's Toxins (Streptococcus pyogenes and Serratia marcescens)* (New York: Cancer Research Institute, Inc., Monograph n° 18, 1991).

R. Oakley, "Reactions of Patients to the Injection of *S. marcescens* Polysaccharide in Nine Further Cases of Malignant Disease" (Washington D.C.: American Association for the Advancement of Science, 1947), pp. 277–278.

L.J. Old, D.A. Clarke and B. Benacerraf, "Effects of Bacillus Calmette-Guérin Infection of Transplanted Tumors in Mice" *Nature* (1959), 194: 291–293.

R.K. Oldham and R.V. Smalley, "Immunotherapy: The Old and the New" *Journal of Biological Response Modifiers* (1983), 2: 1–37.

S. Panem, *The Interferon Crusade* (Washington, D.C.: The Brookings Institution, 1984).

J. Pickstone, "Ways of Knowing: Towards a Historical Sociology of Science, Technology and Medicine" *British Journal of History of Science* (1993), 26: 433–458.

T. Pieters, "Interferon and its First Clinical Trial: Looking Behind the Scenes" *Medical History* (1993a), 37: 270–295.

T. Pieters, "A Comparative Study of Development of Two Biologicals: The Interferons", 1993b. Paper presented at the workshop *Les biologistes entre l'Etat et l'industrie: Des usages à la production du vivant*, Paris, Cité des Sciences et d'Industrie, 3 December 1993.

T.M. Powledge, "Interferon on Trial" *Biotechnology* (March 1984), 214–228.

G. Schwartzman and N. Michailovsky, "Phenomenon of Local Skin Reactivity to Bacterial Filtrates in the Treatment of Mouse Sarcoma 180" *Proceedings of the Society of Experimental Biology and Medicine* (1932), 29: 737–741.

M.J. Shear, "Chemical Treatment of Tumors. Isolation of the Hemorrhage-Producing Fraction from *Serratia marcescens* (*Bacillus prodigiosus*) Culture Filtrate" *Journal of the National Cancer Institute* (1943), 4: 81–97.

M.J. Shear, "Role of the Chemotherapy Research Laboratory in Clinical Cancer Research" *Journal of the National Cancer Institute* (1951), 12: 569–582.

E. Shorter, *The Health Century* (New York: Doubleday, 1987).

K. Sikora, *Interferon and Cancer* (New York and London: Plenum Press, 1983).

A.M. Silverstein, "Introduction", in D. Jan Bibel (ed.), *Milestones in Immunology: A Historical Exploration* (Berlin and Madison, Wisconsin: Springer Verlag, 1988), pp. xiii–xv.

A.M. Silverstein, *A History of Immunology* (San Diego: Academic Press, 1989).

C.M. Southam, "Relationships of Immunology to Cancer: A Review" *Cancer Research* (1960), 20(3): 271–291.

H. Strander, K. Cantell and S. Ingmarson, "Interferon Treatment of Osteogenic Sarcoma: A Clinical Trial" *Conference on the Modulation of Host Immune Resistance in the Prevention or Treatments of Neoplasms* (Washington D.C.: Government Printing Office, 1974).

A.I. Tauber, *The Immune Self* (Cambridge, New York, Melbourne: Cambridge University Press, 1994).

W.D. Terry, "Concluding Remarks" in W.D. Terry and D. Windhorst (eds), *Immunotherapy of Cancer: Present Status of Trials in Man* (New York: Raven Press, 1978), pp. 669–671.

W.D. Terry and Y. Yamamura (eds), *Immunobiology and Immunotherapy of Cancer* (New York, Amsterdam, Oxford: Elsevier, North Holland, 1979).

D. Windhorst, "International Registry of Tumor Immunotherapy: The Contribution of an Information Service to a New Specialty" in W.D. Terry and D. Windhorst (eds), *Immunotherapy of Cancer: Present Status of Trials in Man* (New York: Raven Press, 1978), pp. 615–620.

M. Worboys, "Vaccine Therapy and Laboratory Medicine in Edwardian Britain" in J.V. Pickstone (ed.), *Medical Innovations in Historical Perspective* (London: Macmillan, 1992), pp. 84–103.

C.G. Zubrod, S. Schepartz and J. Leiter, "History of the Cancer Chemotherapy Program" *Cancer Chemotherapy Reports* (1966), 50(7): 349–382.

C.G. Zubrod, "Origins and Development of Chemotherapy Research at the National Cancer Institute" *Cancer Treatment Reports* (1984), 68(1): 9–19.

CHAPTER 10

Reflections: Molecularization, Standardization and the History of Science

Steve Sturdy

The concept of molecularization, as discussed in the various papers in this volume, might be seen as having two distinct layers of meaning. First, it is a technical process, in which the complex, refractory stuff of biological reality is broken down into discrete molecular constituents, to be characterized, standardized and manipulated using various chemical and other scientific techniques. But secondly, molecularization is also a social and cultural phenomenon, evident in the emergence of a wide range of institutions committed to furthering the process of molecularization in its narrower technical sense. Of course, any such distinction between the technical and the social is ultimately misleading if we want to understand and explain the historical processes of scientific development; science is culture, even when it is at its most technical. This is amply demonstrated by the various papers collected here: the characterization and production of molecules is intimately dependent upon, and deeply implicated in, the formation of networks of collaboration and exchange of materials and techniques, both within biomedical research laboratories, and between laboratories and the various other social contexts in which such molecules are put to use.

In their introduction to this volume, the editors take the indivisibility of the technical and the social as their starting point. They begin by defining molecularization as the formation of networks of strategic alliances orientated towards the characterization and deployment of molecules in science and medicine. And they argue that, by focusing on the formation and transformation of such networks, and especially on the way in which they are mediated by the circulation of molecules and their associated technical practices, we are able to move away from static notions of social structure and social interests to a more fluid understanding of how technical development may be directly implicated in processes of social change.

In my concluding remarks, I want to adopt a somewhat different — but complementary — perspective on the history of molecularization. I take my cue from a recent paper by Steven Shapin (Shapin, 1992). While Shapin agrees that attention to the formation and transformation of scientific networks provides a useful corrective to residual suppositions that any principled distinction can be drawn between the technical and the social in scientific practice, he goes on to suggest that such a perspective may also tend to obscure important pragmatic distinctions between different social sites of scientific and technical activity — between different spheres of professional practice and jurisdiction, and between what we and many of the historical actors involved might regard as the distinct processes of scientific research and the wider social uses of the products of that research. In what follows, I will elaborate a couple of points that follow if we direct our attention, less towards the way in which the circulation of particular molecules and techniques mediated the formation and maintenance of social networks, and more towards the heterogeneities and discontinuities in the networks through which those molecules circulated.

In the first place, it is clear that different groups of actors allied themselves to the project of molecularization at different times, and for different reasons. Concentrating on the circulation of molecules and technologies alone does not explain why this should have been the case. By drawing on the various case studies collected in this volume,[1] I want to attempt to sketch out, in very general terms, some of the specific ways in which the characterization and deployment of molecules may have come to serve the interests of distinct groups of actors at specific historical junctures. In particular, I want to suggest that one of the main factors behind this convergence of interests was the overriding interest of some of the most powerful promoters and funders of molecular science in the development of standardized medical technologies, which they regarded as a means of promoting administrative efficiency within systems of mass health care. In this respect, if we are to explain the growth of molecularization as a shared scientific and technical project, it is important to acknowledge some of the wider social interests which sustained the technical work of characterizing and producing biomedical molecules. Inevitably, my account of these wider interests, and the way they informed the development of molecularization, is sketchy and in large part hypothetical. But hopefully it will provide some useful pointers to questions that other researchers may care to address.

Secondly, I want briefly to consider the fact, made clear in several of the papers in this collection, that the technical definition and characterization of molecules was itself the result of negotiation between different groups of actors, not least over how different techniques might legitimately be used in this work. Consequently, I will suggest that, in any discussion of the formation of scientific networks, attention needs also to be paid to the way in which agreement was reached in such negotiations. Specifically, I will argue that the establishment of agreed methods of standardizing molecules and instruments provided a means of negotiating common interests between different groups of practitioners, and hence of establishing an effective basis for collaboration and exchange. While the circulation of molecules and their associated techniques and practices may have served to tie diverse groups of actors into the developing economy of molecularization, standardization was vital in determining the value and meaning of such currency.

STANDARDIZATION AND THE GROWTH OF MOLECULAR CULTURE

In asking how and why the interests of diverse groups of actors, working in different social settings, came to converge on the characterization, production and use of biological molecules, it is perhaps convenient to begin by looking at developments that took place in the pharmaceutical industry from the 1880s onwards. In both Europe and North America, sales of pharmaceutical products had expanded rapidly through the nineteenth century. Much of this growth had been achieved without any significant changes in manufacturing methods. Pharmaceutical production still involved little more than the preparation and packaging of naturally occurring substances. At its most sophisticated, it might include the extraction of pure active chemical constituents — for instance morphine and colchicine — from natural raw materials. But even this was still largely a craft-based rather than a science-based procedure. From about the 1880s, however, a number of companies began to concentrate on producing pure drugs, and now sought to improve the efficiency and reliability of their extraction and purification processes by adopting more advanced methods of chemical manipulation. At the same time, a handful of chemical manufacturers, particularly in Germany, began to look for ways of synthesizing drugs.

New methods of constructing complex organic molecules from simpler chemical building blocks were being developed in the chemical industry, and were now applied to pharmaceutical production as well. Beginning with the relatively simple salicylates, these companies moved on to the synthesis of increasingly complex molecules, including modifications of naturally occurring substances. The adoption of molecular methods of pharmaceutical production brought further commercial benefits. For one thing, synthesis enabled manufacturers to dispense, in many cases, with scarce natural sources of complex chemicals and to turn instead to more readily available raw materials. For another, synthetic chemical processes could be patented, and their products registered under protected trade names, thus providing an effective means of asserting proprietary rights within a highly competitive pharmaceuticals market (Liebenau, 1987, 1988). Thirdly, as pharmaceutical production was broken down into discrete chemical steps, it became possible to reduce wastage by using by-products and intermediaries from one chemical process as raw materials for another. Such efficiency gains favored the growth of large-scale manufacturing organizations with complex and highly integrated internal divisions of labor, while smaller craft-based companies tended to be edged out of the pharmaceutical market.[2]

Important areas of pharmaceutical production thus became reorganized around increasingly sophisticated molecular and other chemical practices from the late nineteenth century. But this reorganization was not simply a matter of introducing new technical procedures into the manufacturing process; it also depended upon the elaboration of industrial management structures. With the subdivision and integration of pharmaceutical production processes, it became necessary to ensure that each step in the overall process was managed and regulated in order to maintain the integrity and efficiency of the production process as a whole. Of vital importance to this process was the establishment of centralized quality control laboratories, which now came to play an important role in maintaining appropriate standards at each stage of production and in different sections of the plant or company (Liebenau, 1987). Internal standardization, in other words, was important for asserting managerial and technical control over the manufacture of medical molecules.

Moreover, such laboratories could be used for other purposes than just quality control. The pursuit of innovation, in terms of both products and production methods, was rapidly becoming an important factor in maintaining the competitiveness of individual companies, and research and development work was increasingly seen to be vital both for improving existing processes and for identifying new drugs that might be brought into industrial production. Initially, such work was carried out in the same laboratories as conducted quality control work. Subsequently, however, many companies invested in separate research departments. This also had implications for universities and other non-industrial institutions. Industry now provided employment opportunities for trained chemists and medical scientists, and a market in which academics could sell their expertise as consultants. Consequently, many scientists in universities and other academic institutions began to tailor their teaching and research to meet this demand. This was particularly the case in Germany, where the exchange of personnel, information and techniques between industry and academia was actively encouraged by government (Lenoir, 1988; Liebenau, 1990); but similar developments took place in the USA and — less obviously — in Britain from about the end of the nineteenth century (Parascandola, 1992; Swann, 1988; Tansey and Milligan, 1990).

The search for new drugs and new manufacturing techniques thus came to inform the work of scientists both within and beyond the confines of industrial organizations. Of course, scientists in different settings tended to address such issues in rather different ways. While research in industry was clearly directed, by and large, towards innovation in industrial production, that in academia tended to be more speculative and open-ended. Still, we might regard this as at least partly an extension of the division of labor within industry itself. Thus pharmaceutical and other chemical manufacturers not only favored the establishment of research laboratories within their own firms; they also threw their political weight behind the creation of academic laboratories in regional and national institutions, in the expectation that their own enterprises would benefit, either directly or indirectly, from the research carried out there. Here, as de

Chadarevian and Kamminga note in their introduction, lie the beginnings of the process of molecularization, in the sense of the formation of networks of alliances and exchange between industrial and academic scientists and the institutions within which they worked. And a clear driving force behind the establishment of such networks was the pursuit of commercial profit and competitiveness in the national and international pharmaceuticals market.

Pharmaceutical molecularization would have served little purpose, however, if the drugs produced by molecular means had not also been commercially successful. The fact is that they were very successful, and increasingly so as the twentieth century drew on. The reasons for this — and, indeed, for the growth of molecular culture more generally — have much to do with changes in the culture of medical practice. The first such changes we need to consider were in doctors' prescribing habits. For most of the nineteenth century, the pharmaceuticals market had been largely dominated by so-called patent medicines. The constitution of these preparations was a matter of strict commercial secrecy, and their sales were based chiefly on manufacturers' own reports of their efficacy, often backed up by affidavits from doctors and patients who had used them (Richards, 1990; Young, 1961). Large quantities of patent medicines were sold directly to patients, but many medical practitioners also found them convenient to use. However, growing numbers of doctors feared that the advertizing of patent medicines played into the hands of unqualified practitioners, or encouraged patients to practice self-medication. Consequently, these doctors preferred to rely on drugs of known chemical constitution, whose pharmacological effects had been demonstrated by laboratory experiments or clinical evidence, and whose use depended on the kind of esoteric knowledge that could only be acquired in the course of a proper medical training (Rosenberg, 1979; Warner, 1986; and, for the persistence of such concerns into a later period, Cantor, 1992).

Pharmaceutical companies consequently had a further reason for adopting sophisticated chemical and other laboratory-based techniques in the manufacturing process. Besides being used to secure greater managerial control of production, such techniques could also be employed to reinforce manufacturers' claims that their products were pure, safe, reliable and efficacious.[3] This language of safety and efficacy became particularly urgent with the appearance on the market of a new class of drugs produced directly from living organisms. From the 1890s, developments in bacteriology and immunology led to the manufacture of a growing range of anti-bacterial vaccines, sera and antitoxins, and precipitated a second wave of innovation in pharmaceutical production. These preparations often varied enormously in concentration, however, and were prone to contamination by dangerous micro-organisms. Consequently, manufacturers were placed under considerable pressure, and in some countries direct legislative requirements, to standardize different batches for immunological activity and to eliminate contaminated batches (Weindling, 1992; Liebenau, 1988). Again, it should be noted that this did not necessarily imply the adoption of molecular or even chemical techniques. Like many other naturally occurring pharmaceuticals, the new biological drugs could at that time only be tested and standardized using methods derived from the sciences of bacteriology and physiology. Nevertheless, the adoption of such methods reinforced the current medical concern with pharmaceutical purity and standardization, and with it the more specific drive to molecularization. Indeed, many of the manufacturers who profited most successfully from immunological preparations had already begun to adopt molecular techniques for the production of more conventional kinds of drugs.

Moreover, interest in the standardization of pharmaceuticals was not confined to medical practitioners. The late nineteenth and early twentieth century was a period of intensifying competition, both industrial and military, between nation states. In this context, ill health was widely seen to pose a serious threat to national strength, vigor and efficiency. This threat might be countered by, among other things, the provision of more effective systems of medical care. In some countries — for instance Britain, where a scheme of National Health Insurance was established in 1911 — governments went so far as to intervene directly in the organization of the medical market place (Gilbert, 1966; Hennock, 1987). But even where such measures were not enacted, there was widespread discussion among doctors and policy makers about how best to bring about improvements in the standard of medical practice. Patent medicines figured large in these discussions. Despite the growing availability of pure drugs, many doctors continued to rely heavily on patent medicines and other commercially available remedies of doubtful therapeutic efficacy. The use of such remedies, reformers argued, was not simply a waste of money; it might actually exacerbate the decline in patients' health, and thereby weaken the national stock. Consequently, much would be gained if doctors could be induced to limit themselves, as far as possible, to using only such therapies as had been shown to be effective, either by laboratory tests or by authoritative clinical investigations.

Thus, for instance, with the establishment of National Health Insurance in Britain, administrators proposed to draw up a list of approved drugs, and to prohibit participating doctors from using any preparations that were not included on that list. This initiative foundered as doctors reasserted their right to use their own professional judgment in medical matters; while doctors welcomed the scientific standardization of commercially available medications, they were not prepared to accept legislative standardization of their own therapeutic practices.[4] Nevertheless, if compulsion was not the answer, it might still be possible to secure the same end by the production and dissemination of accurate scientific information about the pharmacological and clinical effects of specific drugs. Consequently, a wide range of organizations, including not just pharmaceutical companies and universities, but also charities and government bodies, now began to promote concerted research into various forms of therapeutics, and into the related sciences of bacteriology, immunology and physiology.

In Britain, the Lister Institute was established by charitable bequest, initially to emulate the work of the Pasteur Institute in developing new therapeutic and preventive vaccines; by the early 1900s the work of the Institute had diversified to include research into pathological chemistry and into the nutritional and therapeutic value of various foodstuffs (Kamminga, this volume; see also Chick *et al.*, 1971). The national importance of this work was acknowledged in 1913, when a Medical Research Committee was set up under the auspices of, and with funding from, the National Health Insurance administration. Plans were drawn up to hand over the Lister Institute to the new Committee, but negotiations broke down. Instead, the Committee decided to set up its own National Institute for Medical Research. That the National Institute was expected to conduct research of relevance to therapeutic practice is evident from the fact that the staff of its Department of Biochemistry and Pharmacology was recruited directly from the research laboratories of the drug company Burroughs Wellcome (Harington, 1950). In the United States, meanwhile, the Rockefeller Foundation had also established a research Institute, and later a Hospital, to investigate the causes and treatment of particular diseases (Amsterdamska, this volume; Corner, 1964).

Clearly, the work of these and other research institutions was not informed solely by a concern to standardize therapeutic practice. Medical scientists brought to their research a host of other interests and aims, to do with the establishment of science as a profession, with the consolidation of particular disciplinary projects, and with other more local and more personal concerns. Nevertheless, it was clearly hoped that they would help to establish new technically defined standards of good medical practice, and that these standards would in turn lead to greater efficiency and effectiveness in the provision of medical care. By such means, it might be possible to reduce the protean medico-social problems of national ill health to a series of clearly defined diseases with equally clearly defined therapeutic solutions. At a time when national governments were looking for new ways of managing the health of populations, the ideal of a standardized and routinized system of medical practice had obvious administrative appeal. Moreover, there is good reason to suppose that many medical scientists shared that ideal. After all, much of their own research was aimed at reducing the complexities of living processes to regular, reproducible and — above all — standardizable phenomena. Increasingly, the interests of laboratory-based scientists and medical administrators were converging on the pursuit of more efficient forms of therapeutics and, ultimately, more efficient systems of mass medical care (Sturdy and Cooter, unpublished paper). Insofar as such interests were focused on the characterization and use of specific molecules, the pursuit of medical standardization provided a powerful stimulus to molecularization.

As the various papers in this volume show, medical scientists soon proved themselves adept at generating precisely the kinds of standardized — and ultimately molecularized — forms of medical intervention that doctors and health administrators favored. Thus Kamminga shows how work on the therapeutic and preventive uses of specific foodstuffs, conducted at the Lister Institute, provided a basis for large-scale administrative responses to the nutritional problems posed, first by World War I, then by the depressions that followed it.[5] Initially, scientists could do no more than hypothesize that the beneficial effects of these foods were due to the presence of specific molecules. Nevertheless, considerable effort was devoted, both in government and in the food and pharmaceutical industries, to isolating and analysing these putative "accessory food factors", and by the 1930s had progressed far enough for synthesis of a number of vitamins to proceed on a commercial scale. Governments could now do much to address the profound social problems of inadequate nutrition — and, indeed, of inadequate income — by the simple and cheap expedient of distributing vitamin supplements. The standardization of dietary requirements, and the expression of such standards in terms of specific molecules, had proved to be a boon both to public health administrators and pharmaceutical companies.

Interwar examples of successful molecularization could easily be multiplied. Insulin is a case in point. Laboratory scientists had been interested in the physiological and pharmacological activity of organ extracts since at least the 1890s, and by the early 1900s had succeeded in isolating a number of specific chemicals — hormones — from such extracts. Pharmaceutical companies were quick to exploit such discoveries by manufacturing and selling a wide range of organ extracts or "organotherapies" as treatments for an even wider range of conditions. Academic scientists were increasingly suspicious of this practice, which they feared was based on little more than the inflated claims of quack medicine (Borell, 1985). The demonstration that pancreatic extract could be used to treat diabetes, and the attribution of this therapeutic effect to a single, isolable molecule — insulin — helped to foster renewed scientific interest in the biology and chemistry of hormones. Moreover, both the initial development of insulin and its subsequent validation as an effective therapy depended on close collaboration between pharmaceutical companies, laboratory scientists and hospital doctors, and did much to reinforce the shared belief that molecularization of hormones was a viable medical project (Bliss, 1987; Li, 1992). Elsewhere, industrial and academic scientists alike were chasing a number of other molecular hares, including various forms of chemotherapy for cancer, arsenicals and quinine derivatives for the treatment of syphilis and other microparasites, and later the sulfa drugs to treat a range of microbial diseases (Weatherall, 1990). Though less dramatic in their impact than insulin, these different lines of research all yielded enough new drugs to sustain interest in pharmaceutical molecularization.

Another change of gear came with the outbreak of World War II. As in the Great War before it, the massive mobilization of military and civilian personnel raised serious problems of managing, not just battle injuries, but also ill health more generally. Once again, the need for administrative efficiency placed a premium on the development of standardized and routine forms of therapy and prevention. As Angela Creager shows (this volume), molecular techniques proved effective as a way of devising new treatments for wound shock. It was widely agreed that, under ideal circumstances, whole blood transfusion should be the treatment of choice in such cases. But in a situation where large quantities of transfusion materials were needed, and where there was little time to ensure that patients received blood of the correct immunological type, the use of whole blood was highly problematic. By developing a molecular method of isolating distinct plasma fractions from whole blood, Edwin Cohn was able to provide an effective solution to this urgent administrative problem. In so doing, Cohn reinvented his laboratory, not just as a center for molecular research, but as a key site for the integration of scientific, industrial and government interests. A similar response to the pressing administrative problems of wartime medicine can be seen in the identification of penicillin and subsequently streptomycin as effective, standardizable and clinically simple means of treating a wide variety of infections, and in the mobilization of industrial, academic and government scientists in the successful development of efficient methods of production and distribution (Swann, 1983; Neushul, 1993).

This convergence of interests around the standardization of medical care continued after the war, as national governments made renewed efforts to promote and maintain the health of their populations. In the United States, martial attitudes were carried over into the announcement of a "war against disease", which was intended to mobilize the combined scientific resources of industry, government and corporate charity in the search for new methods of prevention and cure. Meanwhile, in Britain, the establishment of the National Health Service, and the decision to provide free medical treatment to the entire population, reinforced official, medical and scientific interest in the development of efficient methods of diagnosis and treatment. The success of molecularization in meeting the administrative demands of wartime medicine provided an obvious starting point for many postwar initiatives, which commonly revolved around the search for molecular causes and treatments for a variety of diseases, including cancer (Gaudillière, this volume; Löwy, this volume), and sickle cell and other anemias (de Chadarevian, this volume).

It is worth noting, moreover, that both the hemoglobin project and the American program of cancer research reflected a significant broadening of the scope of molecular research. Hitherto, the project of medical molecularization had been led, on the whole, by research into therapeutic substances. Insofar as research had been conducted into identifying specific causes of disease, it had tended to be based either on microbiological techniques, or on the therapy-led discovery of particular hormonal or nutritional deficiencies. A major exception to this rule was the work that had been conducted between the wars into possible molecular carcinogenesis; but this had proved inconclusive, and had eventually been diverted into experimental genetics and into the search for possible cancer vaccines (Goodman, this volume; Gaudillière, this volume). But with the postwar revival of interest in standardizing and especially molecularizing medicine, concerted efforts were now made to investigate the possible molecular basis of disease etiology, even where effective therapeutic measures were not yet available. It was clearly expected that continued research would eventually yield such therapies; indeed, as Paul and Edelson make clear in their study of the growth of screening programs for phenylketonuria (this volume), the merest hint of therapeutic promise was sufficient to mobilize considerable political support for continued research. In the meantime, it was at least possible to characterize certain diseases in molecular terms, and to use this characterization as a basis for establishing new diagnostic standards.

One further development marked the post-war consolidation of the culture of medical standardization, and with it of molecularization. Hitherto, medical practitioners had played a highly ambivalent role in the development of that culture. As we have seen, they were often enthusiastic consumers of new therapeutic molecules and other standardized medications. But they tended to resist efforts to standardize and routinize the ways in which they themselves used such products. By the interwar years, it was widely understood that, in order to generate the kind of knowledge best suited to standardizing therapeutic practice, clinical trials should themselves be conducted using standardized therapeutic regimes and standardized populations of patients (Lilienfeld, 1982). This implied the imposition of strict managerial and administrative controls on experimental forms of clinical practice. Clinicians, however, preferred to rely on their own professional judgment in evaluating new treatment methods and their clinical outcomes, and were reluctant to subordinate this judgment to the demands of administrators.

Consequently, it was only under exceptional circumstances that doctors had agreed to participate in efforts to standardize and molecularize either diagnostic or therapeutic procedures. In the peculiar setting of the Rockefeller Hospital, for instance, a handful of doctors had collaborated in the work of reinterpreting various clinical conditions in molecular terms, though others were plainly uncomfortable with this project (Amsterdamska, this volume). Elsewhere, the dramatic effects of insulin, and the fact that laboratory scientists and medical administrators were able to monopolize early supplies of the drug, had persuaded hospital doctors that they must collaborate in determining appropriate standards for clinical use (Liebenau, 1989). But in general, clinicians continued to resist efforts to standardize therapeutic practice at least until the 1940s. We can see this, for instance, in the fact that, once the crisis of the World War II had passed, military doctors reasserted the superiority of whole blood transfusions — which must be adjusted to suit the immunological constitution and clinical requirements of individual cases, and which thus involved a complex combination of technical skill and clinical judgment — over the routine administration of plasma albumin (Creager, this volume).

In other areas of practice, however, doctors were already becoming assimilated into the administrative culture of medical standardization and molecularization. Again, it was the administrative exigencies of wartime, coupled with the dramatic development of antibiotics, that led them to change their attitudes. Despite the efforts of governments and drug companies to scale up the production of penicillin and streptomycin, demand rapidly outstripped the limited supply available at that time. Consequently, government agencies in the United States and, to a lesser extent in Britain, took steps to ensure that antibiotic treatment was restricted to those cases that it would most benefit (Adams, 1991). And in order to determine what kind of cases those might be, they also sought more standardized knowledge of the efficacy of the new drugs. Official control over the distribution of antibiotics now made it possible to generate such knowledge. Doctors had little choice but to submit to the imposition of increasingly strict experimental protocols (Marks, 1988).

The scene was set for the rapid institutionalization of clinical trial procedures in the postwar years. Drug companies, government agencies and charitable organizations now realized that, by exerting strict controls over the supply of new drugs, they could force clinicians to participate in standardized clinical trials. But at the same time, clinicians came to recognize that they too could benefit from participating in such trials. The development of dramatically effective new drugs, including the antibiotics and subsequently such molecules as cortisone (Cantor, 1992; Marks, 1992), did much to raise public expectations of the power of modern medicine. By acting as gatekeepers to potentially beneficial new therapies, clinicians could do much to enhance their own professional prestige and authority over patients. As a result, large scale clinical trials became one of the defining features of the postwar medical landscape. Drug companies and administrative bodies were now able to conduct large scale clinical experiments to measure the therapeutic effects of a wide range of novel substances, including the many new chemotherapeutic molecules issuing from the highly politicized field of cancer research (Löwy, 1995; also Pieters, 1993), and the equally politically sensitive contraceptive use of sex hormones (Marks, this volume; also Meldrum, 1996, for non-molecular contraceptive methods). Doctors had now ceded much of their clinical autonomy to the administrative demand for standardized forms of medical practice.

By the 1960s, then, a wide range of disparate interests had become bound up in the project of medical standardization and molecularization. Pharmaceutical companies found in it a way to enhance the efficiency of their production processes and to eliminate competing patent medicine manufacturers; doctors benefited by using new criteria of medical efficacy and safety to secure a monopoly over the medical marketplace and to enhance their image as practitioners of unique therapeutic power; government agencies and corporate charities used it to generate routine technical responses to a variety of medico-social problems; and scientists in academia and industry alike found employment for their skills in an expanding field of research. All secured different benefits, but all, in one way or another, saw those benefits as intimately connected to the continuing elucidation of the molecular constituents of biological reality, and the production of new molecular medicines.

STANDARDIZATION AND THE ARTICULATION OF TECHNICAL INTERESTS

In telling this story of an overall convergence of interests on a common project of medical standardization and molecularization, I have skated, in rather cavalier fashion, over the more narrowly focused and more detailed accounts of particular technical and social developments provided by other contributors to this volume. Indeed, at the level of resolution offered by these case studies, the picture becomes vastly more complicated. Thus it is plain that the molecularization of biological reality did not proceed unproblematically through the advancement of a single coherent research program. Of the many different lines of molecular research adopted at one time or another, some did indeed lead to the isolation and characterization of particular molecules, but many simply petered out, while others were diverted towards different scientific ends. Thus, for instance, Löwy shows (this volume) that the search for new cancer therapies sometimes revolved around the chemical and pharmacological properties of particular molecules, but at other times came to focus on the properties of the immune system. Within this shifting network of research projects, earlier experimental results could assume a very different significance in the light of new theoretical or methodological assumptions. Molecules themselves might appear in different guises: as problematic objects to be characterized and analysed or as taken-for-granted tools to be used in the analysis of other molecular processes (Goodman, this volume); as indicators of underlying clinical phenomena or as autonomous and interesting clinical phenomena in their own right (Amsterdamska, this volume). Indeed, some molecules might even cease to exist, as happened to the viral particles that were regarded, for a while, as probable agents in the causation of cancer (Gaudillière, this volume).

Much the same view can be taken of the social aspects of molecularization. Seen close up, it is difficult to discern any straightforward convergence of interests of the kind I have discussed above. Finer-grained analysis makes clear that such bodies as the pharmaceutical industry, the medical profession, research scientists and government administrators, far from being homogeneous in social constitution and in their social interests, all consisted of heterogeneous groups of actors, each pursuing their own local interests and often in competition or otherwise at odds with one another. The history of molecularization thus becomes a much more complex story of a multitude of local negotiations and alliances between different groups of actors, some of them relatively stable and long-lasting, others quickly breaking down in mutual indifference or hostility.

Viewed in this way, it is clear that biomedical science in the twentieth century might just as plausibly have fragmented into a chaos of divergent research programs and centrifugal social interests, as that large sections of it should have coalesced around a common core of molecular aims and objects. Nevertheless, the fact remains that, in the long run, such a convergence of interests and research projects did occur, and a relatively coherent and integrated molecular culture did emerge from the chaos of local social interactions and research programs. Consequently, we are justified in asking what larger influences might have structured the social field of the biomedical sciences such that collaboration in research tended, overall, to be favored over conflict.

The obvious answer would seem to lie in the fact that standardized medical technologies, be they therapeutic molecules, other therapeutic substances, or other techniques for standardizing medical practice, were generally valued, both in the medical market place and by the administrators of large-scale medical services. The development of molecular methods which contributed, or which looked likely to contribute, to the production of such technologies consequently attracted funding and other social benefits from a variety of sources outwith the research community itself. That is not to say that all research was necessarily conducted with a clear sense of how the results might be applied in medical or other forms of practice. Nevertheless, as the papers in this volume make clear, much molecular and other biomedical research did indeed lead, either directly or indirectly, to new methods of medical standardization. Moreover, such innovations commonly resulted when one research program drew on insights or techniques developed in another field, or when different teams of researchers combined resources to make a concerted attack on a particular problem. In other words, there were often direct benefits to be gained by interdisciplinary or interspecialism collaboration and cross-fertilization, and by the formation of heterogeneous networks of investigators.

Consequently, though biomedical research was often riven by intellectual and methodological disagreements, and while such links as existed between researchers in different specialisms were often temporary and always provisional, we should not regard the field as a whole as nothing more than a chaos of competing interests. Rather, it makes just as much sense to think of it as a highly fluid division of labor, loosely held together by the financial, material and social gains that ultimately resulted from the production of standardized medical technologies, of which therapeutic molecules were a prime example. Of course, the wider social value that such technologies commanded was itself contingent, as we have seen, upon the social restructuring of medical provision through the emergence of an organised medical profession and the growth of large-scale medical services. And we must acknowledge the likelihood that this restructuring in turn depended, at least to an extent, on the availability of new and standardized medical technologies.[6] Medical technology and medical services, in other words, might best be seen as co-evolving elements within a larger economy of medical practice. Nevertheless, the point stands: within this economy, value accrued through the production and distribution of standardized methods of diagnosis and treatment. It was this that ensured that different groups of manufacturers, researchers and practitioners came to collaborate, in their various ways, in the characterization, production and use of medical molecules.

This raises a further problem, however. Early in my discussion of the development of molecularization, I stressed that the successful establishment of an integrated division of technical labor in pharmaceutical production depended upon the imposition of appropriate forms of managerial control. Specifically, if the overall process was to result in the production of the desired molecules, it was necessary to ensure that different steps in the manufacturing process were all adequately standardized, regulated and coordinated. A similar point can be made about the integration of molecular research and production more generally.

If effective collaboration and division of labor between heterogeneous groups of researchers was to be possible, there had to be some way of ensuring that these groups addressed the same technical questions and worked towards the characterization and production of the same molecular objects. In some cases, this might involve overt managerial intervention.

This is clear, for instance, in Gaudillière's discussion of the organization of large-scale collaborative research into the viral genetics of cancer in the 1960s and 1970s (this volume). In that instance, explicit standards were imposed through the establishment of a strict system of contractual controls on the work conducted in different laboratories. A central managerial body specified the processes to be carried out, and the instruments and the reagents that were to be employed. Likewise, in the case of clinical trials, the managerial imposition of strict experimental protocols was necessary to ensure that the results obtained from different patients in different settings could nevertheless be seen as generating standardized knowledge of the properties of particular molecules (Marks, this volume). But the kind of "big science" discussed by Gaudillière, and the organization of large-scale clinical trials, represent a late formalization of a process of convergence and integration that had been taking place for three quarters of a century. Consequently, in examining the earlier stages of molecularization, we must look for more informal and more consensual ways in which common standards might have been agreed. It is here that we need to take up the point made by the editors and several of the contributors to this volume, and look at how integration might have been effected by the exchange and circulation of technical products — be they molecules, instruments or skilled personnel (see Gaudillière and Creager on various instruments; de Chadarevian on hemoglobin molecules and Lehmann's movements between laboratories).

What needs to be stressed, however, is that such circulation and exchange alone could not have been enough to ensure that the work of different research groups would be directed towards the characterization and production of the same molecular objects. For as we have seen, molecules could and sometimes did change their chemical identities and properties as they moved from one technical or theoretical context to another. Much the same can be said of instruments or of embodied knowledge and skills: it cannot be assumed that these would necessarily yield equivalent observations or objects in different settings. Consequently, if networks of collaboration and division of labor were to remain coherent — if, in effect, the circulation of molecules and technologies was to constitute a shared molecular economy — what was needed was some way of establishing common standards against which various molecules, instruments and facts could all be compared and calibrated (*cf.* Fujimura, 1992).

This is indeed exactly what we see taking place in the development of molecularization as a strategic orientation shared by a growing community of researchers. From the early years of the twentieth century onwards, numerous national and international committees were appointed to agree appropriate standards for a growing variety of biological and chemical entities. By no means all of these standards were molecular in nature. Reference collections of pathogenic and other bacteria were set up, for instance, while many pharmacologically active substances were standardized in terms of their biological effects. But many molecular standards were also created, insulin being an early example (Liebenau, 1989), Lehmann's collection of abnormal hemoglobins a much later one (de Chadarevian, this volume). That is not to say that such standards were incapable of revision or adjustment. On the contrary, they were themselves constituted and characterized through measurements and observations, which had in turn to be calibrated against other standard objects (see Jardine, 1991). There could be no final court of appeal in all this: all measurements and standards were ultimately provisional and subject to adjustment or modification. Nevertheless, if meaningful scientific collaboration was to take place, it was a necessary condition of such collaboration that researchers in different localities or conducting different experiments must have agreed, at any one time, on certain fixed points of empirical and technical reference.[7]

In many cases, such agreement was presumably achieved on a purely local basis. But as wider networks of interest converged on the shared project of molecularization, so researchers came to accept and endorse more generally applicable technical standards, including those embodied in reference collections of standard molecules and in standardised and commercially available instrumentation. Indeed, it was by expanding and elaborating this network of mutually agreed standards and observations that biomedical scientists were able to extend their own research networks, and ultimately to advance their common aim of molecularizing ever larger fields of biological and medical reality. In effect, negotiation and agreement of shared standards provided the means by which different researchers in different centers were able to negotiate and agree common scientific interests.

Moreover, as we have seen, the establishment of molecular standards did not serve solely as a means of organizing social relations and aligning social interests within the research community. It was also the means whereby researchers engaged the interests of different groups in the wider community of medical providers. Indeed, much of the impetus to establish national and international agreement on molecular standards came from administrators' demands for ways of regulating the use of therapeutic substances and diagnostic techniques in medical practice. Consequently, by examining how scientists negotiated and agreed on appropriate molecular standards, we can begin to understand, not only how they managed to organize their own research activities, but also how they contributed to the ordering and organization of social practices and social relations in the world beyond their immediate community (*cf.* Schaffer, 1992). Standardization was thus a vital element in the development of molecularization on two counts: it made possible the continuing characterization and production of new molecules by scientists working within the research community; but equally importantly, it provided the means of making those products interesting and valuable to administrators and others working in the wider society, and hence of generating funds and support for the continuation of the molecular enterprise.

Once we acknowledge the centrality of standardization, it becomes clear that molecularization was not effected simply through the circulation of molecules and their associated technologies and practices around an increasingly wide network of actors and social sites. Such circulation alone would have been meaningless and valueless. Rather, it is perhaps more illuminating to think in terms of the growth of what was in effect a molecular economy, in which the value and meaning of molecules, technologies and practices was established in relation to precise technical standards negotiated and agreed — sometimes formally, more often informally — between diverse groups of actors both within and outwith the research community. And as with any other economy, we need not suppose that the free circulation of goods necessarily dissolved the divergent interests that informed different actors' participation in that economy. On the contrary, we might also ask how different groups sought to maximize the value that accrued to their own activities within that economy, not least by endeavouring to establish their own products and practices as standards against which other activities might be evaluated.

Within the burgeoning project of molecularization, then, the technical was indeed profoundly social. This has been made abundantly clear by the various contributors to this volume. Thus it is plain that the growth of a molecular orientation within biology and medicine cannot be understood simply in terms of the characterization of new molecules and the invention of new techniques. Rather, we must also understand how the circulation of molecules and techniques served to link together networks of actors around a common strategic commitment to elaborating further molecular knowledge and practice. By tracing some of these shifting networks, the papers in this volume have gone a long way towards mapping the social development of this technical project, and towards showing how the characterization and circulation of molecular goods might in effect be regarded as forms of production and exchange within an evolving molecular economy.

Moreover, as many of the contributors are well aware, the social world of molecularization was constituted by more than just technical activities. Though participation in the molecular economy may have brought disparate groups of actors into agreement over a common set of technical and social aims, those actors also had other social interests that they sought to pursue through their molecular work. Consequently, if we are to acquire a satisfactory understanding of how molecularization developed, and why it took the forms and directions it did, we must also pay attention to the different actors' divergent interests, and how those interests informed negotiations over the shared technical standards that were so crucial both for the characterization of molecular reality and for securing wider social support for the enterprise as a whole. The various case studies collected in this volume do much to indicate some of the particular interests that need to be taken into account in constructing a synoptic account of the success of molecularization, and the part it has played in more general processes of social change, in the course of the twentieth century.

NOTES

1 I am well aware of the parasitic nature of my own remarks, and of the danger that I may misrepresent both other contributors' views and the detailed historical picture they are laboring to construct. As someone whose research does not focus primarily on the history of molecules, I am grateful to the editors for inviting me to expatiate on the work of better qualified historians, first at the conference where a number of these papers were initially presented, and now in this volume.

2 Integration was an important feature of the development of the chemical industry as a whole, and led ultimately to wholesale reorganization of national industries with the formation of IG Farben in Germany and Imperial Chemicals in Britain (Haber, 1971; Reader, 1970). Integration of specifically pharmaceutical firms did not occur to the same extent. The diversity and complexity of pharmaceutical chemicals, and the fact that many processes continued to depend on natural sources of pharmacologically active substances, meant that there were fewer points of overlap between different production pathways. Nevertheless, with the growth of science-based production methods, and the need to invest in new plant and laboratory facilities, even pharmaceutical companies benefitted from economies of scale, and were able to pursue a degree of internal division of labor and integration that would not otherwise have been possible.

3 See Liebenau (1986), for how such claims were incorporated into drug manufacturers' promotional materials. The same companies also sometimes turned their chemical laboratories to the work of analysing patent medicines, in order to demonstrate that such preparations contained little in the way of pharmacologically active chemicals, or included dangerous adulterants. In Britain, Burroughs Wellcome adopted this strategy, collaborating with the British Medical Association to publish two volumes of analyses aimed primarily at medical practitioners (British Medical Association, 1909, 1912).

4 See "Report of Proceedings of Conference in regard to Medical Benefit with the Medical Members of the Advisory Committees", 11 June 1912, Public Record Office, PIN 2/18.

5 Smith (1995) and Smith and Nicolson (1996) discuss interwar attempts to establish appropriate dietary standards, though primarily in terms of calorific intake rather than vitamins, and they place these discussions in the context of debates over poverty.

6 In this respect, it would be helpful to know more about just how and why molecular technologies came to be favoured over other possible responses to medico-social problems, as for instance in the case of vitamins, where alternative and less obviously standardized schemes of dietary reform were also proposed (Smith and Nicolson, 1989).

7 Mary Hesse's network theory of meaning might usefully be invoked here (Hesse, 1974). Hesse argues that the theoretical significance of any

observation is defined relative to an inferential network of inter-related facts and postulates. All the various terms in the network, and the relationships between them, are provisional and subject to adjustment, though any such adjustment will have implications for other terms in the network. But unless the whole network of knowledge and meaning is to collapse into incoherence, certain terms and relationships must remain fixed, by agreement or convention, at any one time.

REFERENCES

D.P. Adams, *"The Greatest Good to the Greatest Number": Penicillin Rationing on the American Home Front, 1940–1945* (New York: Peter Lang, 1991).

M. Bliss, *The Discovery of Insulin* (Toronto: McLelland and Stewart, 1982).

M. Borell, "Organotherapy and the Emergence of Reproductive Endocrinology" *Journal of the History of Biology* (1985), 18: 1–30.

British Medical Association, *Secret Remedies: What They Cost and What They Contain* (London: British Medical Association, 1909).

British Medical Association, *More Secret Remedies: What They Cost and What They Contain* (London: British Medical Association, 1912).

D. Cantor, "Cortisone and the Politics of Drama, 1949–55" in John V. Pickstone (ed.), *Medical Innovations in Historical Perspective* (Basingstoke: Macmillan, 1992), pp. 165–184.

H. Chick, M. Hume, M. Macfarlane, *War on Disease: A History of the Lister Institute* (London: André Deutsch, 1971).

G.W. Corner, *History of the Rockefeller Institute, 1901–1953* (New York: Rockefeller Institute Press, 1964).

J.H. Fujimura, "Crafting Science: Standardized Packages, Boundary Objects, and 'Translation'" in A. Pickering (ed.), *Science as Practice and Culture* (Chicago: University of Chicago Press, 1992), pp. 168–211.

B.B. Gilbert, *The Evolution of National Insurance in Great Britain: The Origins of the Welfare State* (London: Michael Joseph, 1966).

L.F. Haber, *The Chemical Industry 1900–1930: Growth and Technical Change* (Oxford: Clarendon Press, 1971).

C. Harington, "The Work of the National Institute for Medical Research" *Proceedings of the Royal Society of London, B* (1950), 136: 333–348.

E.P. Hennock, *British Social Reform and German Precedents: The Case of Social Insurance 1880–1914* (Oxford: Clarendon Press, 1987).

M.B. Hesse, *The Structure of Scientific Inference* (London: Macmillan, 1974).

T. Lenoir, "A Magic Bullet: Research for Profit and the Growth of Knowledge in Germany Around 1900" *Minerva* (1988), 26: 66–88.

A. Li, "J.B. Collip, A.M. Hanson and the Isolation of the Parathyroid Hormone, or Endocrines and Enterprise" *Journal of the History of Medicine and Allied Sciences* (1992), 47: 405–438.

J. Liebenau, "Marketing High Technology: Educating Physicians to Use Innovative Medicines" in R.P.T. Davenport-Hines (ed.) *Markets and Bagmen: Studies in the History of Marketing and British Industrial Performance 1830–1939* (Aldershot: Gower, 1986), pp. 82–101.

J. Liebenau, *Medical Science and Medical Industry: The Formation of the American Pharmaceutical Industry* (Basingstoke: Macmillan, 1987).

J. Liebenau, "Ethical Business: The Formation of the Pharmaceutical Industry in Britain, Germany, and the United States Before 1914" *Business History* (1988), 30: 116–129.

J. Liebenau, "The MRC and the Pharmaceutical Industry: The Model of Insulin" in J. Austoker and L. Bryder (eds), *Historical Perspectives on the Role of the MRC: Essays in the History of the Medical Research Council of the United Kingdom and its Predecessor, the Medical Research Committee, 1913–1953* (Oxford: Oxford University Press, 1989), pp. 163–180.

J. Liebenau, "Paul Ehrlich as a Commercial Scientist and Research Administrator" *Medical History* (1990), 34: 65–78.

A.M. Lilienfeld, "'Ceteris Paribus': The Evolution of the Clinical Trial" *Bulletin of the History of Medicine* (1982), 56: 1–18.

I. Löwy, "Nothing More to be Done: Palliative Care Versus Experimental Therapy" *Science in Context* (1995), 8: 209–229.

H.M. Marks, "Notes From the Underground: The Social Organization of Therapeutic Research" in R.C. Maulitz and D.E. Long (eds), *Grand Rounds: One Hundred Years of Internal Medicine* (Philadelphia: University of Pennsylvania Press, 1988), pp. 297–336.

H.M. Marks, "Cortisone, 1949: A Year in the Political Life of a Drug" *Bulletin of the History of Medicine* (1992), 66: 419–439.

M.L. Meldrum, "'Simple Methods' and 'Determined Contraceptors': The Statistical Evaluation of Fertility Control, 1957–1968" *Bulletin of the History of Medicine* (1996), 70: 266–295.

P. Neushul, "Science, Government, and the Mass Production of Penicillin" *Journal of the History of Medicine and Allied Sciences* (1993) 48: 371–395.

J. Parascandola, *The Development of American Pharmacology: John J. Abel and the Shaping of a Discipline* (Baltimore: Johns Hopkins University Press, 1992).

T. Pieters, "Interferon and its First Clinical Trial: Looking Behind the Scenes" *Medical History* (1993), 37: 270–295.

W.J. Reader, *Imperial Chemical Industries: A History*, Vol. 1: *The Forerunners, 1870–1926* (London: Oxford University Press, 1970).

T. Richards, *The Commodity Culture of Victorian England: Advertising and Spectacle, 1851–1914* (Stanford: Stanford University Press, 1990).

C.E. Rosenberg, "The Therapeutic Revolution: Medicine, Meaning, and Social Change in Nineteenth-Century America" in M.J. Vogel and C.E. Rosenberg (eds), *The Therapeutic Revolution: Essays in the Social History of American Medicine* (Philadelphia: University of Pennsylvania Press, 1979), pp. 3–25.

S. Schaffer, "Late Victorian Metrology: A Manufactory of Ohms" in R. Budd and S.E. Cozzens (eds), *Invisible Connections: Instruments, Institutions and Science* (London: SPIE Optical Engineering Press, 1991).

S. Shapin, "Discipline and Bounding: The History and Sociology of Science as Seen Through the Externalism-Internalism Debate" *History of Science* (1992), 30: 333–369.

D. Smith, "The Social Construction of Dietary Standards: The British Medical Association-Ministry of Health Advisory Committee on Nutrition Report of 1934" in D. Maurer and J. Sobal (eds), *Eating Agendas: Food and Nutrition as Social Problems* (New York: Aldine De Gruyter, 1995), pp. 279–303.

D. Smith and M. Nicolson, "The 'Glasgow School' of Paton, Findlay and Cathcart: Conservative Thought in Chemical Physiology, Nutrition and Public Health" *Social Studies of Science* (1989), 19: 195–238.

D. Smith and M. Nicolson, "Nutrition, Education, Ignorance and Income: A Twentieth-Century Debate" in H. Kamminga and A. Cunningham (eds), *The Science and Culture of Nutrition, 1840–1940* (Amsterdam: Rodopi, 1995), pp. 288–318.

S. Sturdy and R. Cooter, "Science, Scientific Management and the Transformation of Medicine in Britain c.1870–1950" unpublished paper.

J.P. Swann, "The Search for Synthetic Penicillin During World War II" *British Journal for the History of Science* (1983), 16: 154–190.

J.P. Swann, *Academic Scientists and the Pharmaceutical Industry: Collaborative Research in Twentieth-Century America* (Baltimore: Johns Hopkins University Press, 1988).

E.M. Tansey and R.C.E. Milligan, "The Early History of the Wellcome Research Laboratories, 1894–1914" in G.J. Higby and E.C. Stroud (eds), *Pill Peddlers: Essays on the History of the Pharmaceutical Industry* (Madison: American Institute of the History of Pharmacy, 1990), pp. 91–106.

J.H. Warner, *The Therapeutic Perspective: Medical Practice, Knowledge, and Identity in America, 1820–1885* (Cambridge, MA/London: Harvard University Press, 1986).

M. Weatherall, *In Search of a Cure: A History of the Pharmaceutical Industry* (Oxford: Oxford University Press, 1990).

P. Weindling, "From Medical Research to Clinical Practice: Serum Therapy for Diphtheria in the 1890s" in J.V. Pickstone (ed.), *Medical Innovations in Historical Perspective* (Basingstoke: Macmillan, 1992), pp. 72–83.

J.H. Young, *The Toadstool Millionaires: A Social History of Patent Medicine in America Before Federal Regulation* (Princeton: Princeton University Press, 1961).

Index